Multinationals and Economic Geography

Multinationals and Economic Geography

Location, Technology and Innovation

Simona Iammarino

Professor of Economic Geography, London School of Economics and Political Science, UK

Philip McCann

Professor of Economic Geography, University of Groningen, The Netherlands and Professor of Economics, University of Waikato, New Zealand

Edward Elgar

Cheltenham, UK • Northampton, MA, USA

Published by
Edward Elgar Publishing Limited
The Lypiatts
15 Lansdown Road
Cheltenham
Glos GL50 2JA
UK

Edward Elgar Publishing, Inc.
William Pratt House
9 Dewey Court
Northampton
Massachusetts 01060
USA

A catalogue record for this book
is available from the British Library

Library of Congress Control Number: 2012949719

This book is available electronically in the ElgarOnline.com
Economics Subject Collection, E-ISBN 978 1 78195 479 9

MIX
Paper from
responsible sources
FSC www.fsc.org FSC® C018575

ISBN 978 1 78195 478 2 (cased)
978 1 78195 487 4 (paperback)

Typeset by Servis Filmsetting Ltd, Stockport, Cheshire
Printed and bound by MPG Books Group, UK

Contents

1. Introduction: multinational enterprises, innovation and geography in todays' globalized world

1.1 WHAT THIS BOOK IS (AND IS NOT) ABOUT

The existence and behaviour of the multinational enterprise (MNE) has now been analysed for over half a century by economists, international business and management researchers, behavioural and organizational theorists, as well as business historians and sociologists. In spite of the diversified theoretical roots and backgrounds of these different traditions, two main sets of questions have been commonly addressed by social scientists. The first set relates to the behavioural determinants, motives, and strategies of multinational firms, and includes questions such as: *why* do firms become multinational? *where* do they locate their cross-border activities? *how* do they undertake and organize their multinational production and transactions? The second set of questions relates to the wider effects of a multinational presence, and includes questions such as: what is the *impact* on the home economy? what is the *effect* on the host economy? and what is the *effect* on the multinational firm itself?

These two broad lines of inquiry have been further decomposed into a variety of related, but differentiated, sub-questions. What is apparent, after more than fifty years of systematic research on multinationality, is that there is as yet no unified or dominant theory of the MNE. A plethora of theoretical approaches, conceptual frameworks and speculative explanations have been put forward, each of which is developed within the boundaries and assumptions inherent to these diverse analytical backgrounds. Such a multidisciplinary scholarly attention is not surprising, as multinationality is per se a multidimensional, multifaceted phenomenon, and the heterogeneity of facts, activities and behaviours to be explained are not easily fitted within a single conceptual framework. In this sense, the study of MNEs is no different from that of business organizations generally: the study of the firm as an economic organization and a social institution is not, and cannot be, encompassed in a single governing theoretical scheme.

The aim of this book is not to propose a new theory of the MNE or an alternative set of explanations of MNE behaviour. Rather, our objective is to bring into focus one particular dimension of MNE behaviour and activity which has been under-researched in comparison with other aspects of the MNE; namely the *geography* of the multinational enterprise. Indeed, the intersections of the study of MNEs with both geography and innovation are nowadays widely recognized as being very important, but they are still seriously underexplored.

In this book we work with a set of selected theories and conceptual frameworks which, in our view, together help explain much of the spatial behaviour of the modern MNE. These theories and frameworks explain the spatial behaviour of the MNE with reference to both the internal processes of the firm's decision-making and operating structures, and also with reference to its connections and interactions with the outside world, and thus with geography. In the international business and management literature the giant, multidivisional, highly integrated multinational *firm* has been put at the centre of the analysis, and this is often treated as an abstract type of economic agent of unique importance. However, it is more useful, for our purpose here, to follow the view of Edith Penrose and treat the MNE as just a special case of the general firm, or even better, as a stage in its evolution, with the stage – *being* multinational – itself subject to change over time. Indeed, the incredible pace with which the number of internationalized and network-connected firms has grown in the recent years seems to indicate that many large, and even medium and small sized companies, are going through a metamorphosis that will lead to different paths of openness and interdependence, and eventually to different forms of multinationality itself (Box 1.1).

In marked contrast to the international business and management literatures, other fields such as innovation studies and the economics of technological change have mostly considered the uni-national, unilocated, mono-activity *firm* as the main object of the analysis, thereby marginalizing the MNE in the investigation of causes and effects of learning and innovation processes. Likewise, traditional economic geography and location theory, and even the most recent developments in both new trade theory and new economic geography, have tended to concentrate on the *activities* carried out by the multinational firm, mostly defined, if not exclusively so, in terms of foreign direct investment (FDI). Analyses here have tended to emphasize the firm's location choices for plants or establishments in the same area, or alternatively the firm's decisions to replicate production capacity in other far away sites, without paying much atten-

BOX 1.1 SOME FACTS AND FIGURES: THE GROWING NUMBERS OF MNEs

At the end of the 1960s there were approximately 7000 MNEs, which were accounted for almost entirely by just fifteen countries. By 2006, there were an estimated 78,000 MNEs in the global economy with some 780,000 foreign affiliates (UNCTAD 2007). Over the last few years the number of MNEs in the global economy has been increasing at a rate of 1000–2000 per annum, while the number of foreign affiliates has risen by 10,000–20,000 per year. The numbers of MNEs in 2006 represented a 23 per cent increase with respect to the same number in 2000, and a 13 per cent increase in the number of foreign affiliates (UNCTAD 2007).

As well as employment, over recent decades the levels of output and trade which are associated with multinational firms have also increased much more rapidly than the growth of global trade. Foreign direct investment has been growing at twice the speed of world trade, which itself has grown at twice the rate of world income. The result is that foreign direct investment grew by almost six-fold between 1970 and 1999 (Bobonis and Shatz 2007), with 30–40 per cent of US trade currently accounted for by intra-firm trade flows (Lai and Zhu 2006). The 78,000 multinational firms operating in the global economy in 2006 accounted for an estimated $4.8 trillion in value-added and $4.7 trillion in exports (UNCTAD 2007). In order to get a sense of the relative importance of multinational investment, in 2006 the global GDP was over $45 trillion (World Bank 2007) and global exports were $14.1 trillion (UNCTAD 2007): MNEs therefore accounted for over 10 per cent of global GDP, and approximately one third of global exports, as well as 12.6 per cent of global domestic fixed capital formation (UNCTAD 2007). The 780,000 foreign affiliates of MNEs also employed an estimated 73 million workers. This number has not only tripled since 1990 (UNCTAD 2007), but increased by some 20 million from 53 million (UNCTAD 2003) just since 2002. The total number of workers employed in foreign affiliates now represents around 3 per cent of the global workforce (UNCTAD 2007). Recent estimates (Spence 2011) suggest that 31 per cent of US economic growth since 1990 is accounted for by MNEs.

In spite of the negative impact of the global economic crisis on foreign direct investment flows, the internationalization of firms worldwide has continued to grow: the estimated number of MNEs in 2008 was around 82,000, with foreign affiliates' share in global GDP peaking at 11 per cent in 2009 (it was approximately 5 per cent of global GDP in 1980), representing one-third of world exports, and foreign employment of MNEs reaching 80 million in the same year (UNCTAD 2010). MNEs, in their operations at the worldwide level, generated a value added of approximately 16 trillion US dollars in 2010, representing more than a quarter of global GDP (UNCTAD 2011). MNEs' international production (i.e. the value added generated by foreign affiliates) accounts for approximately 40 per cent of MNEs' total value added: despite the economic crisis and the drop in foreign direct investment (FDI) flows since 2008, international production has thus grown from the 35 per cent share registered in 2005 (UNCTAD 2011).

The increasing internationalization of the world economy is strongly reflected in the shares of FDI stocks, both inward and outward, with respect to GDP, which, as is shown in Tables 1.1 and 1.2 has tripled at the global level between 1990 and 2009 (UNCTAD 2010).

Table 1.1 FDI inward stock as a percentage of GDP, 1990–2009 (per cent)

Region / economy	1990	1995	2000	2005	2009
World	9.8	11.4	23.3	25.4	30.7
Developed economies	9.0	10.8	23.0	25.5	31.5
Europe	10.7	12.7	27.5	34.2	46.5
European Union (27)	10.6	12.5	27.5	34.1	45.5
Other developed Europe	13.0	16.1	27.6	36.7	66.4
North America	10.2	14.2	28.6	23.4	23.4
Other developed countries[a]	2.8	2.9	4.1	7.8	10.5
Developing economies	13.6	14.6	25.0	25.2	29.1
Africa	12.1	17.0	26.0	27.3	34.6
North Africa	12.6	16.2	17.4	22.9	32.5
Other Africa	11.8	17.5	32.7	29.8	36.0
Latin America and the Caribbean	9.9	10.5	24.3	29.7	36.5
South and Central America	9.7	10.3	21.5	26.2	29.8
Caribbean	13.4	14.2	81.5	99.8	187.5
Asia	15.8	18.3	25.2	23.1	25.7

Table 1.1 (continued)

Region / economy	1990	1995	2000	2005	2009
West Asia	8.8	8.5	8.8	15.6	25.6
South, East and South-East Asia	17.4	17.6	28.4	24.7	25.8
East Asia	25.9	21.0	31.7	26.0	25.4
South Asia	1.3	2.6	4.2	6.1	11.2
South-East Asia	18.2	22.6	44.5	44.7	46.3
Oceania	24.9	22.5	29.9	26.3	44.1
South-East Europe and the CIS	..	2.1	15.6	25.3	27.5

Source: UNCTAD, FDI/TNC database (www.unctad.org/fdistatistics).
[a] Includes Australia, Bermuda, Israel, Japan, New Zealand.

Table 1.2 FDI outward stock as a percentage of GDP, 1990–2009 (per cent)

Region / economy	1990	1995	2000	2005	2009
World	10.0	12.2	25.2	27.6	33.2
Developed economies	11.2	14.0	28.8	32.7	40.8
Europe	11.8	15.5	42.4	43.6	57.8
European Union (27)	11.3	14.5	41.4	41.9	55.0
Other developed Europe	21.4	35.1	62.7	78.2	110.2
North America	12.8	18.7	27.9	29.8	31.2
Other developed countries[a]	6.9	5.2	7.5	11.4	18.2
Developing economies	4.1	5.9	12.9	12.5	16.5
Africa	4.6	6.8	8.3	5.9	7.9
North Africa	1.0	0.9	1.3	1.2	3.9
Other Africa	7.1	11.0	14.4	8.8	10.5
Latin America and the Caribbean	5.4	5.1	10.2	13.3	16.4
South and Central America	5.3	4.3	5.9	7.4	9.0
Caribbean	11.5	67.8	293.9	371.6	564.2
Asia	3.3	6.1	14.8	13.1	17.6
West Asia	2.1	1.9	2.5	4.2	9.7
South, East and South-East Asia	3.6	6.8	17.1	14.9	19.0
East Asia	5.4	8.8	22.9	18.5	22.2
South Asia	0.1	0.1	0.4	1.0	4.3
South-East Asia	2.8	7.7	15.1	19.7	25.2
Oceania	6.4	7.4	10.1	5.6	5.2
South-East Europe and the CIS	..	0.9	6.0	14.6	16.1

Source: UNCTAD, FDI/TNC database (www.unctad.org/fdistatistics).
[a] Includes Australia, Bermuda, Israel, Japan, New Zealand.

tion to the internal transformation of the firm's structure and organization as such. Finally, there are two other growing bodies of literature we must consider. On the one hand there is the literature looking at industrial and geographical clustering. On the other hand there is the literature looking at the formation of networks as alternative modes of organizing international production, along with the new value-chain divisions of labour implied by these international networks. Both of these literatures focus primarily on the role and nature of *linkages* of both a spatial and a non-spatial nature, and as such both literatures largely overlook the heterogeneity of firms and of the activities and functions that they carry out.

There are, of course, numerous exceptions to these generalizations which we will try to identify and use as bridges between the different scholarly fields. However, our efforts here are aimed at providing the pieces of the puzzle such that the picture hidden in the jigsaw will emerge. As such, a *new* theory is not what we are looking for, nor is what we think is needed, but rather the integration of existing theoretical strands that largely do not talk to each other. We do not therefore aim to provide an exhaustive review of what we know about multinational behaviour or locational patterns or their links with innovation and growth. We leave that to the many excellent existing publications on the different aspects of MNE which have been sources of inspiration for the present work. Our aim is rather to show that these important aspects of the real world need to be analysed in conjunction, and we offer a guide as to how thinking on the spatial dimension of the MNE phenomena could fruitfully evolve. What we want to point out is that, by integrating different theoretical approaches, and by addressing simultaneously different levels of analysis, it is possible to identify the *linkages* between different types of *firms* and *activities*, and the way these linkages are shaped on the one hand by geography and space, on the other hand by innovation, technology and learning, and also by the interactions between these different dimensions and phenomena.

The two-way relationship between multinational corporations and geography will be the focal point here. We claim that geography is increasingly important for MNEs, and in turn MNEs are progressively more important for geography. The pivot on which this relationship turns is the creation, diffusion and management of new knowledge. Our effort is aimed at showing that such a bilateral and mediated relationship, which in principle is valid for any kind of business or firm, is *particularly* valid in the case of MNEs and *particularly* important now for MNEs. Mark Casson has argued that 'the modern theory of the MNE has the potential to become a general theory of the enterprise in space, and as such, to embrace theories of the multi-regional and multi-plant firm' (Casson 1987, p. 1). We fully agree with Casson and, although what we present in this book has a far more modest scope, we hope it can be considered as a small step forward in that direction.

For this purpose, a clarification regarding the interpretative lenses adopted in this book is necessary. Evolutionary perspectives have found extensive applications both in the theory of the MNE and in the economic geography arena. This underlies our choice to adopt such a perspective here, particularly in considering the interdependence between the micro level of the firm, and the meso level of the spatial system. This choice should not be taken as implying a claim that the evolutionary approach to analysis is uniquely or universally superior, but rather that we make use of it because of its particular strengths in interpreting change and evolution. On the other hand, we make use of many concepts which have been developed by different theoretical and conceptual streams, some of which have even been blended. For example, it is hard, if not impossible, to accredit the introduction of the term 'technological competence' to any specific research line, and to determine whether and to what extent this differs from that of 'capability' in the general literature. Yet, to discuss the parenthood of terms is not our aim here, but rather to indicate the building blocks of an analytical framework for interpreting the relationship between space and innovation sources, and their relevance for MNE location and investment choices and evolution.

1.2 THE CHANGING GLOBAL CONTEXT

1.2.1 The Earth, a Round Planet with a Bumpy Surface

Our integrative view features the MNE-knowledge-space nexus in a context of increasing globalization of the world economy. Could this book have been written, or would it have been applicable, prior to the end of the 1980s? As we will see, we think not, because the difference between then and now has been the astounding speed and scale of globalization in the last two decades which has transformed all aspects of multinationality and of being multinational.

The notion of economic globalization, in the sense of the increasing economic openness of national economies and the increasing integration of the world economy, is now widely discussed and well established (Box 1.2). Scholars, governments, international organizations, and individuals have attempted to assess the changes that have occurred as a consequence of an ever increasingly globalized economy and society. In geographical or spatial terms, we may reduce discussions on globalization to two broad questions: is distance becoming less important? and, are different places in

BOX 1.2 SOME FACTS AND FIGURES: GLOBALIZATION OR TRIADIZATION OF THE WORLD'S ECONOMIC ACTIVITY?

As we will see also in later chapters, inward FDI flows into developing and transition economies combined were, until recently, less than half the level of all inward FDI into developed economies. In 2005 and 2006, the rate of growth of FDI inflows to developed economies, at 34 per cent and 45 per cent respectively, was actually more than twice that of FDI inflows (21 per cent) into developing economies (UNCTAD 2006, 2007). As such, the total value of inward FDI into developing and transition economies reached \$379bn in 2006 (UNCTAD 2007), thereby accounting for 29 per cent of global FDI inflows (UNCTAD 2006), with 5–6 per cent being accounted for by transition economies.

The most important macro-region for FDI inflows in recent years is South, East and South-East Asia, which accounts for 53 per cent of total FDI inflows into developing or transition economies. In terms of individual countries, the largest FDI recipient amongst developing countries is China, with \$69.5bn, and amongst transitions economies is Russia, with \$28.7bn (UNCTAD 2007). Although the scale of FDI inflows to India in 2006, at \$17bn, is much smaller than that of China, this was equivalent to the total inward FDI in India during the three previous years combined, suggesting a rapid increase in India's attractiveness for foreign investors (UNCTAD 2007). In comparison, Latin America accounts for 6.4 per cent of FDI inflows, while the whole of Africa receives just 2.7 per cent of global FDI inflows (UNCTAD 2007).

The relative dominance of the developed economies in FDI also emerges when we consider FDI outflows. FDI outflows from Western developed economies plus Japan grew by 45 per cent between 2005 and 2006, accounting for 84 per cent of global outward FDI (UNCTAD 2007). If Taiwan, Korea, Singapore and Hong Kong are also included, the share of global outward FDI increases to 89.6 per cent (UNCTAD 2007). Comparing these shares with those reported below (see Box 1.3), we see that the geographical origins of global FDI outflows very closely reflect the geographical origins of R&D. In terms of the sources of these FDI outflows, the US, Canada, Japan, Australia and Hong Kong, plus ten European economies have been the top 15 sources of investment. Outflows from the European Union (EU) account for 47 per cent of global outflows, rising to 55 per cent when including the four members of the European Free Trade Association (EFTA), today composed by four members – Norway, Iceland, Switzerland and Liechtenstein (UNCTAD 2007). Outflows from the US and Canada together account for 21.5 per cent of the total (UNCTAD 2007). For a single country, the US exhibits by far the largest FDI outflows: annual investment abroad from the US is typically more than twice that of any other country, with the total foreign employment of US MNEs currently running at approximately 9 million employees (UNCTAD 2007).

As we will see later, each of the three super-regions – namely the European Union (EU), North America Free Trade Area (NAFTA) and South and East Asia – is dominated by a core of countries: in the case of NAFTA the dominant element by far is the US; in that of the EU the core is the EU-15, or the members pre-existing the enlargements to Eastern and Central Europe in 2004 and 2007; and in the case of South and East Asia, it is East Asia and particularly Japan, China and Korea. The share of global trade accounted for by these leading countries has increased in line with their share of global output: by 2005 NAFTA accounted for 15 per cent of world trade, the EU-15 for 36 per cent, and East Asia for 22 per cent (UN Comtrade data; see also Fujita 2007a). In terms of foreign direct investment, globalization confirms to be overwhelmingly a rich country phenomenon. In 2006, FDI inflows to developed economies amounted to $857bn (UNCTAD 2007), which is almost two thirds of total FDI flows, and up from 59 per cent in 2005 (UNCTAD 2006). In terms of individual countries, in 2006, with an inflow of $175.4bn, the US recovered its top position as the largest destination for FDI (UNCTAD 2007), still occupied, and even strengthened, in 2010 despite the economic crisis (UNCTAD 2011). The UK continues to maintain its dominant position in terms of the world's top 50 financial MNEs, still hosting the largest number of these particular MNE affiliates (UNCTAD 2007). Meanwhile, across all sectors, the EU alone accounts for 40 per cent of global FDI inflows, rising to over 43 per cent if we include the EFTA countries (UNCTAD 2006, 2007). The US and Canada combined account for just under 19 per cent (UNCTAD 2007). Over the last decades, therefore, as the three super-regions have increasingly controlled global economic activities, the concentration of trade in these blocks has also been increasing consistently, to the point that many scholars prefer to use the term 'Triadization', instead of Globalization, of the world economic activity.

FDI figures for the most recent years are rather volatile, due to the uncertainties brought about by the economic crisis since 2008 and the still rather fragile recovery at the world-wide level. However, it is worth pointing out that the most recent trends have shown a strong resilience of developing and transition economies, that have registered an enormous increase in their relative weight in international production networks, in contrast with the contraction or very slow pace of foreign investment from and into developed countries, particularly the European core countries (UNCTAD 2010). In 2010, developing and transition economies increased remarkably their importance both as FDI recipients – absorbing more than half of global FDI inflows and representing half of the top-20 host economies – and as investors – accounting for 29 per cent of global FDI outflows and being increasingly represented in the top-20 of the world's investors (UNCTAD 2011). These positive trends, however, mask a high variation in the geographical distribution of the involvement in foreign production, with East and South-East Asia, and Latin America increasing their shares at the expenses of other weaker developing economics (UNCTAD 2011). Some of the players in the Triad may lose ground, but a really 'global' convergence remains to be confirmed.

the world becoming more alike? Some have argued that both of these are so, and that the world is becoming in some economic and social sense flat (e.g., O'Brien 1992; Ohmae 1995; Cairncross 1997; Friedman 2005). It is a claim that advances in technology and business organization have enabled actors anywhere in the world, be they consumers, investors, entrepreneurs, or firms, to link up and do business with others anywhere with such ease and accessibility that obviates the advantages that once were accrued to those who were attached to favoured localities, regions or nations. The 'flat world' thesis implies that economic activities can be done just as well in India as America, by small firms as by large firms, because we are in the age of the network organization, and these networks are global.

The story of flatness rests on two big claims, one of which is geographical and one of which is organizational. Many have claimed that distance is becoming unimportant, and many others have argued that big multinational enterprises are dinosaurs while networked small and medium-sized enterprises (SMEs) are the mammals and flowering plants that will take their place. Let us take a quick look at each of these claims on its own.

Some Victorian observers saw the railway as the agent of the obliteration of time and distance. Similarly, in the early 1990s, a new generation of observers made the same claim for the Internet and low cost air travel. Commentators argued that technological improvements meant that geography and distance had become of little or no importance, and that the likely impacts of the emergent information and communication technologies would be to promote dispersal away from areas of population concentration (Warf 1995; Gaspar and Glaeser 1998). These arguments were increasingly bolstered by widespread contemporaneous observations of a process of urban-rural shift, whereby increasing numbers of people and firms in advanced economies appeared to be moving away from major cities and into smaller cities (Fothergill et al. 1985). The death of distance applied at any scale: at the level of a single city, a region, a nation state, or the world. In the last case, it becomes part of a view that has been called 'hyper-globalization' (Amin 1997; Ohmae 1995) that can be applied to many areas of human activity, from popular culture to politics, and there have been several variants of this line of reasoning.

For the purposes of this book we are particularly interested in the application of such claims to the study of multinational enterprises. In this respect, in 1983 Theodore Levitt (1983) heralded the arrival of the global corporation as one which sells the same thing in the same way everywhere. From a slightly different perspective, Kenichi Ohmae (1999) sees the problem facing MNEs as being the achievement of the 'Anchorage perspective'; in other words seeing the world – or rather the parts that matter such as the markets of Europe, North America, and East Asia – as if from Anchorage, Alaska, equidistant from the three. Finally, the 1990s saw the emergence of a small fan-club or cult in some business schools and some parts of the business press which admired and was devoted to the model of ABB, a multinational corporation which was Swedish in origin, headquartered in Switzerland, and which constituted of acquired businesses on every continent while using English as its working language. To these advocates, this stateless entity seemed to represent the ideal shape of the future global company.

What these visions of the global corporation share is that in all of them distance is seen as an obstacle to the homogenization of the world economy, and that improvements in information and communications technologies and also in transportation technologies, have the effect of reducing (and perhaps eliminating) this obstacle. National borders are likewise seen as barriers to homogenization, but the productive opportunities and information flows associated with the death of distance render political barriers increasingly costly and ultimately unsustainable (Strange 1992). The belief that the characteristic business form of this small and flat world will be the decentralized and flat stateless network, rather than the corporate leviathan, therefore also comes from a parallel aspatial discourse on business organization, in addition to a primarily spatial discourse. We are, without doubt, in an age of outsourcing, offshoring, alliances, partnerships, networks, core capabilities and competencies, and clusters. Organizations which once strove to internalize functions – up and down the supply chain – now strive to source them. Leanness, agility, flatness, flexibility, focus, specialization and even virtuality are honoured, and there is a tradition, exemplified by Michael Piore and Charles Sabel (1984) and Hirst and Zeitlin (1992) that holds that these new virtues are best reflected in SMEs. This tradition sees the origins of the large MNE, which is presumed to be fat, rigid, and otherwise antithetical to the new virtues, as emerging from sources. One is in inflexible production technologies which militate against flexible organizational systems, and the other is in the relatively high costs of organizing production between firms via market transactions, rather than within firms via administrative coordination (Williamson 1985). In this view two of the standard claims are that in a post-mass production world, networks of SMEs have many inherent advantages over corporate giants in that they are perceived as being better at innovation and also better able to adjust output levels. The realization of these advantages depends on the ability to cut transactions costs. In the 'flat world' vision two sorts of barriers are perceived as having been removed, namely the barriers to operating at a distance and across borders, and the transactions costs between firms.

That flat world, of course, is a fiction. One might be tempted to say that it is a hypothetical idealized state, one that we may never attain but can hope to glimpse on the far horizon as the barriers of distance, borders, and transactions costs are reduced by advances in technology. Even this, though, would be incorrect. At the simplest level, we can say that there is ample evidence that the removal of barriers to international trade, investment and communication can actually make geographical proximity more, rather than less, important. Many economic activities benefit from co-location, whether for reasons of simple logistics, shared inputs, or benefits of face-to-face contact (Arita and McCann 2000). When barriers to cross-border or long-distance trade are reduced, some agglomerations – local, regional, or even national – may find that their comparative advantages produces growth. The integrated world becomes a more differentiated one, and proximity to places such as Silicon Valley, the City of

London, or Hong Kong becomes more important than ever. Rather than becoming flatter, in many aspects globalization is nowadays making the world more uneven, more curved (McCann 2008), or more spiky (Florida 2005a, b; Rodríguez-Pose and Crescenzi 2008). In formal economics, as discussed later, these changes are represented by the increasing returns models of the new trade theory and the new economic geography (see, for all, Krugman 1991a, b; Fujita, Krugman and Venables 1999).

In the organizational dimension, it is true that networks have grown, as the reduction of transactions costs has made it possible for companies to outsource many functions. Yet, innovation and flexibility can thrive under administrative guidance just as mass production did. Rumours of the death of the large corporation have been greatly exaggerated (Piore 1994a, b).

Another problem with the picture of flat global networks is that flatness and global reach are often in conflict. The global reach of production networks is to a substantial extent an administrative accomplishment: distance has been shrunk not only by improved technology, but also by corporate organization. Managerial hierarchies and standardized procedures are, among their other functions, distance-spanning devices. Conversely, the most robust and innovative *flat* networks are disproportionately local. One role played by large MNEs is to link such local clusters and networks to the larger world, extending them beyond national boundaries (Amin and Thrift 1992; Enright 2000). This is an increasingly important role, as we will see, because the reductions of transactions, communication and transportation costs, where they are evident, are not accessible everywhere and to any firm, while they are relatively readily accessible to large MNEs.

1.2.2 Evolving Dinosaurs and the 'Natural' Selection of Geography

The multinational enterprise can be broadly defined as a multiplant, multi-activity firm that engages in the foreign production of goods and services, as well as in a variety of other businesses carried out across national borders.[1] In so far as MNEs own and control their cross-border activities, the explanations of the determinants and motivations of multinational behaviour have to take into account ownership, control, organization and the location of such operations (Dunning 1993). The first three aspects, namely ownership, control and organization, were almost entirely ignored in the economic theory of the first half of the twentieth century, which had largely assumed that factors of production were brought together by perfect and costless market mechanisms. At the same time, the geographical patterns of multinational activity were broadly and indirectly addressed by classical and neoclassical theories of the international allocation of resources and endowments. Indeed, the international geographical distribution of factor endowments was, and in some respects still is, assumed to be a major determinant of cross-border flows in both old and new theories of international trade and production.

During the decades immediately after the Second World War, MNEs were operating within the Bretton-Woods system of largely closed economies in which currency parities were fixed. Within this framework, neoclassical explanations of trade without any role for geography naturally flourished. The resulting theories of international business and multinational enterprise which emerged as a hybrid agenda largely from Vernon's (1966) product cycle model (and later encompassed in John Dunning's OLI eclectic paradigm, which, as we will see, plays a paramount role in our attempts at providing the theoretical building blocks for interpreting the relationship between MNEs and space) were, not surprisingly, based almost entirely on the nature, organization, objectives and behaviour of mid-twentieth century US manufacturing firms, the dominant form and mode of multinational organization. However, this particular institutional system and its associated analytical framework, started to break down following a series of events that began with the Eurodollar waves of migration of US merchant banks into London in the 1960s and then continuously on through the late 1970s and early 1980s, and was maintained with the rise of the Japanese and East Asian manufacturing MNEs in the 1980s. These modes of multinational activity were quite different from the received frameworks, either because of sectoral differences (services versus manufacturing), because of different organizational behaviour and production philosophies (Just-in-Time JIT and Total Quality Management TQM), or because of the different and wider arrays of functions included in multinational expansion.

Since the 1960s, issues other than factor endowments therefore became increasingly important to international trade and industrial economists, most notably, market structures, economies of scale, knowledge and technology. However, in geographical terms, as these new elements were progressively incorporated into a variety of theoretical approaches devoting attention to the MNE, they were treated along the lines of the unspecified macro-geography inherited from the international trade theory. By and large, in this schema, the competitive advantages of firms coincide with those of the country of origin, and the attraction forces pulling MNE operations were those of the country of destination. National and even supranational geographies have been the obvious level of analysis when considering both the determinants and effects of MNE activity. Cities, sub-national regions and differentiation *within* nation states were largely absent.

Institutional and technological environments have changed radically since the late 1980s, and multinational corporations, the supposedly inflexible dinosaurs, have changed accordingly, and to a much larger extent and rather more quickly than many other types of firms. The modes of international investment, the organization and management of intra-firm vertical and horizontal relationships, the types of affiliation linkages, the diversification and distribution of functions, the integration of subsidiary objectives into the overall goals and strategy of the MNE, have all gone through substantial and rapid changes. These new organizational modes have occurred within MNEs, and are also seen in terms of their external relationships. Increasingly complex and systemic forms of integration of international operations have involved external firms, often SMEs, which are connected through contractual relations (outsourcing) to the global production networks led by large corporations. Also, the growing degree of complexity in MNE forms and organization has made clear that the boundaries and definitions of multiproduct and multi-technology MNEs with respect to individual industrial sectors and technological fields are increasingly blurred.

As we argue in later chapters, the identification of the MNE's spatial location, and its characteristics with respect to geography, industry structure, knowledge and technology flows, is a result of complex interactions between firm, industry, organizational and knowledge characteristics. The simple host-home country dichotomy becomes insufficient. This is particularly so in the developed parts of the world, where host and home may actually overlap to a great extent, and where the locations which are host to a major presence of MNEs are also those which are most generally connected by outward linkages. The location of corporate headquarters of large MNEs nowadays has little geographical connection with the location base for specific business units and operations. An MNE's multiple locations are best understood as a specific sub-national area – a region, city or industrial cluster – where a firm locates its core functions, including strategic decision-making, research and development, and often some core manufacturing or services activities. Each function (or activity) of an MNE tends to favour different spatial characteristics. Yet, opposing this dispersion force are vertical linkages between stages of production which can encourage firms to co-locate different activities in the same location (Defever 2006).

Flows, and particularly flows of knowledge, are increasingly understood as being bi-directional or multi-directional, with concepts such as 'openness' and 'connectivity' coming to replace terms such as 'inward' and 'outward' flows. At the same time, the stable relationship between ownership and control, which has long been understood as being problematic when looked at in terms of the divide between investor and executives, has been disturbed both along the supply chain and within corporate hierarchies (Ietto-Gillies 2005). In outsourcing and offshoring strategies, ownership changes but control of the value chain activities is largely retained through various means of pressure on the suppliers and their competitive bidding (e.g., narrow transfers of technology, strict specifications of the product, tight supplying schedules, etc.). Conversely, in vertical integration strategies ownership is not altered, but the distribution of control within the MNE can vary greatly, with different degrees of autonomy of the affiliates and subsidiaries that can lead to intra-firm competition and even to a certain degree of restraint in the powers of the central headquarters of the MNE. Such changes have had huge implications for the location choices and advantages of MNEs, which are increasingly dependent on the balance between technological competencies and capabilities *within* and *outside* the firm or, as we will describe it later, on the integration of various sources of knowledge which are internal and external to the firm. Also, different geographies have emerged in relation to different types of multinational vertical integration, such as more centralized/decentralized control within the firm, the unbundling of headquarters and core functions, and hierarchical versus heterarchical structural forms and internal organization (Desai 2009).

Thus, as inward and outward flows have largely ceased to describe what happens at the sites of MNE activity, also the dichotomy between 'hierarchies versus networks', which itself is a recent substitute for the traditional poles of the market and hierarchy, is losing its analytical usefulness. Neither large vertically integrated MNEs nor small specialized producers' networks represent the ideal institutional structure to cope with the rapidly changing competitive advantages. In the words of Robertson and Langlois (1995, p. 543): 'The relative desirability of the various structures, then, depends on the nature and scope of technological change in the industry and on the effects of various product life-cycle patterns'.

MNEs have changed more than other firm types, whether SMEs or large multiplant uni-national firms, because of the more intense interaction between MNEs and global institutional, organizational and technological changes. The causality nexus between MNE evolution and globalization processes is not straightforward. Globalization exerts pressure on MNEs, but MNEs are one of the leading agents of globalization.

One of the crucial aspects of the current phase of economic globalization lies in the new modes of creating and diffusing new knowledge and technology (Box 1.3). The central role played by contemporary multinational corporations in such processes has been described in the evolutionary economics literature, which stresses the metamorphosis of the MNE

BOX 1.3 SOME FACTS AND FIGURES: SKEWNESS IN GLOBAL R&D AND TOP MNEs IN THE WORLD

Global R&D is overwhelmingly accounted for by the advanced economies. Global public plus private sector R&D expenditure grew rapidly during the 1990s to some $677bn in 2002, of which some $450bn is in the form of global private R&D expenditure (UNCTAD 2005). In 2002, the advanced economies accounted for over 90 per cent of this global R&D expenditure, an even higher proportion than the share of global outward FDI which is accounted for by the same group of countries. In 2002, just the top nine origin countries for outward FDI accounted for 87 per cent of both total global R&D and global private sector R&D (UNCTAD 2005).

Yet, even allowing for these impressive figures, the role of multinationals in the generation of world R&D has grown markedly. The major multinational firms account for an ever-increasing share of global R&D. By 2005 the 700 largest R&D expenditure MNEs were estimated to account for over $310bn in R&D investment, which represents around 46 per cent of all global R&D and 69 per cent of the $450bn of global business sector R&D expenditure (UNCTAD 2005). As such, almost half of global R&D and over two thirds of global private sector R&D is concentrated in less than 1 per cent of the world's 78,000 MNEs. Multinational R&D is therefore extremely skewed according to firm size when measured at the global scale.

As well as being concentrated in a tiny number of firms, such a high degree of concentration or skewness in global multinational R&D is in fact also manifested in terms of both sector and geographical origin. As for the former, of the 700 MNEs dominating global R&D more than half are in just three sectors, namely IT hardware, the automotive industry, and the pharmaceuticals and biotechnology industry (UNCTAD 2007). In terms of the nationality of origin, we have already seen that all forms of R&D activity are dominated by the most advanced economies and it is therefore not surprising that MNE activity closely reflects these overall knowledge patterns. In fact, the US is the individual largest single source of R&D expenditure, while over 80 per cent of the 700 largest R&D expenditure MNEs come from only five countries, namely US, Japan, Germany, UK and France. These firm concentration patterns are therefore even more skewed than for FDI flows in general. This is reflected in the fact that, in 2001, the top 100 non-financial MNEs were all from developed economies, including Hong Kong, Singapore and Korea. In 2009, only four out of the world's top 100 non-financial MNEs were from developing economies (two from China, one from Malaysia and one from Mexico), and these countries counted none of the world's top 50 financial MNEs (UNCTAD 2010). Tables 1.3 and 1.5 report the top-20 non-financial and financial MNEs ranked in terms of the geographical spread of their foreign affiliates. As we can see, only one multinational corporation from a developing economy, namely Mexico, appears among the top-20 non-financial MNEs. On average, we see from Table 1.4, the top non-financial MNEs from developing countries are still dominated by those from the highly advanced economy of Hong Kong in China, and even these are far smaller and far less internationalized in general than those from developed countries (UNCTAD 2010).

Table 1.3 The world's top 20 non-financial TNCs, ranked by Transnationality Index, 2008[a]

Ranking by: Foreign assets	Ranking by: TNI [b]	Corporation	Home economy	Industry[c]	TNI[b] (per cent)
37	1	Xstrata PLC	United Kingdom	Mining & quarrying	93.2
87	2	ABB Ltd.	Switzerland	Engineering services	90.4
40	3	Nokia	Finland	Electrical & electronic equipment	90.3
66	4	Pernod Ricard SA	France	Food, beverages and tobacco	89.1
67	5	WPP Group Plc	United Kingdom	Business services	88.9
3	6	Vodafone Group Plc	United Kingdom	Telecommunications	88.6
72	7	Linde AG	Germany	Chemicals	88.3
13	8	Anheuser-Busch Inbev SA	Netherlands	Food, beverages and tobacco	87.9

Table 1.3 (continued)

Ranking by: Foreign assets	TNI [b]	Corporation	Home economy	Industry[c]	TNI[b] (per cent)
46	9	Anglo American	United Kingdom	Mining & quarrying	87.5
10	10	ArcelorMittal	Luxembourg	Metal and metal products	87.2
27	11	Nestlé SA	Switzerland	Food, beverages and tobacco	87.1
88	12	Air Liquide	France	Chemical/Non-metallic mineral products	86.9
61	13	Liberty Global Inc	United States	Telecommunications	86.2
58	14	Astrazeneca Plc	United Kingdom	Pharmaceuticals	85.4
100	15	Teva Pharma-ceutical Industries Limited	Israel	Pharmaceuticals	84.4
41	16	Lafarge SA	France	Non-metallic mineral products	84.2
57	17	Volvo AB	Sweden	Motor vehicles	82.3
25	18	Hutchison Whampoa Limited	Hong Kong, China	Diversified	82.0
55	19	Cemex S.A.	Mexico	Non-metallic mineral products	81.6
4	20	BP PLC	United Kingdom	Petroleum expl. /ref./ distr.	81.0

Source: *UNCTAD/Erasmus Universities database.*

[a] All data are based on the companies' annual reports unless otherwise stated.

[b] TNI is calculated as the average of the following three ratios: foreign assets to total assets, foreign sales to total sales, foreign employment to total employment.

[c] Industry classification for companies follows the United States Standard Industrial Classification Commission (SEC).

[d] In some cases foreign employment was calculated as the share of foreign employment in total employment of the previous year to total employment of 2008.

Table 1.4 The top 20 non-financial TNCs from developing and transition economies, ranked by Transnationality Index, 2008[a]

Ranking by: Foreign assets	Ranking by: TNI[b]	Corporation	Home economy	Industry[c]	TNI[b] (per cent)
39	1	First Pacific Company Limited	Hong Kong, China	Electrical & electronic equipment	99.0
35	2	China Merchants Holdings international	Hong Kong, China	Diversified	96.8
73	3	Guangdong Investment Limited	Hong Kong, China	Diversified	95.1
93	4	Road King Infrastructure Limited	Hong Kong, China	Transport and storage	90.4
60	5	Li & Fung Limited	Hong Kong, China	Wholesale trade	90.3
87	6	Shougang Concord International	Hong Kong, China	Metal and metal products	89.1
34	7	China Resources Enterprises	Hong Kong, China	Petroleum expl./ref./ distr.	89.0
49	8	Sappi Limited	South Africa	Wood and paper products	85.2
1	9	Hutchison Whampoa Limited	Hong Kong, China	Diversified	82.0
95	10	Techtronic Industries Company Limited	Hong Kong, China	Other equipments goods	81.8
3	11	Cemex S.A.	Mexico	Non-metalic mineral products	81.6
63	12	Acer Inc.	Taiwan Province of China	Electrical & electronic equipment	79.9
59	13	Medi Clinic Corp. Limited	South Africa	Other consumer services	78.7

Table 1.4 (continued)

Ranking by: Foreign assets	Ranking by: TNI [b]	Corporation	Home economy	Industry[c]	TNI[b] (per cent)
75	14	Beijing Enterprises Holdings Ltd.	China	Diversified	77.0
54	15	Suzlon Energy Limited	India	Diversified	75.7
62	16	Pou Chen Corp	Taiwan Province of China	Other consumer goods	71.6
29	17	Hindalco Industries Limited	India	Diversified	71.6
15	18	Tata Steel Ltd.	India	Metal and metal products	69.8
98	19	TPV Technology Limited	China	Wholesale trade	69.8
24	20	Qatar Telecom	Qatar	Telecommunications	69.7

Source: *UNCTAD/Erasmus Universities database.*

[a] All data are based on the companies' annual reports unless otherwise stated.

[b] TNI is calculated as the average of the following three ratios: foreign assets to total assets, foreign sales to total sales, foreign employment to total employment.

[c] Industry classification for companies follows the United States Standard Industrial Classification Commission (SEC).

[d] In some cases foreign employment was calculated as the share of foreign employment in total employment of the previous year to total employment of 2008.

However, to explain more recent trends, in 2006 74 per cent of the international policy measures instituted to facilitate FDI came from developing countries (UNCTAD 2007). These policy changes focus on issues such as reducing corporate taxes, structural liberalization and deregulation. As a result, developing countries are rapidly increasing their shares of FDI in other developing countries: this is particularly noticeable in the case of Asian countries (UNCTAD 2010). Yet, even though FDI from

Table 1.5 The Top 20 financial TNCs ranked by Geographical Spread Index (GSI), 2009[a]

Rank 2009	GSI[b]	Financial TNCs	Home economy	Affiliates			
				Total	Number of foreign affiliates	I.I.[c]	Number of host countries
1	75.3	Citigroup Inc	United States	796	601	75.5	75
2	69.4	BNP Paribas	France	755	596	78.9	61
3	68.7	Allianz SE	Germany	546	444	81.3	58
4	65.0	Generali Spa	Italy	278	261	93.9	45
5	64.5	Societe Generate	France	380	277	72.9	57
6	61.0	UBS AG	Switzerland	615	602	97.9	38
7	60.8	Unicredito Italiano Spa	Italy	853	829	97.2	38
8	59.5	HSBC Holdings PLC	United Kingdom	741	485	65.5	54
9	58.3	Axa	France	542	485	89.5	38
10	57.5	Deutsche Bank AG	Germany	949	804	84.7	39
11	53.2	Zurich Financial Services	Switzerland	265	259	97.7	29
12	53.1	Credit Agricole SA	France	312	191	61.2	46
13	51.5	Credit Suisse Group AG	Switzerland	209	179	85.6	31
14	50.8	ING Group NV	Netherlands	884	506	57.2	45
15	48.2	Morgan Stanley	United States	147	118	80.3	29
16	47.9	Swiss Reinsurance Company	Switzerland	91	87	95.6	24
17	47.4	The Bank of Nova Scotia	Canada	118	106	89.8	25
18	47.0	Banco Sant-ander SA	Spain	390	308	79.0	28

Table 1.5 (continued)

Rank 2009	GSI[b]	Financial TNCs	Home economy	Affiliates Total	Number of foreign affiliates	I.I.[c]	Number of host countries
19	47.0	Natixis	France	217	137	63.1	35
20	46.4	Nomura Holdings Inc	Japan	104	86	82.7	26

Source: UNCTAD/HEC Montréal.
[a] Data on affiliates is based on the Dun and Bradstreet 'Who owns Whom' database.
[b] GSI, the "Geographical Spread Index", is calculated as the square root of the internationalization index multiplied by the number of host countries.
[c] II, the "Internationalization Index", is calculated as the number of foreign affiliates divided by the number of all affiliates.

developing countries into other developing countries is growing quickly, by far the most popular location for the affiliates of the top one hundred MNEs from developing and transition countries is the UK, followed by the US (UNCTAD 2007). These two economies account for more than twice the number of affiliates of these developing economies' MNEs than any other country.

from mere 'vehicle' of technical knowledge to 'creator' of new technology (e.g., Cantwell 1994; Cantwell and Iammarino 2003). MNEs are today the largest source of technology generation, transfer and diffusion in the world. In comparison with all other firm types, the share of new technologies produced globally by MNEs is increasing; they are associated with a higher probability of entry into new and more distantly related technological fields; they account for the great bulk of expenditure on knowledge-creating and skill-enhancing activities, and of trade in technology and technology-intensive products (UNCTAD 2009). MNEs' access to a broad variety of sources of new knowledge, both intra- and inter-firm, provides massive opportunities to acquire new competitive advantages and their internal and external sources of knowledge and information have obvious geographical manifestations.

Recent years have also seen economics direct new attention to the phenomena of spatial concentration, and one result has been an increasing body of evidence of the spatially bounded character of knowledge creation and spillovers. This spatial boundedness is a major explanation for the emergence and reinforcement of core-periphery forces and regional divergence, especially in the face of processes of economic integration and globalization (e.g., Fagerberg et al. 1997; Rodríguez-Pose 1994, 1998; Caniëls 2000; Verspagen and Caniëls 2001; Rodríguez-Pose and Crescenzi 2008; Crescenzi and Rodríguez-Pose 2009; Paci and Usai 2009). Indeed, the spatial proximity of firms and people within individual urban areas becomes over time relatively more important as a source of knowledge spillovers and agglomeration economies. Moreover, dominant agglomerations also appear to become more specialized within the global economic system (Gordon 2002), such that different localities are likely to be progressively more differentiated in terms of their competitive advantages.

The importance of specific geographies also renders problematic the traditional scholarly separation between the questions regarding the causes and impacts of multinationality (Ietto-Gillies 2005). The valuation of MNE effects is inherently connected to the determinants of the MNE existence, the why, how and where the multinational firm operates. As we will see later, in spite of some ground-breaking insights provided by Vernon and, in particular, by Hymer, the international business theory has not explicitly distinguished between the different geographies of MNEs activity, whether with respect to the process of becoming multinational, or with reference to the effects of foreign investment (McCann and Mudambi 2004). Moreover, the national geographical units of study which are still the exclusive focus of international data and statistics have tended far too often to be too general for any meaningful level of spatial analysis. Such a lack of geographical specificity has also therefore precluded any coherent analysis of the changing dynamics of spatial industrial configurations under the pressures of globalization. Yet, these dynamics narrow the scope of the relationships between MNEs and technological and innovation processes, and many aspects of these relationships which, by definition, are often highly embedded in specific geographical settings. The specificity of MNE geography-knowledge-technology interactions is increasing, while new developments in theory and data lag far behind.

At the same time, these developments also provide for the possibility of more sophisticated, subtle and imaginative solutions on the part of MNE strategies, as well as for places. Geography can thus be viewed as a powerful selection mechanism that may (or not) provide conditions favourable to meet the requirements of institutional, organizational and technological change under globalization (Cohen and Levinthal 1990; Lambooy and Boschma 2001; Boschma 2005). Geography will become ever more important as a competitive consideration for international businesses, and all the available evidence suggests that in many ways MNEs will increasingly contribute to making the world a more uneven and more differentiated place (McCann 2008).

1.3 THE STRUCTURE OF THE BOOK

This book proposes an integrated perspective in which knowledge and innovation are regarded as the driving forces underlying the competitive advantage of firms, and location and space provide the platforms to access the requisite knowledge and technology generating that competitive advantage. This perspective is based on a synthesis and integration of rapidly growing and disparate literatures in the fields of economic geography and location theory, innovation and technological change, and international business and management.

The book is divided into two Parts. Part I – Multinationals, location and innovation: foundations and extensions – includes Chapters 2 to 4, and is intended to provide an overview of the theoretical and conceptual building blocks that, combined together, can offer a sound analytical framework for looking at the economic geography of the multinational enterprise.

Chapter 2 briefly reviews some of the main interpretations provided by both economics and international business studies on the determinants and effects of multinational enterprises' activities, with a specific and highly selective focus on those that, directly or indirectly, offer more insights to the issues of geography and location. Here we argue that although the Ownership-Location-Internalization (OLI) eclectic paradigm proposed by John Dunning has demonstrated great usefulness and versatility for over almost four decades since its original formulation in the mid-1970s, its Achilles heel has lain in the weakness of the 'L'. The theoretical roots of the L advantage in the OLI have been identified either in the trade domain (e.g., Hirsch 1979; Beckman and Thisse 1986; Brainard 1993), or in the new economic geography (e.g., Krugman 1991b; Fujita et al. 1999). In these approaches, as well as in the OLI, the geographical element is more virtual than real, with space occupying a sort of 'black box' which is rarely helpful for elucidating the variety of spatial industrial configurations. More importantly, innovation and technological processes, which are central to the transformation of MNE's structure and behaviour over time, are left somehow outside the L, even though from widespread empirical evidence we know that many aspects of knowledge, information and technology are highly geographically specific. The models discussed in Chapter 2 are all well-known to international business and management scholars, but much less so to many economic geographers or economists analysing urban and regional phenomena. While the primary aim of this chapter is to bring these MNE knowledge-internalization and organizational issues to the attention of economic geographers, regional scientists and urban economists who in general are largely unschooled in these matters, our secondary aim is also to demonstrate to both international business and management scholars as well as to economic geographers and regional economists that many of the theories of MNE behaviour originated from observations within urban economics. One of the key themes of this book is that these original observations from economic geography which subsequently were adopted and adapted in international business literature are still basically valid and as such warrant much further investigation.

Chapter 3 reviews traditional location theory with the aim of highlighting the relevance to multiplant and multinational enterprises of very orthodox and standard approaches to firm location behaviour. We will consider the relationships between firm location choices, factor costs, and firm profitability from the perspective of three traditional types of firm location models, namely the Weber, the Hotelling and the Salop models, each of which will provide us with different insights into the nature of firm location strategies. In addition, by looking at leader-follower and clustering behaviour, and the possible shifts from hierarchical versus heterarchical organizational structures in multiplant and multinational firms, in this chapter we will start to take on board the issues of local knowledge accessibility, thereby allowing for an extension of the analysis to more comprehensive and modern-looking MNE types. The models presented in this chapter are well-known to economic geographers, regional scientists and urban economists, but very much less so to international business or management scholars. While the aim of this chapter is primarily to bring to the attention of international business and management scholars the value in applying microeconomic location theory to the spatial behaviour of multiplant and multinational enterprises, the secondary aim of the chapter is also to demonstrate to economic geographers and regional scientists the fact that many elements of their standard toolkit are applicable for analysing the behaviour of firm types which are well outside of their typical domain.

Taken together, Chapters 2 and 3 are intended to demonstrate that the international business and management literatures analysing multinationals and the economic geography and regional science literatures share far more in common than is typically understood by either literature. Moreover, the common ground that these literatures all share centres on the issues of knowledge – and its various manifestations in the form of technology and innovation – and the treatment of the transactions costs faced by firms striving to acquire, generate and exploit this knowledge.

For the purposes of this book these are the commonly-shared avenues which require much deeper exploration.

Chapter 4 therefore firstly reviews the main tenets of the literatures surrounding the economics of the firm growth and of technological change, in order to clarify their treatment of innovation sources, competencies, capabilities and learning processes. This chapter aims at bringing the micro-level of analysis, as represented by the firm, up to the meso-level of analysis as represented by the industrial cluster and the regional economic and innovation system. Following on from Chapter 3 we know that in terms of microeconomic location theory the MNE is seen to be simply a special case of the firm in general, and in particular of the multiplant firm. However, Chapter 4 emphasizes the importance from the literature of understanding the critical role played by knowledge acquisition and exploitation in MNE behaviour. Therefore, an examination of the growth impacts of an MNE's location behaviour on the MNE itself due to its locating in a particular region is seen to be a function of the number and variety of the sources of knowledge accumulation which the MNE derives from the region. Similarly, the growth impact on the region in which an MNE locates depends on the number and variety of the knowledge outflows from the MNE to the surrounding territory. Mutually reinforcing firm-environment knowledge flows must be two-way in order to foster sustained local learning and innovation processes. Understanding how, when and where these strong local complementarities arise between learning processes and sources of innovation, irrespective of whether they are internal or external to the firm, therefore calls for the identification of those commonalities which are displayed by *groups* of firms in the same industry or location. It is the innovation systems literature which explores these issues and based on the earlier chapters, our argument here is that it is uniquely the *regional* innovations systems literature, rather than other innovations systems literatures, that best captures these commonalities.

Part II of the book – Multinationals and the changing economic geography of globalization – comprises Chapters 5 to 8, and aims to provide a more ground-based view of the evolutionary changes in firms' structure and organization, and the associated changes in the spatial systems at different levels of analysis which modern globalization induces.

Chapter 5 builds on the micro foundations of firm location and innovative behaviour examined in the earlier chapters, focusing in particular on MNEs' technological capabilities and strategies, and integrates these with the meso perspectives of regional and local industrial systems. This allows us to explore the new combinations of internal and external sources of innovation which are accessible to MNEs in the modern era of globalization. By taking into account the ways in which MNEs can interact with the local industrial environment, and the degrees of openness and attractiveness of regions and clusters, we are able to move forward our understanding of their dynamics and their likely directions of global competition. The chapter presents a spatial classification scheme based on the transactions costs relationships evident among co-located firms, and then expands this into a broader and more comprehensive taxonomy defined according to the knowledge, innovation and transactions costs characteristics of the cluster or region. The spatial and industrial categories, which are also illustrated with empirical examples, provide steps forward in uncovering the evolutionary trajectories of different spatial economic configurations, all of which may exhibit a major presence of multinational investment. The objective of this chapter is to provide some new analytical tools for assessing the degree of fit between the MNE and its local environment. This taxonometric schema allows us to better evaluate the likelihood of the development of local innovative networks and the direction of the co-evolutionary processes linking firms and their spatial-industrial structures. The schema described in this chapter builds on earlier widely-published research. However, the extended taxonomy described here is both more detailed and also more comprehensive than previous frameworks, and explicitly positions issues of knowledge and innovation at the very centre of the analysis.

In order to appreciate the importance of the extended taxonomy provided in Chapter 5 and also to situate these discussions in a real-world context, a historical overview of globalization is essential in order to understand how firms, activities and their geographical location have co-evolved over time and continue to do so. Chapter 6 therefore expands the discussion developed in the previous chapters in the context of historical trends in globalization. From this perspective which integrates knowledge, technology, innovation and geography, the MNE is seen to be the most important actor in modern globalization processes. The chapter shows how in order to keep abreast with global competition the interdependence between institutional, organizational and geographical changes reflects the increasing pressures on firms to exploit the complementarities resulting from the systemic nature of knowledge and technology. However, what emerges is also that such changes, and the potential benefits that they trigger, are disproportionately associated with particular types of places and regions, both at supra-regional and at sub-national regional levels. Importantly, following the arguments in this book we see that MNEs play an ever increasing role in determining the distribution of these costs and benefits across these different levels of geography.

In terms of economic geography, one of the dominant features of the current wave of globalization is the emergence of a new set of relationships between the location of knowledge assets, the role of particular city-regions as knowledge sources, and the location behaviour of MNEs. Chapter 7 shows how current competitive advantages are increasingly associated with cities or city-regions being the major open centres of knowledge and exchanges. The examples provided illustrate the difference between simple 'connectedness', defined in terms of the architecture of transport and communications infrastructure, and the much broader concept of 'connectivity', which is a behavioural concept incorporating the capability of individuals, firms, organizations and institutions to interact and engage across different locations and within networks. Connectivity emphasizes the substantial degree of two-way inward and outward openness of certain regions and also the firms embedded in these regions, in terms of many behavioural and organizational dimensions of knowledge connectedness. As we demonstrate in this chapter, many features of the current stage of globalization are seen to be making the world more uneven in a variety of different ways, whereby global networks and local agglomerations can act as complementary forces strengthening each other in determining the growth and productivity 'spikes' of the world economy.

Finally, Chapter 8 focuses on new trends emerging with respect to the combined importance of MNEs and developing and transition economies, with particular reference to the BRIICS countries. The increasing role in R&D played by the foreign affiliates of MNEs, on the one hand, and the growing attractiveness of developing economies as locations for MNE foreign affiliates, on the other hand, have sharply accelerated their pace of development and integration in the global economy in the last two decades. These trends provide significant opportunities for developing economies to improve and upgrade their competencies and capabilities, and their capacity to accumulate knowledge and technological assets via multinational conduits. However, agglomeration phenomena and sub-national differentials are also increasing at an impressive speed than was previously anticipated. The major cities in these emerging economies are expected to play an ever more prominent role in the global city rankings, and their surrounding regions will progressively acquire the status of core regional locations in the global economy. The distinction between winner and losers, both within and across national boundaries, is therefore likely to be further exacerbated by these new drifts.

NOTE

1. See Ietto-Gillies (2005) for a comprehensive review of FDI forms and alternative modalities of MNE business operations.

PART I

Multinationals, location and innovation: foundations and extensions

2. Old and new(er) theories of multinational enterprises: selected perspectives and the search for location

2.1 INTRODUCTION

As we stated in the introductory chapter, the major questions about multinationality have typically focussed on the determinants – i.e. the *why*, *where* and *how* firms become multinational – and the effects – i.e. the change on the economies invested in by the MNE operations and also the impacts on the MNE firm itself which are induced by the MNE's expansion and activity. The focus of the present chapter is on the geography of the multinational enterprise as discussed in the theoretical literature. Even though one of the core questions regarding MNEs is precisely that of where their different operations take place, the location of MNEs has so far been largely neglected.

The theoretical approaches discussed below are not meant to provide an exhaustive survey of the literature on the determinants and impacts of multinational activity, but rather to explain in broad terms how geography and space have been dealt with implicitly or explicitly in scholarly thinking on MNEs. Beyond those reported in the present chapter, many other attempts to theorize multinationality find their roots in economics and finance (e.g., Aliber 1970; Rugman 1979; Culem 1988; Kojima 1973, 1982b, 1990), as well as in other disciplines such as political economics (e.g., Cowling and Sugden 1987a) or organizational theory (e.g., Aharoni 1966). A general aspect, however, which is often indicated as a problem with the vast literature on MNEs, is the lack of a unifying or holistic approach to multinational behaviour.

In the presentation of selected theoretical perspectives which are particularly interesting for our purposes here, we will make use of a well-known broad and comprehensive framework which brings together different theoretical building-blocks and explanations of a wide range of aspects of multinationality. John Dunning's framework, also known as the *eclectic*

OLI paradigm, has proved not only to be rather helpful for the purpose of dealing analytically with the MNE phenomenon, but also to be flexible enough to adapt to change over time. Thus, it will be used throughout the chapter as a guideline to the discussion and as a benchmark for assessing the state of the art of the theoretical thinking on the spatial dimension of MNEs.

2.2 THE ECLECTIC OLI PARADIGM

The Ownership-Location-Internalization (OLI) paradigm was originally formulated by John Dunning (1977, 1979, 1981, 1988a) and subsequently updated and adapted by Dunning himself (e.g., 2000, 2003a, b, 2009b) and a number of other scholars. The eclectic OLI paradigm has had the capacity to accommodate and compare different major economic, business and managerial theories aimed at explaining the Ownership (O) advantages, i.e. *why* firms become multinational; the Location (L) advantages, i.e. *where* firms go to internationalize their activities; and the Internalization (I) advantages, i.e. *how* firms carry out their multinational experience. The necessary and sufficient conditions for MNE international operations are thus subsumed in these three analytical categories:

- *Ownership advantages:* the firm must possess some specific tangible or intangible assets vis-à-vis its domestic competitors, as a result of which the firm could achieve either lower costs or higher product quality. The possession of unique technological competence, or of reputation for product excellence, for instance, is a source of market power: such assets can be replicated easily across different locations, and the marginal cost of transferring them within the firm is usually very low. Although O advantages are seen as firm-specific, they are often intended to reflect those of the MNE home country.
- *Location advantages*: the control of distinctive assets per se, however, is not sufficient: to be able to cope with the costs of doing business abroad, the firm must combine its O advantages with some incentive to locate at least part of its operations outside the home economy. Locational advantages are inputs, intermediates, services or other tangible and intangible assets and characteristics specific to some foreign location. Natural resources, capital and labour inputs, skills and capabilities, as well as cost differentials, market size and agglomeration economies, are all examples of the location benefits available in the host economy.

- *Internalization advantages*: finally, even the combination of O and L advantages is not sufficient to explain why the firm does not exploit them by simply using arm-length markets (i.e. through licencing, franchising or establishing alliances with other firms in the host location). Internalization advantages are market-replacement activities that, through the direct control over their assets, confer their own hierarchical advantages to the MNE firm. This category highlights the crucial role of market imperfections in the exploitation of O and L advantages (Dunning 2001).

The simultaneous combination of the O, L and I advantages has to be seen as the necessary and sufficient condition for an MNE's existence and its dynamics over time. In order to be in the position to exploit the benefits associated with these O, L and I advantages the multinational firm must have the financial, managerial and organizational corporate resources that allow it to maintain and adapt its international multi-located operations to changes over time. Thus, dynamic ownership advantages are required in order to modify locational and organizational strategies, given a change in the other types of advantages (Dunning 1995, 1998, 2000). In the words of John Dunning (1998 and 2009b, p. 5): 'The OLI triad of variables . . . may be likened to a three-legged stool: each leg is supportive of the other, and the stool is only functional if the three legs are evenly balanced.' The interdependence among the O, the L and the I, as we will see, is a critical argument for the analysis of the geography of MNEs. Borrowing again from Dunning (1998 and 2009b, p. 5): 'the geography of international business activity is not independent of its entry mode; nor, indeed, of the competitive advantages of the investing firms.'

Thus, beyond the claims of a lack of a holistic approach to MNEs, the OLI eclectic paradigm offers a general and simple framework for interpreting MNE activity by combining both micro- and macroeconomic perspectives, by allowing multiple levels and units of analysis, and by accommodating different theoretical strands under a unifying umbrella. One of the most remarkable properties of the OLI paradigm lies precisely in the balance between micro- and macro-foundations. The approach neither proposes itself as an analytical tool for examining the firm behaviour per se, nor does it point to particular foreign production modes. Indeed, it considers both those activities carried out in foreign locations by a country's own firms, and also foreign-owned operations carried out in the domestic economy. However, in this framework each particular activity performed across international boundaries is seen, at least potentially, as a consequence of specific advantages of the micro-units, which are the multinational firms themselves.

Also associated with the OLI paradigm, and central to our arguments here, is Dunning's classification of the different types of MNE activities (1993, 1994), which is based on his motivation to understand the establishment of foreign affiliates and subsidiaries according to the locational advantages offered by the host economy. Building on the earlier FDI typology of Behrman (1972), the classification identifies four main types of MNEs (see also Dunning and Lundan 2008):

- *Resource or asset seekers*: MNEs investing in particular locations abroad in order to have access to tangible or intangible resources and assets, such as raw materials, labour and skills, which are either unavailable in the home location or available in the host location at a lower cost than at home. It is critical, in this case, to draw a clear distinction between MNEs aiming at host locations' physical resources and cheap labour, which we will call *general-asset seekers*, and MNEs which are prompted to establish affiliates abroad specifically in response to the need to acquire particular technological capabilities, management and marketing expertise or organizational skills, here called *specific-asset seekers*.
- *Market seekers*: MNEs investing in order to supply local or adjacent markets via proximity to demand. Such MNEs are mostly motivated either by the emergence or growth of new or existing leading markets; or by the need to adapt products to local tastes; or by the existence of high transactions costs or trade barriers favouring 'tariff-jumping' local production in preference to exporting.
- *Efficiency seekers*: MNEs investing in order to rationalize and restructure previous investments which are either resource- or market-led. It is useful to distinguish two main types also in this category: *global value chain efficiency seekers*, which are the MNEs taking advantage of international differences in factor costs and endowments among locations typically at different stages of economic development; and *scale and scope economies seekers*, which are the MNEs looking for rationalization across different institutional settings, market and industrial structures, and polices, among locations with fairly similar levels of economic development.
- *Strategic-asset seekers*: MNEs which are engaging in foreign operations usually by acquiring the assets of foreign firms, and aiming primarily at advancing their long-term strategic objectives in terms of global capabilities and competitiveness. By purchasing overseas assets – and therefore localized knowledge, competencies and capabilities – MNEs in this category strengthen their own O advantages and/or weaken those of their competitors.

As said above, such a classification of MNEs is particularly significant here as it is based primarily on the L advantages. These different types of MNEs, which are not mutually exclusive as they can, and increasingly do, belong simultaneously to all four categories and also as we will see more in detail in Chapters 3 and 5, imply very different types of geographies.

In what follows we review some of the main theories of the MNE accommodated by the OLI paradigm, paying particular attention to the interactions between the triumvirate of the determinants of multinationality, and the associated geographical issues.

2.3 THE DETERMINANTS OF MULTINATIONAL ACTIVITY: O-ADVANTAGES

2.3.1 Hymer: Control, the 'Correspondence Principle' and the Locational Pyramid

Among the very first contributions which are considered to be the major pioneers in the recognition of the importance of ownership as an essential aspect of multinationality, is the work of Stephen Hymer (1960).[1] He focuses in particular on the identification of the nation-specific ownership advantages of multinational firms. Hymer's analysis is rooted in Joe Bain's industrial organization work, which aims at explaining differences among selected US industries on the basis of firms' entry and exit dynamics (Bain 1956). Bain shows that different industry-specific entry barriers, and thereby different kinds of protection in certain industrial sectors, provide different competitive advantages to firms in terms of ownership, increasing returns to scale, costs and market access.

Hymer applies Bain's approach to the study of various types of foreign direct investment (FDI)[2] activities of US firms. Hymer points out that multinational behaviour implies the existence of international flows of groups of activities and resources, including technology, capabilities, skills, and entrepreneurship. In contrast to the typical cross-border transactions analysed by economic theory up to that point,[3] Hymer demonstrates that such flows also require the same ownership and control across national boundaries. Indeed, his argument is that common ownership and control confer specific advantages to the firm, making it able to counterbalance the disadvantages of operating in foreign environments. Thus, the MNE emerges from market failure. As such, Hymer breaks with traditional economics, putting at the centre of his analysis the nature of the firm and its expansion process, rather than the market exchange activities carried out by representative Marshallian-type firms.

Hymer's thinking later develops through different stages in various directions, such as the growth of the firm, the internal integration of the firm, and the transactions costs faced by the firm (Hymer 1968, 1970a), and his thinking is particularly influenced by the seminal article *The Nature of the Firm* written by Ronald Coase (1937).[4]

In the first Hymer (1960) argument, the focus of the analysis is on the investor's control afforded by a common ownership and organizational system, which ensures the exploitation of advantages. The analytical approach makes no serious attempt to explain the geographical spread of production across many sites. However, a relevant implication for location is found in Hymer's pioneering recognition that foreign direct investment tends to be concentrated in certain industries across countries, rather than in specific countries across industries. In his 1968 article, which is considered by many to be path-breaking for its anticipation of the internalization and transactions costs approaches to the existence and growth of MNEs, Hymer links more explicitly his ideas on control with geography by suggesting that different segmented geographical markets can be treated as 'spheres of influence'.

Drawing both on the historical account given by Alfred Chandler on the evolution and surge of 'big business' in nineteenth and early twentieth century American industry (Chandler 1959, 1962), and also on the 'creative destruction' arguments of Josef Schumpeter (1934), Hymer formulates his 'law of increasing firm size' (Hymer 1970a). He observes three main stages of corporate evolution: from the competitive Marshallian firm, with a single function, industry and ownership, to the large national oligopolistic corporation, vertically integrated with dispersed ownership, to the modern conglomerate enterprise, multidivisional with cross-country located affiliates all coordinated by a centralized 'corporate brain' (Hymer 1970a, p. 442). Rehearsing Chandler and his co-author Fritz Redlich (1962), Hymer explains that during the evolutionary process of the firm towards the worldwide enterprise – the modern MNE – a pyramidal structure of three levels of business administration and decision-making could be detected. Level III, the lowest and widest level, is the level dealing with ordinary daily operations which are connected to production activities. Level II, which is the intermediate level, is generated by the separation of the head office functions from the field offices of the firm, and it controls Level III as well as coordinating the management at that lower level. The highest is Level I, the top level of management, which is responsible for the setting of the goals and strategies for the whole firm. The multinational firm is therefore rather different from other concepts of the firm. In the representative Marshallian type of firm typical in economic theory all three levels are embodied in the owner-entrepreneur. In contrast, in the case of the national corporation, Level II and I are split off from Level III, whereas in the modern MNE, Level I and II become completely separated, with the top Level I acting as the corporate brain, responsible for the firm's overall strategy.

This hierarchy of corporate decision-making is further discussed in subsequent work, in which the 'law of increasing firm size' is merged with the 'law of uneven development' (Pitelis 1991, 2005). In particular, Hymer (1972) suggests a 'correspondence principle', that is the existence of a direct relationship between the centralization of power and control within the MNE and the centralization of power and control in the world economy. The 'correspondence principle' thereby highlights the correlation between corporate growth and evolution, on the one hand, and uneven development across different parts of the world, on the other.[5] It is here, in the 'correspondence principle', where the critical connection between the MNE's ownership advantages and its strategic behaviour, and its spatial location and economic geography lies.

Hymer suggests that by applying the location theory to the Chandler-Redlich model, the 'spatial dimension of the corporate hierarchy' could be effectively analysed (Hymer 1972, p. 122). According to Hymer, the pyramidal structure of corporate control centralization translates directly into a *hierarchical structure of geographical locations*. Some of these locations are heavily dependent on others and it is this dependence relation which underlies the uneven spatial structure of economic development. Therefore, Level III activities will tend to be relatively evenly spread and distributed world-wide, according to the attractiveness of resources such as labour, markets and raw materials. It is this phenomenon which diffuses industrialization across less advanced economies. In contrast, Level II activities will tend to be concentrated in large cities, as they require primarily white-collar labour, as well as sufficient communication and information systems. At this level, MNEs from different industries will locate in the same cities, thereby giving rise to a strong spatial and functional concentration at the sub-national level, far stronger than that of Level III. According to the Hymer schema, Level I activities will be even more geographically clustered than those of Level II, as they need to be close to sufficient supplies of highly specialized services, capital markets, media and governments. The provision and exploitation of these high profile service relationships require high levels of connectivity, both locally and with the rest of the world. Hence, according to Hymer, the highest level functions of the leading MNEs will almost all be located in the world's major *global cities*, which themselves will be 'surrounded by regional sub-capitals' (Hymer 1970a, p. 446). Hymer (1972, p. 124) argues that, eventually, the major global cities which will be the home to the core headquarter functions and high level strategic planning of the world's MNEs will be New York, London, Paris, Bonn, Tokyo, Moscow and Beijing. From these highest levels, Hymer envisages that a geographical hierarchy will characterize the spread of MNE operations, with 'intermediate' and 'lower-level' activities distributed across lower tier urban centres and regions. In his words: 'geographical specialization will come to reflect the hierarchy of corporate decision making, and the occupational distribution of labour in a city or region will depend upon its function in the international economic system' (Hymer 1972, p. 124). Taking into account what has happened in the last two decades of globalization, as will be discussed in Chapter 7, Hymer's foresight is all the more remarkable given that he was writing during the height of the Cold War. An important qualification made by Hymer, however, and also pre-empting Krugman (1991a, b), is that while growth in the hinterland or periphery means growth in the centre, the reverse does not necessarily hold.

Hymer is therefore one of the very first scholars to understand the limitations of the traditional international economics approach which prevailed at the time. As he notes (Hymer 1968, p. 9), 'the theory of the firm and oligopoly theory have not been applied to the problems of trade and international investment yet'. Much of the later international business literature has explicitly and implicitly borrowed from Hymer's analysis (see the comprehensive assessment in Dunning and Pitelis 2008). Hymer's arguments have also been somewhat influential in certain types of geography and location studies (e.g., Dicken 1977; Norton 1979, 1987). Indeed, his 'correspondence principle' connects, by means of location theory, the centralization of control within the modern corporation and the spatial concentration of control in contemporary economic systems.[6] His work emphasizes the critical role of MNEs in spurring uneven regional development. Curiously, however, in spite of a few studies (e.g., Young et al. 1994; Bailey and Driffield 2002; Pitelis 2002a, b; Kottaridi 2005a) explicitly acknowledging Hymer's key importance in explaining the relationship between multinational activity and sub-national regional development, the international business literature itself has largely overlooked the *ante litteram* relevance attributed by Hymer to the interplay between spatial, and in particular urban, organizational and industrial structures, all of which underlie the locational choice and the economic geography of MNEs.

2.3.2 Vernon: Product Life Cycle and the Comparative Advantages of Nations (and the Metropolis)

The other seminal contribution to the economic theory of MNEs that is particularly relevant for our purposes here is the product life cycle (PLC) model of Raymond Vernon (1966). Adopting a microeconomic perspective, Vernon extends the traditional trade theory into one of international production, while acknowledging market imperfections. His focus is, on the one hand, on the sources of nation-specific ownership advantages afforded by factor endowments and market structures and, on the other hand, on the strategic behaviour arising from oligopolistic markets. Influenced by the emerging trade theories of that time which focussed on technology gaps (Posner 1961; Hufbauer 1966) and the life cycle of products (Kutznets 1953; Hirsch 1965), Vernon links innovation-based ownership advantages to features such as market size, income elasticities of demand and wage levels. Vernon's approach therefore departs from the prevailing neoclassical tradition in that he treats innovation as an endogenous phenomenon conditioned on structural issues, rather than as an exogenous phenomenon dependent on the efficiency of market clearing.

In the first stage of the PLC the innovating firm produces and sells its product in the home market. New products hinge on applications of scientific advances. Innovations are thus seen as country-specific, as they depend on the existence of a large local market to bear the high cost and risk involved in R&D. The new product starts to be exported abroad to countries with similar income and demand conditions as the home economy. In the second stage production gradually shifts to foreign countries with the largest domestic markets to supply the local demand directly through FDI. As innovations in Vernon's scheme are demand-driven, the catching up of demand through technology diffusion allows foreign competitors to gain the potential to develop their own similar technological capabilities, eventually eroding the innovating country firms' technological leadership. With rival firms continuing to enter the market, competition increases and drives down both price and profits. The cycle progresses in the third stage, with the standardization of production processes hindering further reductions in production costs. Firms gain a competitive edge by establishing production facilities in those countries where labour costs are least expensive, and the home country market is now predominantly supplied by imports of the product from foreign plants.

In Vernon's product life cycle analysis, and also to some extent as in Hymer's, the geographical location of international production follows a hierarchical pattern which is strictly linked to the ownership advantages that characterize each stage of the oligopolistic structure (Ietto-Gillies 2005). In the innovation-based oligopoly, the location of production is naturally placed in the economy where the innovation process firstly initiated. In this early stage, when technology is unstable, the demand for the new product is uncertain and technological capacity is vital. Location occurs in the metropolitan centres of advanced countries, especially the

US. This observation stems from the earlier work of Vernon on agglomeration economies in metropolitan regions such as New York (Vernon 1957, 1960). In this work he stresses that among the main causes underlying the growth of the core urban areas is the availability of a variety of business services, which previously were provided internally by the firm's business functions (Vernon 1957, p. 17). In mature oligopolies, advantages based on economies of scale, transport and marketing lead to a strong concentration of investment either in the home country of the MNE, or in similarly advanced economies. During this growth stage, when the number of competitors is increasing and firms are concerned with controlling their production costs, the location of the firm's investments spreads out of the initial agglomerations. Foreign production starts in countries where demand conditions are similar, and the restructuring process itself can lead to improvements in the original innovative product. In price competitive oligopolies, low cost locations in less developed countries become particularly attractive (Vernon 1966, 1974, 1979b). Thus, the mature or standardized stage of the product cycle, which is characterized by stable technology, long production runs, strong price competition and an intensity of unskilled labour, offers less developed countries the opportunity to successfully enter the international competitive environment by attracting MNEs of advanced economies into their developing local industries (see also Hirsch 1967).

The product cycle model has also seen wide applications in regional and urban analysis, basically predicting that the concentration of early stages of product development will be in urban areas, where high-skilled labour, external economies and top management are easily available, followed by subsequent shifts of standardized production to more rural and lower costs regions (Norton and Rees 1979). However, in this literature the structural and technological aspects of Vernon's arguments as they relate to MNEs are largely ignored. On the other hand, the same simplistic skilled–unskilled dichotomy of the international product cycle model is frequently applied also at the sub-national scale, thereby attracting criticism that these approaches have mainly failed to explain the processes of historical change in industry organization and spatial behaviour (e.g., Storper 1985; Taylor 1986, 1987; Mack and Schaeffer 1993). Among the criticisms of the PLC type of behaviour, two in particular seem to be most relevant for our purpose here. Firstly, the models tend to display a simplified and deterministic view of the innovation process, which is assumed to be a linear and uniform sequence from invention to a marketable product to standardization. Secondly, the locational issues arising in each of the stages of the product cycle model following the introduction of the new product have been disregarded. In many PLC models the spatial determinants of foreign and local firms' behaviour are reduced to mainly labour cost differences, thereby largely ignoring the firm's internal and external relationships and its spatial economic environment. This 'ambiguous enterprise context' (Taylor 1987, p. 82) does not allow us to distinguish between global and local competition; nor does it shed light on the intra-firm versus inter-firm relationships which are crucial to identify a specific spatial dimension of both industry and technological cycles.

Yet, although the product cycle model has been given a space-specific characterization by authors who have followed Vernon's own extensive work on location and urban and regional economics (e.g., Vernon 1957, 1959, 1960, 1966, 1991a), such aspects have been surprisingly overlooked in the international trade and business literature. In fact, in spite of Vernon's path-breaking analysis of the causes of the urban concentration and then the geographical spread of multiplant activities which takes place across boundaries of product lines, the subsequent Vernon-inspired literature in international business still tended to treat MNE location behaviour using the highly unspecific geographical dichotomy between skilled capital-abundant advanced countries and unskilled labour-abundant developing economies. In terms of economic geography the Vernon tradition therefore largely fails to move beyond the typical insights of the Heckscher-Ohlin neoclassical model, relying on its implicit assumptions of catching up effects through technology diffusion and imitation.[7] Vernon's emphasis on the spatial shifts involved in industry development cycles, and the changing relevance of external economies during such cycles, should have called for an explicit incorporation of locational issues in the explanation of MNEs operations. This, however, is still largely missing in the international business literature. Meanwhile, on the other hand, a rigorous examination of ownership and organizational issues is largely absent in regional and urban economics.

The major criticisms of the PLC-based explanations of international production, at least as related to the geographical dimension, can thus be split into two themes.[8] On the one hand, on the basis of the evolution of the contemporary world economic environment, ownership advantages are to be attributed to firms, rather than countries, thus making the *geographical origin* of MNEs much less predetermined by the national level. This point, which was seemingly captured in Vernon's original framework, has been highlighted by both regional economics (e.g., Taylor 1986, 1987) and international business (e.g., Cantwell 1995) scholars who attribute the limitations of the PLC model to an inadequate conceptualization of both the firm and technological progress. On the other hand, observation suggests that agglomeration forces have attracted MNE activities to particular locations in both advanced as well as emerging economies, thus making the *geographical destination* of MNEs far less dependent on purely cost-based and intrinsically static hierarchical patterns. Indeed, this latter observation seems to point to the opposite side of the globalization argument. Skills formation and knowledge accumulation processes are ever more dependent on sources that are external to any single firm (however large and multinational it may be) and are highly spatially situated (Storper 1997). Some of these limitations are acknowledged by Vernon himself (1979a, 1991b), who recognizes that the expansion of MNEs' multi-product networks into world-wide production networks would render locational issues much more complex than what was depicted in the original formulation of his model.

2.3.3 Caves: Vertical and Horizontal Integration

Following Hymer in his industrial organization approach to international production, Richard Caves's work (1971) was the first to introduce the ground-breaking distinction between horizontal and vertical integration of MNE operations across national boundaries. Horizontal FDI (hereinafter HFDI) implies the production of the same good or service produced at home in a new foreign location, thus replicating identical production processes across countries. Vertical FDI (hereinafter VFDI) instead involves the shift abroad of some stages of the production process, either backward (upstream) or forward (downstream) or both, thus fragmenting the MNE production process vertically across countries. Interestingly, Caves also allows for the possibility of MNEs carrying out foreign production that is neither horizontally nor vertically integrated: this is said by Caves to be of a 'conglomerate diversification' type (Caves 1971, p. 3).

Hence, the advantages offered by FDI can be related to either vertical or horizontal determinants of foreign production, while taking into account industry characteristics. In terms of HFDI, the ownership advantages are those possessed by MNEs operating in industries characterized by oligopolistic structure and substantial product differentiation, and particularly those with the most considerable knowledge- or research-intensive activities. The major advantages of VFDI relate to the structural features of the markets in which the MNE is active. In particular, VFDI advantages relate to where there are incentives to eliminate oligopolistic uncertainty regarding input supplies and to raise barriers to entry by vertically integrating the stages of production across national boundaries. In Caves' own words (1971, p. 1): 'foreign direct investment occurs mainly in industries characterized by certain market structures in both the 'lending' (or home) and 'borrowing' (or host) countries'.

From this standpoint, the main issues addressed by Caves relate to the variability of foreign production displayed by different industries even from the same country on the basis of different endowments of particular, and combined, intangible assets, such as research and technology, differentiation capacity, managerial and marketing skills, social capital and entrepreneurship. The emphasis on industrial, structural and institutional features – which was rather novel for its time – is critical also for its geographical dimension, though this is still at a macro level. A *propos* of vertical FDI, for example, Caves explains (1971, p. 10): 'The corporation may integrate backwards to produce raw materials in less-developed countries where they might otherwise not be forthcoming, due to shortages of local social overhead capital and entrepreneurship; but this line of argument fails to explain extensive vertical direct investments in countries like Canada and Norway which seem well endowed with railroads and local entrepreneurs.'

2.3.4 The Knowledge-Capital Model

During the 1980s and 1990s, developments in new trade theory (NTT) lead to the construction of what becomes known as the knowledge-capital model (KCM). The knowledge-capital model was first termed as such by one of the its most prominent authors, James Markusen (1984, 1995, 2002), but such a model can be thought of as a research programme which includes the work of a variety of leading scholars including, among others, Ethier (1986), Helpman (1984, 1985), Markusen (1984), Helpman and Krugman (1985), Horstmann and Markusen (1987a, b), Ethier and Markusen (1996), Markusen and Venables (1996, 1998, 2000), Venables (1999a), Carr et al. (2001). The research agenda set out by the knowledge capital model is to develop the study of the determinants and effects of multinational activity in line with the analytical approach of mainstream economics. These models are developed within a formalized general equilibrium framework, and attempt to integrate MNE activity into the mathematical modelling of imperfect competition and increasing returns, while at the same time borrowing some elements from previous traditions including the work of Vernon, Caves, and also some of the internalization theories.

The basic KCM approach rests on the main idea that MNEs are intensive in the use of knowledge-based assets, and the approach combines both horizontal integration associated with the proximity to demand, and vertical integration associated with the search for lower costs, as determinants of MNE location and investment activities. Drawing extensively upon Caves' work, although, curiously, seldom acknowledging it, the KCM splits multinational firms into two types of firms, namely horizontally integrated firms known as type-h firms, and vertically-integrated firms known as type-v firms (Markusen 2002). Type-h firms, which correspond to those which in Caves' schema are denoted as carrying out HFDI, are MNEs which have production establishments at similar levels in the production chain in more than one country. The KCM model is based on the assumption that the firm's knowledge assets are basically a public good within the firm, whose costs of supply to the firm's foreign plants are very low. Vertically-integrated type-v firms, which in the work of Caves are those undertaking VFDI, are the multinational firms which have different production establishments in different international locations. The concept of type-v firms arises from the assumption that firms can split different types of activities along the value chain and locate them in various places on the basis of international differences in factor endowments, and in particular on the basis of the knowledge and skill intensities of the local labour force.

Whether the firm decides to supply foreign and overseas markets directly via exports from the home country or via local supply from foreign affiliates depends on the balance between local production economies of scale and international trade or transport costs (e.g., Markusen 2002; Lankhuizen 2009; but see also Caves 1971, 1982a). In general, high trade and transport costs encourage FDI, specifically HFDI, as firms seek to gain easier access to a foreign market, while low trade costs encourage domestic production and exporting. Similarly, high economies of scale encourage single site production and exporting, whereas low economies of scale encourage the establishment of different facilities in different countries.

Regarding the patterns of FDI, the KCM predicts that type-h multinational firms will tend to dominate when the markets in both the home and host countries are large and similar in terms of their local labour skills endowments, whereas type-v multinational firms will tend to dominate when the markets in both countries are very different in terms of their size and local labour skills endowments. It should be noted that most KCM conclusions are derived from simulations rather than closed form analytical solutions because of the multiple dimension and inequalities embedded in these complex analytical formulations (Markusen 2002; Barba et al. 2004; Lankhuizen 2009).

Most recently, however, an apparent conflict between the KCM explanation of MNE activity and the trends observed in current globalization has been pointed out by trade theorists. As Neary (2009) argues, if transport and trade costs fall, which, as we will see in subsequent chapters has indeed been the case in recent decades, then according to the KCM one would expect that horizontal FDI will decrease, as falling transport and trade costs imply that exporting from a domestic location becomes more attractive. However, HFDI and multinationalism have increased dramatically over recent decades, thereby producing outcomes which appear to be counterintuitive according to the KCM.

One way of reconciling these observations is to assume that the set-up costs of individual foreign establishments have fallen over time. Indeed, the building of plants and the establishing of new 'turnkey' production facilities is becoming increasingly sophisticated, thereby pointing to the conclusion that set-up costs of overseas establishments are falling. However, as we will see in later chapters, plant set-up costs may well also involve issues related to labour knowledge, skills and training, and there is very little evidence which suggests that these costs have fallen over time. As such, the overall evidence on this point is inconclusive.

Alternatively we could assume that the location-specific economies of scale have actually become less important over recent years, and the decline in the importance of economies of scale has been even greater than the decline in transport and trade costs. Indeed, there is growing evidence that the geography of many production systems and input-output chains are actually becoming more spatially fragmented (e.g., Parr et al. 2002; Klier and Rubenstein 2008). However, the suggestion that location-specific economies of scale have fallen over time appears to be at odds with the fundamental assumptions of the new economic geography[9] and also with the wealth of emerging empirical evidence (OECD 2006; World Bank 2009), the emphasis of which is on the increasing worldwide importance of agglomeration effects, at least during the 1990s.

A third attempt at reconciling theory and observation is to suggest that as transport costs fall the potential profits of foreign-firm acquisition systematically favour the acquiring firms, thereby promoting outward FDI based on mergers and acquisitions (Neary 2008). This insight is obviously consistent with much of the available evidence. Moreover, the establishment of export 'platforms' can become even more attractive under conditions of foreign economic integration, as the size of the market that can be served from that establishment becomes more important than simply the size of the domestic market (Neary 2009).

Thus, the dichotomy between *horizontal versus vertical* integration seems to be no longer suitable for reflecting the main organizational forms of international operations implemented by MNEs. The complexity of MNEs' integration strategies has been acknowledged also by the KCM scholars, who have recognized that MNEs actually follow *complex integration strategies* rather than those in one or either category (e.g., Grossman, et al. 2003; Yeaple 2003; Neary 2009). As such, MNEs are both horizontally and vertically integrated, mixing up different strategies including that of the international diversification across products and space firstly described by Caves (1971, 1982).

2.3.5 Resources, Technological Competence and Capabilities

The increasing attention devoted to the internationalization of one of the major ownership advantages, that is, technological competence and innovative activity, has resulted in a renewed questioning of the rather narrow role conferred upon the MNE subsidiary by Vernon (1966). The aim to redefine MNE units as key creators of innovation and technological knowledge was originally suggested by Dunning (1970) and later developed by Chesnais (1988), Cantwell (1994) and Fors (1998) among others.[10] Building on the seminal work of Edith Penrose (1959) and on evolutionary views of technological progress, the MNE is seen as a 'bundle of productive resources and competencies', physical, human and technological, which are idiosyncratic to each specific company, and which represent the firm's major competitive advantage. Interestingly, as discussed in Chapter 4, in spite of the huge relevance of Penrose's work for the conceptualization of the modern competence-based, specific-asset and strategic-asset seeking MNE, her treatment of multinationality as a stage in the growth of the firm, rather than as an archetype itself, has somehow borne only a limited influence on much of the international business literature (Pitelis, 2007a, b). It has, however, provided a huge inspiration for the study of the internationalization of MNE innovation and its technology-based ownership advantages.

In the competence-based view, the geographic sources of MNE competitiveness are seen to be associated with the capacity of foreign affiliates to generate innovation by tapping into local expertise and industrial clustering specific advantages (Birkinshaw and Hood 2000; Mudambi et al. 2007). Attention has therefore been increasingly paid towards understanding the new roles played by affiliates in the MNEs' global innovation efforts (Pearce 1999), the development of the affiliate's strategy, and the emergence of corporate centres of excellence (Birkinshaw and Hood 1998; Birkinshaw 2000a; Frost et al. 2002).

Whereas the traditional subsidiary was viewed as exploiting the technological advantages (known also as capabilities and competencies) of its parent MNE, more and more modern subsidiaries have been developing competencies of their own, that can be leveraged to advantage throughout the whole MNE network. Growing empirical evidence suggests that the rising importance of networked production systems has progressed along with major changes in the role played by modern MNE subsidiaries (e.g., Pine et al. 1993; Birkinshaw et al. 1998, 2002; Johnston and Menguc 2007;

Rabbiosi 2008; Marin and Giuliani 2011) and an increased role for strategic decision making at the subsidiary level focussed on knowledge-based activities (e.g., Cantwell 1987; Pearce 1999; Frost et al. 2002; Young and Tavares 2004). This implies that competencies created in one location can be used in another, so that there may be a multi-directional flow of knowledge, information and goods between relatively autonomous subsidiaries.[11] In order to realize these advantages, MNEs have adopted a more sophisticated means of coordination so as to continually maintain their local and global knowledge advantage.

As a consequence, MNEs have become increasingly regarded as evolving organizations which strongly interact with socio-economic contexts in both the home and host locations (Teece 1977; Dunning 2000; Janne 2002a, b; Frenz and Ietto-Gillies 2007). The evolution of their organizational features is central to explaining the multi-way relationship between multinational expansion and innovation processes. The nature of these interactions is seen to vary across countries, regions, sectors and firms, with often increasing levels of interrelatedness at the different levels. Dunning and Wymbs (1999) demonstrate that the degree of multinationality is nowadays significantly associated with the perception that increased global technological advantage has foreign sources. Multinational firms pursue this goal by establishing integrated networks of affiliates, in order to build up a sustained competitive advantage which is based much more on capability building and dynamic improvements than on static efficiency criteria (Malmberg et al. 1996; Zanfei 2000; Frost 2001; Ietto-Gillies 2001; Veugelers and Cassiman 2004, Narula and Zanfei 2005).

Thus, for these diversified approaches to the study of MNE, which variously group together perspectives from business and management, sociology and evolutionary economics perspectives, the empirical observation that FDI, and multinational operations more broadly intended, have increased over recent years while transport and trade costs have fallen, is not at all problematic. The reason is that such approaches do not treat knowledge as a generic public good within the firm, but rather interpret MNE location behaviour as being the result of the interdependence between the knowledge, organizational, and transactions characteristics of both the firms themselves and also of the potential regions in which they might locate. Multinational firms are thus seen as complex networked systems of knowledge and technology and many of these interrelationships are regarded as being location specific. Therefore, in these approaches to MNE location behaviour, finding the best 'fit' between a multinational firm and its local economic environment is regarded as being the crucial issue.

Yet, in spite of these many analytical and observational developments (see also Cantwell 1989; Kogut and Chang 1991; Dunning 1995; Frost 1996; Almeida 1996; Kuemmerle 1999; Cantwell and Janne 1999; Pearce 1999; Zander 1999a, b; Frost 2001; Mudambi 2002) the mechanisms of these technological linkages have, until recently, tended to be depicted in terms of intra-firm and inter-firm interactions, while the geographical space itself has still been treated mostly as an exogenous variable whose structure and dynamics are hardly explained at all (Rugman and Verbeke 2002). As we shall see in the following chapters, only very recently have numerous empirical studies been undertaken in the direction of defining the actual 'space' of MNE innovative activity. This qualitative and quantitative evidence is regarded as an extremely valuable and fruitful extension of the study of the MNE phenomenon, whose feedbacks have yet to be incorporated in the wider analytical framework of the OLI.

2.4 THE DETERMINANTS OF MULTINATIONAL ACTIVITY: I ADVANTAGES

2.4.1 From O to I Advantages

Building on the ground laid by Hymer, Caves and Vernon,[12] further work in economics explores the ownership advantages of MNEs, addressing the role which both market imperfections and strategic behaviour play in the promotion of international production. In terms of the geography of MNE operations the tradition which follows Hymer and Caves[13] is similar to that of Vernon and his followers.[14] However, in the discussion of country-specific ownership advantages the Hymer-inspired literature puts relatively more emphasis on the relevance of the structural characteristics of the investor's economy, rather than on Vernon's focus on the strategic behaviour of oligopolistic players. From a spatial perspective this is an important point, because in the Hymer–Caves inspired framework the (macro) locational features of the MNE origin, such as the institutional and social environment, take on a crucial role as determinants of comparative advantages, and thus of multinational expansion. This is quite different to the Vernon framework which views locational issues in terms of the strategic responses by firms to product life cycles and market distortions. However, in each of these theoretical approaches – particularly during the 1960s and the 1970s – the main objective of the analysis is still the question of *why* firms become multinationals, with relatively little attention paid to the question of *how* multinationals organize their activities across multiple sites. The focus is therefore primarily on ownership and comparative advantages, on markets and price mechanisms, and thus on the transactions costs which are external to the firm. Market imperfections are only considered as long as they are structural, with little or no discussion regarding the role played by knowledge imperfections and information asymmetries.

2.4.2 Internalization and Transactions Costs

The most influential set of contributions on the question of 'how' multinational activities are organized is based on the transactions costs theory of the firm. The antecedents of this literature are to be found in Coase (1937) and Penrose (1959), and the literature shifts the focus on the relationship between the *ownership* and the *organization* of multinationality. Transactions costs approaches assume microeconomic and organizational views of the firm (Buckley and Casson 1976; Teece 1977; Rugman 1981; Hennart 1982). As international markets, and particularly those for intermediates, technology and knowledge, are all characterized by considerable transactional imperfections, the need arises for the creation of internal markets. MNEs are perceived as being the result of the internalization of cross-border operations (Buckley and Casson 1976). Internalization is thus seen as the outcome of the comparison of the costs and benefits of organizing transactions internally within the firm relative to conducting external market transactions. The decision to internalize activities therefore provides a rationale for both the horizontal and vertical integration of cross-border activities,[15] and the development and organization of international production occurs via the creation of hierarchies rather than markets. Moreover, a variety of outcomes are expected because hierarchical organizational costs relative to market transactions are seen to be specific to the MNE, its industry and country of origin. The hierarchy consequently implies that ownership, control and organization are considered as an unbundled set of motives for multinationality.

Transactions costs economics in many ways provides a much richer understanding of the multinational firm in terms of its behaviour, contracts and asset specificity (Williamson 1985) than other approaches. Its strength lies in the focus on the overall organizational capability of the firm, and on the intra-firm linkages between different kinds of value-added activities and economies of scope. As already mentioned, internalization and integration become strictly intertwined. Teece (1980) expands the vertical integration argument of Williamson (1975) to explain the multi-product diversification, highlighting the fact that both vertical integration and diversification are ultimately related to the particular types of cross-border transactions which are to be internalized.

As we shall see later in Chapter 4, however, one of the weaknesses of transactions costs based explanation of the existence of the MNE is the exclusive emphasis on the firm as an alternative to the market, rather than as an institution for learning in its own right, irrespective of how well markets work. Furthermore, as we will see later, the effects of globalization and the growth of alliance capitalism, strategic collaborations and partnerships, outsourcing and offshoring, production and distribution networking, and asset-augmenting foreign investment, have all greatly affected the nature and scope of firm internalization processes. As Cantwell and Narula put it: 'Internalization issues have received the most attention, as the conventional belief that full internalization is the preferred mode of MNEs has been challenged by the growing use of strategic alliances' (Cantwell and Narula 2001, p. 169).

For our purposes, what needs to be highlighted at this point is that spatial patterns and locational variables are virtually non-existent in the internalization theory of MNEs. In part, this is the consequence of the emphasis on transactional imperfections, which tend to overlook the structural imperfections related to socio-economic and institutional geographical contexts. At the same time, however, the importance attributed by the transactions costs theory to organizational issues, technology, know-how and R&D, as well as the opportunity costs associated with internal transactions and external exchanges, makes the transactions costs approach very suitable for identifying and distinguishing between the different types of economic geographies in which MNEs may choose to locate. These different economic geographies will be the specific focus of Chapters 4 and 5, where we will build on transactions costs approaches to examine the relationship between multinationality and innovation.[16] This relationship is a critical feature of the internalization theory, and the transactions costs perspective allows us to look at the relationship between internalization and the exchange of knowledge in a specific spatial perspective.

2.5 THE DETERMINANTS OF MULTINATIONAL ACTIVITY: L ADVANTAGES

2.5.1 From I to L Advantages

As we have seen, economic and business theory has managed to successfully integrate O advantages and I advantages, and to consider issues of ownership, control, organization and integration, as well as transactions costs both internal and external to the firm, as pooled determinants for multinationality. Yet, the question of *where* MNEs locate is left rather undefined and indeterminate, and locational advantages are still resting on broad categorizations of national geographies. The distinction between HFDI and VFDI in studying MNE operations, which as we have seen is central to different theoretical views, implies some significant implications for the spatial configuration of MNE activities. In particular, the two types of firm integration, as stylized as they may be, follow rather distinct locational patterns. In the case of HFDI, this type of integration is more prone to urban (or metropolitan) concentration; in the case of VFDI, one more oriented towards sub-urban or industrial regions. Such critical insights, derived by combining different forms of firm integration with market and industry structure, have been uncovered by the scholars of the New Economic Geography (see, for examples, Markusen and Venables, 1996; Venables, 2006).

2.5.2 New Economic Geography: Dispersion Versus Concentration

New economic geography (NEG) models first raised the possibility that the locational determinants of MNE activities may be understood in terms of various tensions. These are captured in terms of forces of dispersion versus concentration on the basis of firm and industry-specific trade-offs between scale economies and market access (HFDI), and between scale of integration and factor costs differentials (VFDI), all of which turn out to be critical for understanding the geographical patterns of MNE activity. HFDI largely explains foreign investment among advanced economies, while VFDI largely accounts for flows of FDI from developed to developing countries. Among the determinants of FDI,[17] both supra-national regional economic integration and sub-national agglomeration economies have been incorporated in KCM-type models to determine the locational choices of MNEs (Barba et al. 2004).

The arguments so far tend to be based on two country models. However, when we start to take account of geography in a more explicit manner, then the issue becomes more complex again. With regards to the geography of international production, a central issue discussed by NEG approaches is that of the geographical *dispersion* versus *concentration* of multinational activity. The new economic geography develops from the seminal work of Paul Krugman (1991a, b), which shares the same broad theoretical background of the KCM. NEG builds on the earlier developments in new trade theory, which focussed on the reasons why most international trade and FDI were actually observed to be between advanced countries, primarily taking the form of horizontal rather than vertical integration. In particular, mergers and acquisitions dominate many aspects of foreign production. The key insight of these theories is that such trade and FDI patterns are driven by economies of scale at the level of a firm, an industry, and the domestic market. The importance of the size of the domestic market for the multinational firms is known as the 'home market effect'. New economic geography approaches then translate these arguments into the case of cities and sub-national agglomerations, and demonstrate that the location behaviour of MNEs depends on the balance of the input supply and scale economy advantages of geographical concentration versus the trade cost and market access advantages of geographical dispersion. In the NEG terminology such advantages are referred to as centripetal forces and centrifugal forces, respectively. Dispersion forces are strengthened to the extent that wages and land prices fall with distance, although this benefit may not be applicable to technology-intensive activities (Barba et al. 2004).[18]

Combining these various insights with issues of firm integration, it becomes clear that the distinction between HFDI and VFDI has important implications for the spatial configuration of multinational activities. Research in urban economics (Glaeser et al. 1992; Henderson et al. 1995) demonstrates that the growth and productivity advantages for firms in particular locations depend on the degree of industrial diversity or specialization of the local economic environments in which the firms operate. Certain types of firms appear to benefit from being located in a region which exhibits industrial diversity, while others appear to benefit more from being located in a region which exhibits industrial specialization.

Adopting the lines of enquiry developed by NTT and NEG, most notably characterized by the work of Markusen and Venables (1996), Fujita et al. (1999), and Venables (2006), it is possible to argue that the attractiveness – or L advantages – of particular locations for MNE investment will depend on both the variety and the nature of the local linkages which the firm wishes to access. The location choice of MNEs which are primarily focussed on HFDI will tend to be oriented towards areas of agglomeration which are characterized by industrial diversity, whereas the location behaviour of firms focussed on VFDI will tend to be more oriented towards specialized locations. In general, in situations where final producers and also intermediate suppliers both exhibit economies of scale, they will tend to locate in the same region (Venables 1996). Where economies of scale operate in the same sector (i.e. localization economies are at work), then VFDI will dominate and the region will be highly specialized. In contrast, where the economies of scale operate in different sectors (i.e. urbanization economies are at work), then HFDI will dominate and the region will be highly diversified.[19] Even more complex patterns are possible, in that differences in the size of the market and local skills endowments can also play a role. As such, these two different types of firm integration, namely HFDI and VFDI, are likely in many cases to follow rather distinct locational patterns.

In many ways, some of the most important recent insights for our understanding of the L advantages of MNEs have therefore come about from the microeconomic approaches represented by the NEG, whereas most of the breakthroughs on the O and I advantages have come from the international business and management literature. However, in terms of the explicitly spatial behaviour of multinational firms at the sub-national level, even the NEG has also shown some serious limitations. The reason is that neither framework deals simultaneously and explicitly with the geographical, organizational and institutional variations which are typical of sub-national contexts. While the NTT/KCM tradition tends to focus on either two-country or three-country home versus foreign market type cases, in which the internal structure of each country is treated as being effectively homogeneous, the NEG models often prefer to subsume all such differences in terms of an 'iceberg' type specification of transport-distance costs. At the same time, in cases where NEG models attempt to analyse spatial variations in the internal structure of countries, they do not account for the variations in MNE multiplant multi-function characteristics.

2.6 THE EFFECTS OF MULTINATIONAL ACTIVITY: EXTERNALITIES, SPILLOVERS AND DEVELOPMENT

As highlighted above, along with the interest in understanding the determinants of MNE activity, many questions regarding the effects of multinational presence on host economies have always been at the core of the attention of economists and international business theorists, as well as of other social science scholars. In order to be able to assess if MNEs are overall advantageous or disadvantageous to host economies, different theoretical perspectives have been adopted which analyse the issue from the point of view of either the home or the host economy or that of the MNE itself. One of the major issues discussed is that of the externalities and, particularly, potential 'spillovers' generated and diffused by the MNEs in the host economy. To go into depth in the understanding of when and where spillovers may or may not occur it is necessary to have a thorough discussion on the concepts of location strategies, knowledge and information sources, technological change, and their spatial dimensions. These discussions will be the focus of Chapters 3 to 5 in this book. In what follows here some of the remaining conclusions in the literature regarding the impact of MNE activity are reported.

2.6.1 Spillovers, 'Home' and 'Host' Economies

It is a difficult problem to analyse externality issues because the impacts of MNEs can operate simultaneously at different levels, such as that of the firm, of the industry, of the region, of the country and also of the global level: trying to disentangle them has proved to be extremely tricky, if not impossible. As Ietto-Gillies (2005) points out, the assessment of MNE effects is closely associated with the explanations of why, how and where the multinational firm operates, thus making it rather problematic to separate the questions on the causes from those on the impacts of multinationality. What seems to be the case, however, is that the traditional lack of an effective spatial perspective in the MNE analysis has greatly limited our understanding of some of the consequences and implications of MNE activity. In particular, what are known as *spillover effects* – which are almost universally acknowledged by all different theoretical strands as the largest beneficial outcome of multinational operations – are virtually impossible to be grasped without an explicit spatial perspective.

The geographical distinction between *home* and *host economies* has been generally viewed across the different theoretical approaches reviewed above, as the country of origin and that of destination of the MNE, respectively. Although the home-host dichotomy, which is based mainly on the direction of FDI, is rather simplistic and imprecise (Ietto-Gillies 2005), the bulk of the literature has found it convenient in providing a rationale for classifying MNE effects. Leaving aside social, cultural and political issues, the home and host dichotomy allows us to classify MNEs according to differences in factor intensities of production structures, volumes and directions of international trade, levels of wages and employment, market structures, resources and capabilities, innovation and technology creation and diffusion, as well as productivity and growth. In terms of the home-host dichotomy, such issues can be seen as either mutually exclusive – i.e. the benefits of the home country are the costs of the host country and vice versa – or as complements – i.e. there may be benefits or even costs in both the home and the host economies.

The issue of spillovers is strictly linked with that of knowledge flows and technology transfer. In the work of Caves (1974), three main benefits stemming from an MNE's presence in the host location are distinguished: increased market competition, which may improve allocative efficiency; demonstration effects, spurring domestic firms to increase their level of technical efficiency, or X-efficiency; and transfer and diffusion of technology, encouraging local technical change. Already in Caves' work, thus, the issue of knowledge spillovers is discussed by assuming mostly a host economy viewpoint, as is the case of much subsequent work on spillovers effects (e.g., Blomström and Persson 1983; Kokko 1994, 1996; Blomström and Kokko 1996, 1997, 1998; Lipsey 2001a, b; Javorcik 2004; Lin and Saggi 2004; Javorcik and Spatareanu 2008). The benefits of an MNE's presence for host countries are broadly classified into two types: productivity spillovers and market access spillovers. The former kind of spillovers are the result of tougher competition following foreign entry. This may create incentives for local firms to introduce new technologies and organizational practices in order to compete with the new entrants. In addition, MNEs allow local firms access to new technologies and skills from backward and forward linkages, as well as personnel exchanges with foreign affiliates. Market access spillovers therefore come from the experience and knowledge that MNEs have of international marketing, distribution networks, and lobbying power. As a result of their own operations, MNEs may therefore pave the way for local firms to enter the same export markets, either because of the infrastructure created or because of the dissemination of information.

The literature on FDI spillovers generally assumes that the positive effects of MNEs on the production and technological capacity of firms in the host countries are largely unintended, even if they do indeed arise from formal interactions and linkages between local firms and MNEs. For our purposes here, some theoretical limitations of such an assumption are worth mentioning (see also Padilla-Perez 2006). First, it often confuses two forms of positive externality arising from the interactions between co-located firms: linkages (backward, forward and horizontal) coming from the division of production between firms on the basis either of specialization or of matching excess demand; and knowledge sharing and spillovers (where the former is deliberate and the latter are unintentional) between firms (Kamnungwut and Guy 2012). Second, and related, this assumption does not take into account any systemic interactions, such as the relationships between industry structures and resulting interdependencies with institutions and organizations in the location in which production and technological capabilities are created. Third, the focus is on increases in productivity levels as the main and sole output of indigenous technical change, whereas the outcomes may be more varied (e.g., in terms of employment, innovation opportunities and overall growth). Fourth, technical change is treated as an exogenous variable without identifying its sources such as learning-by-doing, learning-by-using, learning-by-interacting, or the types and functions of innovation processes, such as product-, process- or organizational-oriented, as well as adaptive, incremental, radical innovation, etc. Last, but not least, it assumes that technology *spills over* freely to everybody, implying no real distinction between information and knowledge, thereby treating local firms as passive technological recipients in the process of technology transfer, which in turn is seen as strictly unidirectional. All these issues point to the critical importance of introducing specific geographies and spatial structures in analysing externalities and spillovers, as we will see in Chapter 4.

In his highly cited NBER working paper which reviews the evidence on the distribution between home and host economy of the benefits and costs of FDI, Lipsey (2002, p. 1) concludes that 'Much of the impact is from the transfer of knowledge of world markets and of ways of fitting into worldwide production networks, not visible in standard productivity measurements.' Following Hymer's 'correspondence principle', the effects of MNEs may therefore overall be either beneficial or detrimental in the sense that they may preserve or even reinforce existing inequalities between core and peripheral areas, *across and within countries*. A similar argument is also found in Vernon (1957), who argues that MNEs spur an increasing share of employment in that 'mixed bag of business services' (Vernon 1957, pp. 17–18) in a large metropolis, thereby stimulating growth, rising factor prices and also, as a consequence, interregional intra-national differences. With an increase in spatial agglomeration at the functional level, these differences will be evident even if traditional employment in manufacturing following the presence of MNEs is likely to expand more equally between cores and peripheries. Therefore, it may well be that the benefits from MNE location between the core and the hinterland regions will never be evenly distributed (Pitelis, 2002b; Tolentino, 2002; Kottaridi, 2005a, b). The spatial effects are seen here as a direct consequence of MNEs' locational choices, both in the location of origin and in that of destination.

2.6.2 Multinationals and Economic Development

The search for often unspecified spillovers in the economic literature has somewhat hidden the effects of MNEs on capabilities development, particularly in developing economies. One of the most effective applications of the OLI paradigm has been on development issues, through the concept of *investment development path* (IDP) (e.g., Dunning 1981, 1988a, 1993, 2001; Dunning and Narula 1996; Narula 1996). The main tenet of the IDP is that as a country develops, the configuration of the OLI advantages facing both MNEs and local firms changes, as do their interactions, eventually reversing even the ambiguous role of 'home' and 'host' economy. The IDP identifies various development stages, and also the conditions providing for the change, and their effects on the path of the country's development. Such effects take place through three main mechanisms. The first is via backward linkages of the MNEs with the host economy; that is, the linkages between domestic firms supplying intermediate inputs or services and their multinational customers. Through such contacts, local suppliers obtain technical assistance, incentives to improve product quality and/or cost reductions, and opportunities to reap economies of scale and scope. The second mechanism is via imitation and adaptation; local firms may replicate and adapt to their environment innovations of either a technological and organizational nature, generated by the presence of foreign firms. The third mechanism triggering positive effects is represented by skills upgrading and human capital formation, ranging from basic learning-by-doing and learning-by-using to formalized advanced technical training courses (see Dunning, 1993; Hobday 1995; Radosevic 1999a, b; Padilla-Perez 2006). Importantly, the learning processes implied by all three mechanisms occur via both informal connections and formal collaboration.

However, although MNEs undoubtedly play a central role in development paths, as we will see more in detail in Chapter 5, learning and capabilities building processes – and therefore the impact of MNEs on host economies – are strongly dependent on the characteristics of local actors and environments. These in turn are highly diversified within national boundaries of developed countries and even more so in developing economies. Thus, the IDP hypothesis has missed one key aspect of the evolving configuration of the OLI advantages during the development trajectory of the host economy; namely, that it is particularly the L that triggers such a change.

Indeed, Dunning and Pitelis (2008, p. 171) remark that 'Hymer's law of uneven development follows only under his very specific assumptions and assertions, namely the dominance of MNEs, the absence of learning by nation-states in developing countries, and the lack of 'nationalism' on their part. Certainly, in today's economic and political scenario these are questionable assumptions.' Yet, it can be argued that in the current phase of globalization, Hymer's basic assumptions actually hold very strongly if a detailed and explicitly spatial perspective is applied to the argument. The Dunning and Pitelis's remark above represents a fundamentally aspatial interpretation of the role that MNEs can play in regional and local development. However, they also contend that 'Hymer also paid little attention to the ways in which capitalist governments might facilitate the positive externalities of inward FDI. One example is the promotion of 'industrial districts' and 'clusters'. As documented inter alia by Porter (1990) and Enright (2000, 2003), agglomerations of interlinked firms, including MNEs that compete and cooperate in a particular activity in a particular location are frequently a potent source of locally based economic development, and a magnet for FDI' (Dunning and Pitelis, 2008, p. 173).

What we argue here, and will also be further clarified in Chapter 5, is that this top-down approach to local economic development needs to be treated with some caution, because the 'promotion' of industrial districts and clusters for attracting foreign investors has often produced rather limited outcomes, over a period of now more than two decades, and thereby to some extent replicating the aspatial view of growth and development mechanisms. If we start to take geography and space seriously, then it becomes very clear that the extent to which local economic growth and development may be spurred by MNEs in 'particular' activities and locations can be determined only by assuming space as an endogenous variable whose L features combine explicitly with both O and I advantages.

2.7 THE ECONOMIC GEOGRAPHY OF MULTINATIONALS: SEARCHING FOR A STRONG L

2.7.1 The Theory of MNE and the Problematic L

For more than three decades now the eclectic OLI paradigm has provided the principal analytical framework for examining both the growth of multinational activity and its changing patterns over time. In spite of the criticisms that the OLI has encountered regarding its explanatory power, it has consistently been demonstrated that the OLI approach is in many ways still more than suitable as a tool for interpreting the complexity of multinational expansion, and particularly in a historical context (e.g., Dunning 1995, 2001), than any other overall schema. That this is so is clear from the extensive array of counter-arguments to such criticisms which the OLI paradigm offers (Cantwell and Narula 2001).

In contrast, however, the OLI paradigm's ability to effectively integrate both micro and macro perspectives does not translate well when we are looking at the issue of location. The reason for this failure does not reside only in the OLI itself, but also in the theories that underpin it.

The theory of the MNE has developed along the lines of a sharp distinction between firms, as organizations subject to centralized control, and markets, treated as environments characterized by independent actors engaged in full arm's length transactions. Although both of these lines have deeply evolved over time and remain theoretically valid and useful in today's analysis of firms and markets, they also still represent the major constraints to the full integration of a spatial dimension within the study of MNE operations.

Although strongly micro-founded, both traditional economic theory, with its focus on international production and product or market dichotomies, and the international business literature, which concentrates mainly on the business firm, treat geography and space at a highly stylized and unspecific macro level. This is typically that of the country, if not even a macro-region or in some cases a continent. Geography is therefore intended as an 'international geography' rather than as a 'subnational space' (McCann and Mudambi 2004). The general lack of locational features of such theories in isolation would appear to be harder to justify in the eclectic framework offered by the OLI paradigm, where the interest in explaining comparative advantages through ownership and organization would make the question of *where* MNEs internationalize their activities in principle inseparable from the questions of *why* and *how*. Locational choices, that are prerogatives of firms and other agents and not of countries, are by definition part of wider strategies that underlie both the comparative and competitive advantages of firms.

From a conceptual viewpoint, some attempts have been made in order to explicitly incorporate location within the international business and management literature, and these are most notably associated with Michael Porter's (1990) work on clusters. However, Porter's most famous and valuable contribution is still about the O-type advantages. The four central determinants of competitive advantage, namely: factor conditions; demand conditions; related and supporting industries; and firm strategy, structure, and rivalry – are all properties of the home base, and particularly the nation-base, of the firm. Yet, by leaving aside proper economic geography perspectives, such attempts have failed to discuss the sub-national regional location behaviour of MNEs with respect to anything more than the very simplest – and aspatial – Porter's notions of industrial clustering.

The absence of any 'real' geography in the international business literature ironically therefore parallels the criticisms made by international business scholars of both new trade theories and NEG with respect to their largely abstract treatment of space (see Dunning 1998; Ietto-Gillies 2002, 2005; Buckley and Ghauri 2004). Such criticisms are to be taken as a first step to acknowledge and address the issue of location in the established explanations of MNEs activities. Yet, NEG has demonstrated the critical tensions which exist between cores and peripheries and revived the academic interest in the atavistic contrast between central and more marginal regions, and the role which firm location plays in these tensions.

In spite of the criticisms attracted over time, the work of both Hymer and Vernon has been considered seminal in the analysis of MNEs for two reasons. Firstly, they both introduce market imperfections and dynamic elements into the analysis, such as changes in demand structures, innovation and technological advantages and information lags. Secondly, they both recognize the importance of ownership comparative advantages in explaining multinational behaviour. As discussed above, both Hymer's and Vernon's pioneer works have strong micro-foundations, and as such provide crucial connections between the structure and behaviour of MNEs and the sub-national spatial organization of their economic activities. However, in spite of the relevance attributed to locational elements in the original work of both authors, the subsequent literature on MNEs has essentially privileged the traditional macroeconomic perspective of international trade theory in treating countries as the main spatial units of observation, where the firm is primarily regarded as merely the expression of a certain nationality.

Thus, while the integration between O and I is so highly effective, the L advantages are essentially attributed to either countries or firms in a rather non-systematic and non-analytical manner by the different theoretical approaches to the MNE (e.g., see Dunning 1993, p. 77). In general, the L aspect of multinationalism has tended to be treated as a kind of exogenous context variable conditioned on national features, and it is really only very recently that the empirical literature in the OLI and economics tradition has begun to actually address the locational strategies of MNEs at a more fine-grained territorial perspective.

2.7.2 The Sources of Knowledge and the Centrality of the L

Part of the explanation for the surprising delay in acknowledging the role of space in MNE behaviour lies in the fact that most of the international economics and business emphasis has been, as we have seen above, on either the macroeconomics of international production or on the characteristics of the processes which are internal to the firm. As we will see in later chapters, prior to the advent of the current era of globalization since the 1990s, it is clear that both of these lines of enquiry were motivated primarily by historical contingency (Dunning 2001), in the sense that the priorities of the foreign investment decisions made by MNEs were perceived of as being far more related to national and international differences rather than to (sub-national) regional differences. This accounts for the fact that, while L is apparently central to the eclectic OLI paradigm, its theoretical foundations and reliance on micro-meso interactions seem to have been substantially understated.

Much of the available real world evidence regarding the current phase of globalization points to a growing importance of new types of structures and new forms of economic coordination, that go well beyond the traditional distinction between the firm and the market. These structures include networks, value chains and quasi-market relationships of various different kinds which have grown exponentially in importance (Guy

2009). The 'new' economic structures and forms of governance pose major challenges to the existing explanations of the determinants and effects of multinational firms' behaviour. In these emerging network structures, international trade, production and knowledge creation occur both *within* the firm across space and *within* networks, some of which are highly spatially concentrated and some of which largely aspatial.

Such massive, profound and ongoing changes have obviously affected both the nature and configuration of the OLI advantages, and their interactions. Although such interdependences are very difficult to disentangle 'bilaterally', we argue that the changes in the global institutional and technological environment have had important repercussions for the balance of the 'three-legged stool' (Dunning 1998 and 2009b, p. 5). This affects in particular the centrality of the L and, as a consequence, its interaction with both O and I advantages.

O advantages have been historically exploited largely via internalization (intra-firm networks), or I advantages. Growing multinationality generates per se new O advantages through experience and capability accumulation, which can then be exploited by both internal and external means (Castellani and Zanfei 2004, 2006), giving rise to a positive cumulative causation mechanism of the reinforcement of such advantages. On the other hand, recent evolutionary views of technological change applied to MNE behaviour and strategy have contemplated the interactions between O and L, providing grounds for some significant advances in the field, at least in terms of our empirical understanding. As we will see, O advantages are increasingly reliant on the ability to explore and select among a wide range of knowledge and quality sources (e.g., Arora and Gambardella 1995, Cantwell and Piscitello 1999). Such mostly intangible L advantages are highly localized and concentrated within specific locations, and contribute to enhancing firm-specific O advantages, which in turn strengthen those of the home and host location at the same time. It is important to note that when competitive advantages are seen through the lens of economic geography and thus are perceived as both firm-specific and region-specific competitive advantages, the effects of the presence of MNEs is much less constrained or conditioned by the traditional host country-home country dichotomy, because the sources of knowledge and the overall openness of the host economy in terms of the inflows and outflows of knowledge resources, become much more relevant issues. Similarly, as seen above, another focus has been on local spillovers coming from a local multinational presence. The widespread lack of any direct observation of spillovers, allied with only a fuzzy notion of the L which is generally conceived primarily in terms of national borders (e.g., Dunning 1981, 1988a; Blomstrom 1989b, c; Ozawa 1992; Kokko 1992a; Lall 1993a), have therefore limited the consideration of externalities to one particular type and to only one direction; namely, spillovers which are transmitted from MNEs to the host economy, disregarding the critical link that goes from region-specific L advantages to the growth of the firm itself.

In addition, it is also rather surprising how the reciprocal interactions between I and L have so far been almost entirely ignored. Instead, a major focus of theoretical and empirical research has been on firm growth as a consequence of the increasing intensity of O and I.

Turning to the different types of MNEs, from the literature the recommendation emerges clearly that it is necessary to move beyond vertical and horizontal FDI/multinational firms. In the words of Neary 'the distinction between horizontal and vertical FDI is useful for pedagogic purposes but otherwise not very helpful.' (Neary 2009, p. 215). Such a distinction reflects neither the organizational forms of multilocated operations implemented by MNEs nor the complexity of their functional and territorial integration strategies, and constrains our understanding of the determinants of MNE activity exclusively to that of the O advantages. Borrowing from a Berhman-Dunning's type of classification, it would instead be more suitable to capture the contemporary variety and complexity of L advantages and MNEs' strategies, which nowadays are primarily, although not exclusively, specific-asset and strategic-asset seekers.

As we will see in later chapters, in this book we adopt a mixed, heterodox and combined approach to MNEs. This allows us to consider MNE activities well beyond that of simply FDI and also allows us to include among other features the role of knowledge and technology creation and diffusion. From this perspective, it becomes clear that while transport and trade costs have indeed fallen significantly over time, the transactions costs related to knowledge and technological inputs have not at all fallen, and in many cases have actually risen over recent decades (McCann 2007, 2008; Boulhol and de Serres 2010). As such, in knowledge terms the importance of accessibility to both markets and inputs has increased over recent years in different ways, thereby encouraging further foreign expansion and a range of new multinational operations, exactly as we observe empirically.

2.8 CONCLUSIONS

The theories involved, combined and compared within the analytical framework provided by the eclectic OLI paradigm have been coherently and powerfully integrated as far as macro- and trade-oriented perspectives are concerned, giving rise to a strong rationale for the O advantages (the *why multinational?*). Similarly, sound underpinnings for our understanding of the internal behaviour and organization of firms, the I advantages (the *how multinational?*), have relied on both internalization theories and on managerial and sociological perspectives. Theories of market power (Hymer 1960), macroeconomic models of international production (Vernon 1966), imperfect competition and international investment (Horstman and Markusen 1987a), transactions costs approach (Buckley and Casson 1976; Rugman 1981; Hennart 2000), resource and competence based views of the firm, as well as evolutionary theories of technological change (Cantwell 1989; Kogut and Zander 1993), have all contributed to our understanding of the nature of O advantages, and have also helped to explain its growing connections and interactions with the I.

While acknowledging the success of the eclectic OLI paradigm in standing the test of time, a major shortcoming has been, and still is, observable in its fundamentally weak conception of the L. The L has always constituted the Achilles' heel of the OLI, both in terms of its initial formulation and also in terms of its subsequent adaptations which gave rise to the resulting major theoretical explanations of multinationality.

Later on in Chapter 5 it will be seen that a transactions costs analysis of the firm, combined with an evolutionary view of technological change and regional industrial structures, provides a novel and systematic way of integrating international business insights with the economic geography literature on firm location behaviour, clusters, innovation and knowledge spillovers. This will then allow us to tease out the key linkages between particular firms and particular types of regions, and to distinguish between which types of regional structural conditions will be favourable to which types of MNE firms. Above all, the variety of potential outcomes in multinational location choices is seen to be much broader than the tensions between HFDI and VFDI or between concentration versus dispersion. Multiple types of firm-cluster interactions are discernable, all of which can involve multinational investment.

However, before we can reach these conclusions, it is necessary to first investigate the major theoretical insights associated with firm location behaviour in general. If the focus is indeed to be on actual location behaviour, one important clarification has to be made with respect to multiplant domestic enterprises (MPDEs). Indeed, as with MNEs, MPDEs carry out their economic activities in multiple geographical sites. However, unlike MNEs, MPDEs' locational choices occur within the same national boundaries. In practice, from the perspective of a spatial analysis, the two categories of firms can be treated in a similar manner. However, as MNEs are always multiplant firms, whilst MPDEs are not by definition involved in international production, there are elements which critically distinguish the two types of firms. In particular

MNEs operate across highly diversified socio-economic environments and national regulatory regimes, and as such MNEs have the ability to access to and combine together a variety of knowledge sources, and therefore OLI advantages. Given that MPDEs are confined to an individual country, their ability to combine and exploit a variety of different knowledge sources is dramatically reduced in comparison with MNEs, as is their ability to access market scale. The main insights of traditional location theories for MPDE, and some implications for MNEs, will be discussed next in Chapter 3. In Chapter 4 we discuss the issues related to the firm, and in particular to the MNE as a special case of the MPDE, whose special nature arises from its unique ability to integrate and exploit profoundly different knowledge sources in a manner which an MPDE is unable to do or is inexperienced in doing. Looking at the nature of the firm via the lens of knowledge and innovation allows us in Chapter 4 for the first time in this book to introduce the concept of an innovation system, of which individual firms, including MPDEs and MNEs, are an integral part. As we will see, there are various types of innovation system concepts, of which the regional innovation system (RIS) concept is by far the most important for the purposes of this book. The arguments, concepts, insights and frameworks discussed in the four chapters comprising Part I of this book provide the various building blocks for us to develop in Part II of the book our analysis of the economic geography of modern multinationals in the context of the current era of globalization.

NOTES

1. Hymer's PhD dissertation at MIT was published only in 1976, after Hymer's death in 1974.
2. As already stated, one of the main, but not the sole, activities of a multinational enterprise (see, for example, Wilkins 2001; Ietto-Gillies 2005). Hymer himself considered joint ventures, licensing, collusion, and FDI.
3. That is, international trade in commodities and services and portfolio investments (the latter motivated by cross-border interest rates differential).
4. As discussed below, Coase's influential essay introduced the concepts of transactions costs in exploring under what circumstances new firms arise and grow. For a thoughtful review of the different stages of Hymer's work see Pitelis (2002a, b).
5. Actually, the 'correspondence principle' appears implicitly already in Hymer's *American Economic Review* article (1970), where he argues that 'The multinational corporation tends to create a world in its own image by creating a division of labour between countries that corresponds to the division of labour between various levels of the corporate hierarchy' (Hymer 1970, p. 446). However, its exact formulation and connection to locational issues are made explicit in the 1972 paper.
6. See also the exposition on Hymer's legacy by Pitelis (2002a, b, 2005) and Kottaridi (2005a).
7. Vernon's approach is also recognized as the trade/*location* approach.

8. The PCM has been one of the most cited and criticized theories of the whole literature on international production. For more extensive views on its limits see, among others, Cantwell (1989, 1995) and Ietto-Gillies (2005).
9. See discussion below.
10. This stream of literature will be discussed more in detail in Chapter 4.
11. Birkinshaw (1996) has discussed the progression of a subsidiary's development in terms of the geographic reach of its responsibilities. Thus, a subsidiary may acquire a 'regional mandate' where it coordinates activities with regard to a particular class of products overseeing other subsidiaries in the same region. Eventually it may obtain a 'global product mandate' where its responsibilities become worldwide.
12. Including, as already mentioned, the contribution of these authors themselves (see Hymer 1970; Vernon 1974, 1979a, b; Caves 1982a, b).
13. See for example Johnson (1970), Horst (1972), Franko (1976), Magee (1977).
14. See for example Knickerbocker (1973), Graham (1975).
15. Indeed, Caves himself in his later work (1982a, b, 1996) reinterpreted the MNE integration across borders in terms of the internalization theory.
16. The crucial developments on multinationality and technological accumulation, and their geographical implications, will be discussed in detail in Chapter 4, as they address more specifically the issue of the internationalization of MNE innovation activities, rather than being concerned with MNE operations in general.
17. Again, used mostly as the best proxy for MNE activity.
18. See Barba et al. (2004) for their effective classification of costs and benefits of FDI by typology.
19. For localization versus diversification economies see also Chapter 5.

3. Firm location behaviour in theory: extensions to multiplant and multinational firms

3.1 INTRODUCTION

As we saw in Chapter 2, issues of geography and location are only included in traditional models of multinational investment and behaviour at a very cursory level, with locations being defined primarily in terms of countries in international economics, and via the inclusion of location 'L' in Dunning's eclectic OLI paradigm. In both of these traditions in general, the various specific local aspects of geography and location which affect and are affected by the presence of MNE investment are not explicitly discussed. However, whenever MNEs make investment decisions, they need to consider exactly *where* such investments are to be located, and the level of geographical specificity that MNE firms need to consider for such purposes is always much more detailed than simply in which country they should invest. The reason is that project investment returns and firm profitability can vary enormously over even very short distances within an individual national economy. As such, firms must consider first where to invest in terms of which country to invest in, but also at which particular location within that country should the investment be situated.

This argument implies that MNE investment decisions tend to exhibit a *hierarchical* structure, depicted in Figure 3.1, in which the first level of the location decision of MNE tends to be in terms of which country, A, B, or C, to invest in. Once this has been decided, the second level choice of location within the individual country must be made, denoted as L_{A1}, L_{A2}, and so on. Both stages of the hierarchical location strategy are essential in order for the MNE to choose the optimal location for each investment project. However, traditionally, only the first level of decision-making – namely the national or country level – has been the level of analysis incorporated into the economics and international business literature, as represented by the 'L' in the OLI paradigm. Until very recently, the lower level of analysis – namely the sub-national geographical level – was mainly

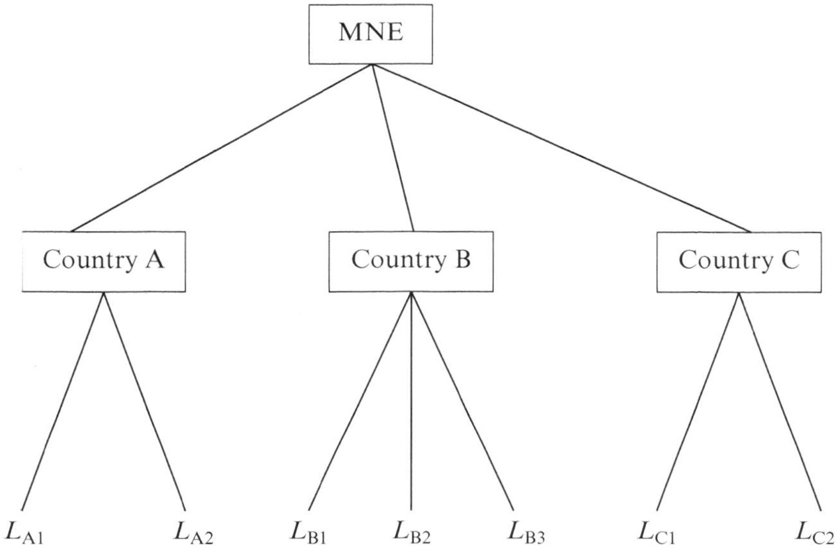

Figure 3.1 Hierarchical locational decision-making

confined to research within the urban and regional economics tradition and also that of economic geography, both of which are research fields which were largely isolated from the main research on MNEs. Today, however, these differences are rapidly disappearing and research in these various fields is converging following a rising and widespread interest in the links between geography, trade and multinational investment. This interest is not confined to any particular part of the world, although the major emphasis of these discussions has tended to take place among OECD countries. As we will see in later chapters, there are three main reasons for the converging interest in the role which geography plays in determining MNE behaviour; and these reasons are related to technological progress, to institutional change, and to transformation in corporate behaviours and strategies, most notably connected to outsourcing and offshoring.

In terms of our hierarchical decision-making model described in Figure 3.1, these technological, institutional and behavioural developments imply that MNE investment choices can and must be considered in much greater detail than simply that of the first level of analysis – that is, the country level – as has traditionally been the case. The two levels of analysis – namely that of the country and then that of the city or regional locality within the country – must be increasingly seen as part of the same location decision problem. As such, international geography and subnational geography are really different spatial dimensions of the same investment decision.

In this chapter we begin to consider these questions in detail. Our framework for analysing these issues will be built around three stages. The first stage explains ways by which we can begin to understand the relationships between firm location behaviour, factor costs, and firm profitability. We will discuss these issues from the perspective of four types of firm location models: the Weber production-location model and its related but more sophisticated *logistics-costs* model, and also the market area location-competition Hotelling and the Salop models. Under certain conditions, the Weber model and the logistics-costs models will allow us to identify the optimal location for a firm's investment project. In addition, this approach will also help identify the conditions under which a firm will also change the geographical location of its investments. Thus, in terms of the Behrman-Dunning classification seen in Chapter 2, the Weber and logistics-costs frameworks can be seen as broadly representative of the *resource* or *asset seeking* approach to MNE location behaviour, and in particular of those MNEs defined as *general-asset seekers*, whereby firms take advantage of location-specific input factors.

The second stage of our analysis is that of the market area models, of which the Hotelling and Salop models are the most important. These analyses are broadly consistent with the *market seeking* type of MNE location behaviour, in that they allow us to identify the ways in which firms use location as a means of capturing market demand. Although none of these models is specifically related to MNEs or MPDEs, taken together, the Weber, logistics-costs, Hotelling and Salop models allow us to understand location decision-making behaviour in terms of corporate investment strategy.

The third stage of our analysis is to briefly reconsider the various models in the light of the issues raised by the problematic interrelationships between locational competition and price competition. This will then provide the base for a more detailed examination of how uncertainty and access to knowledge and information sources critically shape firms location behaviour, often encouraging firms to cluster together in space. Uncertainty and imperfect information issues will themselves imply that firm locational decision-making will display particular characteristics, which overlay and somewhat modify the insights gained from the basic Weber, logistics-costs, Hotelling and Salop-type models, allowing for a much richer understanding of the nature and spatial configuration of modern multiplant (MPDE) and multinational (MNE) firms.[1]

3.2 DETERMINING THE OPTIMUM LOCATION BEHAVIOUR OF THE FIRM

All MNE investment decisions involve location choices. Firms must decide where to situate their investments and, following on from the discussion above, these location decisions must necessarily be taken at a spatial scale which is more specific than simply that of an individual country. Firms wish to make the best investment decisions and therefore it is necessary for us to understand how the optimality of a firm's investment choices is related to economic geography.

Production factor inputs are most simply defined in terms of capital and labour. However, in economic geography models they are usually defined in terms of three broad types, namely capital, labour and land, whereby the factor earnings are profits, wages and rents, respectively. At this stage, in order to keep things very simple, we ignore other potential factor inputs such as technology and entrepreneurship. Later in the book our discussion will also investigate how the additional issues associated with innovation, technology and entrepreneurship also affect and are affected by firm location behaviour. In this chapter we concentrate solely on the determinants of the location of a firm's capital investments, by assuming at this point that both labour and land are immobile; in other words, we assume that labour and land are location-specific while capital is geographically mobile. We will relax these assumptions later on.

In order to understand the level of capital investment in an area it is necessary to ask why a particular firm has chosen to locate there and why the specific levels and types of investment in the area are as they are. These issues can be addressed initially by using the approach set out by the simplest Weber location-production model.

For our analysis to proceed we will also first assume that the firm is a single establishment. This is clearly contradictory to the multiplant nature of MNEs, but once the analysis has been fully developed, we can then also relax this assumption and allow for the multiplant nature of firms, in particular MNEs. In addition, we assume that the firm's decision as to where to locate is a 'one-shot' decision, in that there is no hierarchical nature to the decision-making process, as implied by Figure 3.1. Once again, when the analysis has been fully developed, we can also relax this assumption and allow for the hierarchical nature of multinational location decision-making.

3.2.1 The Weber Location-Production Model: Fixed Input Factor Prices

The Weber location production model (Weber 1909) is conceptually the simplest analytical framework for analysing location decisions, and is therefore frequently used as the starting point for discussions about firm location behaviour. The reason is that the basic Weber argument, which is a purely theoretical construct, is framed within a triangular geographical space, and the triangle is the simplest two-dimensional analytical framework. While this may seem unrealistic for our purposes, in that MPDEs and MNEs often have hundreds of suppliers and hundreds of market points all situated in many different locations, the analytical solutions to the two-dimensional triangular case can also be shown to hold for more complex multidimensional cases (Eswaran et al. 1981), and in particular for spatial forms typical of MNEs.

In order to motivate the models, we begin by adopting the standard microeconomic assumption that the firm aims to maximize its profits. Later on we will partially relax this assumption when distinguishing between different firm types. However, for the moment, by assuming the profit maximizing rationale for the firm, the question of where a firm will locate within the Weber framework becomes the question of at which location a firm will maximize its profits. The optimal location decision is thus related to profitability, and profits maximization means that returns to the capital investment are also maximized.

Figure 3.2 describes a Weber location-production triangle, in which the firm consumes two inputs in order to produce a single output. In this model, we assume that the goods being produced by the firm are physical and transportable commodities, as would be the case for a manufacturing firm.

In the simple Weber location-production model we assume that the firm consumes material inputs 1 and 2, which are then combined by the firm at a location K in order to produce an output commodity 3. K represents the location of the firm's capital investment and, as yet, this is to be determined. The production locations of the input sources of 1 and 2 are defined as S_1 and S_2 and are assumed to be given, as is the location of the output market M_3, at which output good 3 is sold. The prices per tonne of the inputs 1 and 2 are given as p_1 and p_2, at the points of production S_1 and S_2, respectively. The price per tonne of the output good 3 at the market location M_3 is assumed to be given as p_3, such that the firm is a price taker; and we also assume that the firm is able to sell unlimited quantities of output 3 at this given output market price of p_3. The transport rates are denoted as t_1, t_2 and t_3, and they represent the costs of transporting one tonne of each commodity 1, 2 and 3, respectively, over one kilometre. Finally, the distances d_1, d_2 and d_3, represent the distances over which each of the goods 1, 2 and 3 are shipped.

We also assume that the coefficients of production for the firm located at K are fixed, so that there is a fixed relationship between the quantities of

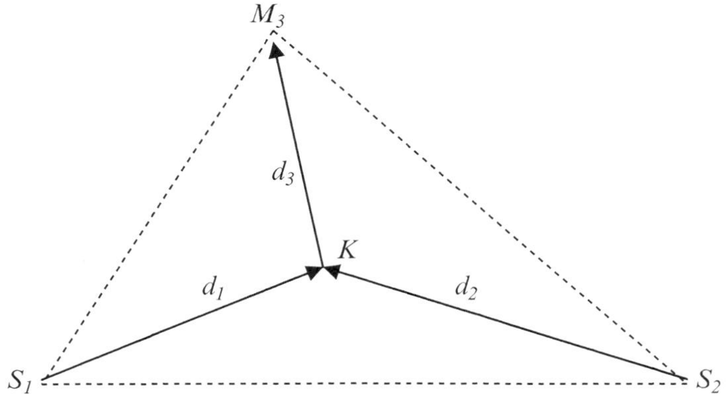

Notes:

m_1, m_2	weight (tonnes) of material of input goods 1 and 2 consumed by the firm.
m_3	weight of output good 3 produced by the firm.
p_1, p_2	prices per ton of the input goods 1 and 2 at their points of production.
p_3	price per ton of the output good 3 at the market location.
S_1, S_2	production locations of input goods 1 and 2.
M_3	market location for the output good 3.
t_1, t_2	transport rates per ton-mile (or per ton-kilometre) for hauling input goods 1 and 2.
t_3	transport rates per ton-mile (or per ton-kilometre) for hauling output goods 3.
K	the location of the firm.

Figure 3.2 Weber location-production triangle

each input required in order to produce a single unit of the output. These coefficients are assumed, for the purposes of this model, to be exogenously given.

Our production function therefore takes the general form:

$$m_3 = f(k_1 m_1, k_2 m_2) \tag{3.1}$$

where k_1 and k_2 are fixed coefficients.

In order to develop the locational analysis we also assume that the labour and capital input production factors are freely available everywhere at factor prices and qualities that do not change with location, and that land is homogeneous. In other words, we assume for the moment that the price and quality of labour is equal everywhere, as is the quality and rental cost of land. This does not imply that the prices of labour and land are equal to each other, rather that all locations exhibit the same attributes in terms of their production factor availability. Geography and space are therefore assumed to be homogeneous.

If the firm is able to choose to locate anywhere it wishes, in order to maximize the returns to its capital investment the firm will therefore locate at whichever location allows it to earn maximum profits. In the model structure, so far we have assumed that the prices of all the input and output goods are exogenously set, and that the prices of all production factors are invariant with respect to space. Therefore, in this particular framework, the only issue which will alter the relative profitability of different locations is the distance of any particular potential production location from the geographically fixed input sources S_1 and S_2 and the geographically fixed output market point M_3. The reason for this is that different production locations will incur different costs associated with transporting inputs from their source production points S_1 and S_2 to the location of the firm K, and transporting the output from the location of the firm K to the market point M_3.

With these simple assumptions it is now possible to write the profit (π) function of the firm as:

$$\pi = m_3(p_3 - t_3d_3) - m_1(p_1 + t_1d_1) - m_2(p_2 + t_2d_2) \qquad (3.2)$$

in which the firm's profit is defined as the sales revenue, calculated at the location of the firm at K, minus the total value of the input purchases minus the transportation costs incurred on both inputs and outputs. The gross sales revenue at the market point M_3 is denoted as p_3m_3, but the net sales revenue accruing to the firm at the location of the firm K is defined as minus the transportation costs of the output shipments. Therefore, the first term in brackets on the right hand side of equation (3.2) represents the net sales revenue received at K per tonne of output after taking out the output transportation costs, and the second two terms in brackets on the right hand side of equation (3.2) represent the delivered prices of the input goods at the location of the firm K.

Rearranging (3.2) now gives:

$$\pi = [m_3p_3 - m_1p_1 - m_2p_2] - \{m_3t_3d_3 - m_1t_1d_1 - m_2t_2d_2\} \qquad (3.3)$$

Here we can now split the purchase costs out from the transportation costs. In equation (3.3) we see that the terms in the square brackets are all assumed to be invariant with respect to location, while the terms in the second set of brackets are all dependent on the location K of the firm. As such, if the price per unit of output p_3 is fixed, the location that ensures that maximum profits are earned by the firm, denoted as K^*, is the location at which, *ceteris paribus*, the total input plus output transport costs are minimized. This is known as the *Weber optimum location* (McCann 2001).

In principle finding the Weber optimum location K^* involves comparing the relative total input plus output transport costs at each location: the actual K^* will be the particular location at which the sum of these total costs (TC) is minimized. The cost condition that determines the Weber optimum location can be described as:

$$TC = Min\sum_{i=1}^{3} m_i t_i d_i \tag{3.4}$$

where the subscript i refers to the particular weights, transport rates, and distances over which goods are shipped to and from each location point K. In the simple Weber location-production model, the value of $i = 3$. However, this general and simple principle can be applied to any situations where $i > 3$, in which the firm has multiple input source and output market locations. With information on the actual values corresponding to each of the spatial and non-spatial parameters, in the simple triangular case described here it is quite straightforward using basic geometry to calculate the total production plus transportation costs incurred by the firm associated with being at any arbitrary location K. The total costs for all locations within the triangular space can then be compared, and the location with the lowest total costs is denoted as the Weber optimum location K^*. For more complex geometrical shapes, in which there are multiple input source output market locations, it is possible to use operations research techniques or Geographical Information Systems techniques in order to compute the total costs associated with any given location K. Once again, by comparing all of the possible locations for K, we are able to identify the Weber optimum K^*. Given our assumptions that the firm will behave so as to maximize its profits we can therefore assume that, as long as the firm has all of the available information at its disposal, the minimum cost location K^* will be the actual chosen location of the firm.

Given our Weber model assumptions in which the firm is assumed to be a price taker for both inputs and outputs, it can be easily demonstrated that the actual chosen Weber optimum K^* will be independent of the values of p_1, p_2 and p_3. Moreover, once the Weber optimum has been found, the optimum location of the firm will remain in the same place irrespective of the output level m_3, and the required inputs quantities m_1 and m_2, as the relationships between these are assumed to all be exogenously fixed.

Although it is always possible to calculate the optimum location of the firm in each specific case, of particular interest to us here is to understand how the location of the Weber optimum will itself be affected by the levels of, and changes in, any of the parameters p_1, p_2, p_3, m_1, m_2, m_3, t_1, t_2, t_3. For the moment, we will ignore the effects of the input and output prices p_1, p_2

and p_3 on the optimum location decision, although we will return to this issue shortly. At this point we will instead concentrate on the effects of differences in the values of m_1, m_2 and m_3 and the impacts that these have on the optimum location K^* of the firm.

In order to explain this, we can adopt a hypothetical set of examples.

We can imagine that the Weber location-production model in Figure 3.2 represents the case of a firm that produces automobiles from inputs of steel and plastic. We can consider that the output good 3 is therefore defined in terms of automobiles and these are sold at the market point M_3. Now we can assume that input 1 is steel and input 2 is plastic, and these are produced at source locations S_1 and S_2, respectively. Let us assume that the automobile manufacturing firm produces a car weighing two tonnes from one tonne of steel and one tonne of plastic. The relationships between the total weight of the final car which is produced and the specific weights of the steel and plastic required in order to produce each car represents the production function coefficients, described in equation (3.1) above. If the fixed transport rate per tonne per kilometre for steel t_1 is half that for plastic t_2 – given that plastic is much less dense and bulkier than steel and therefore more expensive to transport – the automobile manufacturing firm will locate relatively close to the source of the plastics production S_2. The reason is that the firm will wish to reduce the higher transport costs associated with shipping plastic inputs relative to shipping steel inputs. The firm can do this simply by reducing the value of d_2 relative to d_1, the result of which is that the firm moves closer to S_2. On the other hand, if the automobile manufacturing firm had a different production function, such that it produces a car weighing two tonnes from 1.5 tonnes of steel and 0.5 tonnes of plastic, then even with transport rates t_1 for steel which are half that of the transport rate for plastics t_2, the automobile manufacturing firm will now be incurring higher total transport costs associated with steel shipments. The reason for this is that although a tonne of plastic is twice as expensive to transport per kilometre as steel, the total quantity of steel being shipped is three times that of plastic. If the shipment distances for the two inputs are identical, in which case the firm would be equidistant between S_1 and S_2, the total transport costs on steel would be 1.5 times higher than the total transport costs of plastics. In order to reduce its total transport costs, the automobile manufacturing firm can reduce the value of d_1 relative to d_2 by moving closer to the source of the steel inputs S_1. The optimum location of the automobile manufacturing firm will therefore now tend towards the location of production for the steel input S_1. Within the Weber framework, this argument allows us to consider the locational effects on the firm of having different production function relationships.

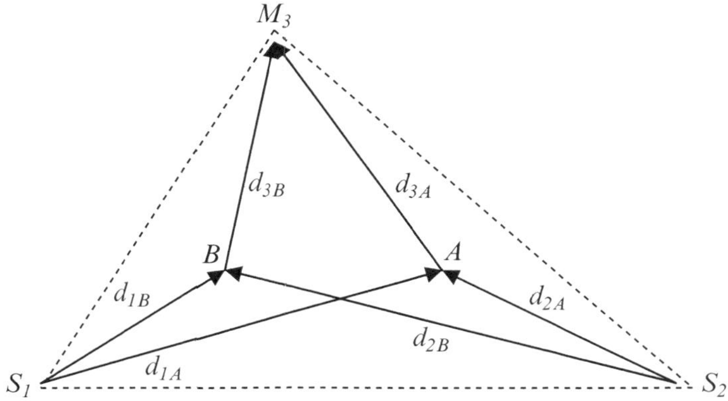

Figure 3.3 *Relative input transport costs and location*

We can also extend this argument to consider how the locational behaviour of different firms will differ according to their production function characteristics. For example, we can imagine two types of automobile-producing firms, one of which has a production function which is relatively plastic intensive, and the other whose production function is relatively steel intensive. As such, for the moment we associate different production function characteristics with representing different firms. Firm A exhibits the plastic intensive production function, while firm B exhibits the steel intensive production function. As we see in Figure 3.3, from the argument above we know that firm A will locate relatively close to S_2, the source of plastic, while firm B will locate relatively close to S_1, the source of steel. If both firms have just a single establishment each, then the actual locational outcomes can be easily predicted. If, however, the firms are multiplant establishments, as is the case with all MNEs, then the issue is far more complex, as we will see shortly.

This basic line of reasoning can also be applied to the case of the relationship between the firm's chosen optimal location and the transportation costs involved in its output shipments. Although so far we have only considered the transport cost pull of the input sources on the location decision of the firm, the same line of reasoning suggest that the market itself will display a 'pull' effect on the location behaviour of the manufacturing firm. Once again, we can explain this issue by using our hypothetical example above of two automobile producer firms, A and B, each consuming inputs of steel and plastic, but in order to simplify the analysis we can rule out the locational effect of inputs described above by imagining a situation in which the production function coefficients for the inputs

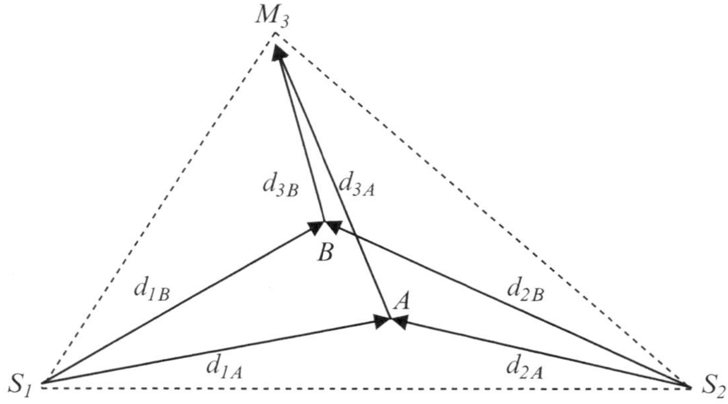

Figure 3.4 Relative output transport costs and location

of both firms are exactly the same. This allows us to concentrate only on the locational effect of output shipments.

Let us assume that firm *A* specializes in the production of small very compact vehicles suited to the traffic conditions in high density urban areas, while firm *B* produces large four-wheel drive vehicles suitable for wide roads and open spaces. In particular, we assume that for every tonne of steel and every tonne of plastic consumed, firm *A* produces two small cars of one tonne each, while firm *B* produces one large vehicle weighing two tonnes. As we have already seen, transport rates depend on the bulk of the product, and products which have a high density will exhibit lower transport costs per tonne than products with a low density. In this situation firm *B* produces goods which are very bulky, whereas firm *A* produces goods which are relatively dense. If a single large four wheel drive vehicle is more bulky than the combined bulk of the two small compact vehicles, then the output of firm *B* will be more expensive to transport than that of firm *A*. This will encourage firm *B* to move closer to the market than firm *A*. As we see in Figure 3.4, in locational terms the result of these differences in the products produced will mean that firm *B* will be more market-oriented than firm *A*.

In all manufacturing production process, inputs are transformed into different outputs. The point of this hypothetical example is to demonstrate that the *change* in density of the products consumed and produced during this transformation process also affects the firm's optimum location. For production processes in which the goods produced increase in density, this tends to encourage the firm to move towards the input sources. In contrast, where the goods produced exhibit a falling density during the production process, then the location will tend towards the market point. Examples here would be the case of microprocessors, whose density tends to increase significantly during the production process and also that of furniture, whose density tends to fall significantly during the production process. These would suggest that within the simple Weber model framework at least, microprocessors will be produced by firms located close to the input sources, whereas furniture will be produced by firms located close to the market.

Such varying locational outcomes can also be generated within the simple Weber framework in situations where firms differ in terms of their technical efficiency. If firm *A* discards 50 per cent of the inputs during the production process, whereas firm *B* discards only 25 per cent of the inputs during the production process, for the same total weight of inputs consumed by each firm, the total output weight m_3 produced by firm *B* is twice as great as that of firm *A*. Following the logic of the model, the greater weight of outputs m_3 produced by firm *B* will also encourage it to move closer towards the market point and further away from the inputs points than firm *A*.

Although our analysis here has so far been developed within just the triangular case of only two input source locations and one output market location, the Weber location-production types of arguments are perfectly applicable to the case of firms with multiple input and output locations (Eswaran et al. 1981; Revelle 1986), as is typical of MPDEs and MNEs. Figure 3.5 depicts the case of a single establishment firm which has multiple input source locations S_1, S_2, S_3, S_4, and S_5, and also multiple output market locations, M_3, M_4, M_5 and M_6. In fact, this type of complex supply chain system involving multiple input-output arrangements and which produces a polygonal form of location-production system, is the normal observed case, and is almost always the case for MNEs. In contrast, while the triangular type of system depicted in Figures 3.2–3.4 is observationally an unusual case, it is not analytically a special case, in that all the optimum location results generated by the triangular case are also applicable to all polygonal cases. Therefore, the reason for employing the triangular case of the two input locations and one output market location is simply pedagogical, in that this particular spatial structure is the easiest two-dimensional model to explain.

The importance of the simple Weber model is that it helps us to understand the advantages which geography confers on particular locations as sites for mobile firm investment. However, the Weber analysis so far has been based on the assumption that the firms have a fixed coefficients production function. More complex models based on the seminal paper of Moses (1958) demonstrate that the basic Weber insights can also be shown

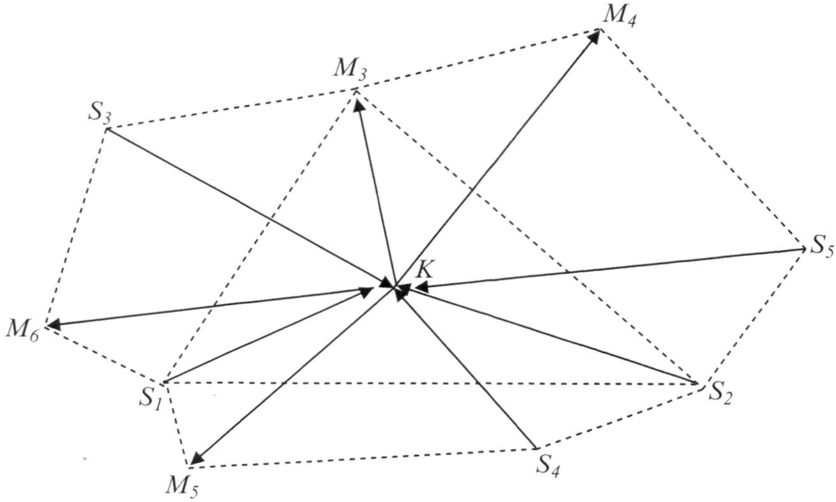

Figure 3.5 Polygonal location-production system

to hold for firms with more orthodox production functions which allow for input substitution between the inputs sources from S_1, S_2, S_3 and so on. (Moses 1958; Khalili et al. 1974; Miller and Jensen 1978; Eswaran et al. 1981). Although the technical analysis of these models is beyond the scope of this book, it is interesting to highlight the main conclusions. In particular, firstly, where the prices of inputs and outputs p_1, p_2, and p_3 are not only exogenous but are also independent of the distance costs t_1, t_2, t_3, and so on, the optimum location of the firm K^* stays exactly the same even as the firm expands its level of output, as long as the firm exhibits constant returns to scale and also firm's expansion path is a straight line from the origin (McCann 2001). This is the case where the firm's production function is either linear, as in a Leontief function, or homogenous of degree one, as in a Cobb-Douglas function. In other words, with these particular types of production functions, once we calculate the firm's Weber optimum K^*, this will always remain the firm's optimum location whatever the firm's level of output, as long as the location-specific labour prices and land prices do not change. This is empirically a very useful insight, because the production functions of many manufacturing firms in reality do have these particular properties (Morroni 1992). In situations where as the firm expands its output it tends to substitute in favour of one input and away from another input, then the optimum location of the firm will tend to move towards the input whose use is increasing faster as the output expands.

On the other hand, it is also possible to argue that input and output prices are often not independent of distance costs t_1, t_2, t_3. Large firms consuming large quantities of inputs and producing large quantities of outputs are often able to achieve economies of scale in transportation either via their own fleet management practices, or by negotiating favourable deals with haulage companies. As such, as the firms become larger their transport rates may fall. Under these conditions, the Weber model's optimum location outcomes become much more complex and depend on the relative changes in the transportation rates of the inputs and the outputs. In general though, we can still conclude that the firm will tend to move away from the shipment direction where the transport costs have fallen the most in order to economize on the relatively higher transport costs.

In the most complex situations where both the production coefficients and the transport rates are changing as output expands, the actual optimum location will depend on the relative directions and rates of change of these various features.

3.2.2 The Weber Location-Production Model: Varying Input Factor Prices

So far in our location theory models we have been considering simply the case of a firm which has a single production establishment. We have also been focusing on the case where a firm is producing physical tangible goods and merchandise. We will keep these assumptions intact during the following sections in order to keep the explanations of the theoretical models as simple as possible. In the final sections of this chapter we will then start to relax these assumptions and allow for the case of firms with multiple establishments, as is typical of MNEs, and also allow for firms producing intangible products and services.

The locational analysis so far has proceeded on the assumption that labour and land prices are identical across all locations, whereas we know that in reality the prices of production factors such as labour and land can vary significantly over space. Yet, it is very straightforward to employ the Weber-type of approach in order to consider how factor price variations across space will affect the location behaviour of the firm. In order to make the models simple, for the moment we will also assume that the costs of relocating a firm's establishment are zero.

Here we can once again assume that the firm is still consuming inputs from S_1 and S_2 and producing an output for the market at M_3. From the above discussion, we know that if all factor prices are equal across space the Weber optimum location K^*, which is the minimum transport cost

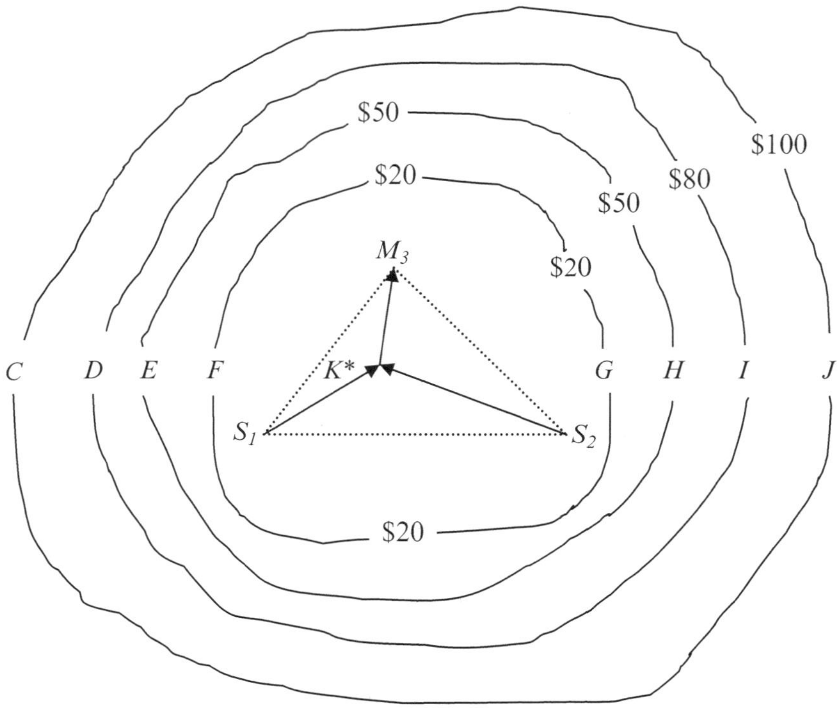

Figure 3.6 Isodapane analysis

location, will be the location of the firm. Moreover, even if we extend the analysis to incorporate all logistics costs elements, the basic principles remain the same. For simplicity, however, we will confine the analysis here to the case of the Weber optimum. Our starting point now is therefore to consider by how much factor prices will need to vary across space, relative to the Weber optimum K^*, so as to encourage a firm to move away from the Weber optimum to another location. In order to do this, it is first necessary for us to construct a contour map on our Weber triangle, as depicted in Figure 3.6, using contours known as *isodapanes*.

On a normal geographical map contours are the lines that link all the locations with the same altitude. On the other hand, in a Weber model analysis, each isodapane links all the locations which exhibit the same increase in total input + output transport costs, per unit of output m_3 produced, relative to the Weber optimum location at K^*. As with contours on a normal geographical map, isodapanes cannot cross each other. Therefore, isodapanes at a greater distance from the Weber optimum K^*

reflect higher values of total input plus output transport costs, per unit of output m_3 produced, relative to the Weber optimum K^*. In other words, if the location of the firm were to be moved away from the Weber optimum in any direction, the firm would be incurring increasing transport costs relative to the Weber optimum. The result of this is that, *ceteris paribus*, if factor prices are equal across space, the firm would exhibit successively lower profits relative to K^*, and therefore locations further away from the Weber optimum will become successively less desirable sites for investment.

This line of argument now allows us to ask directly the question: by how much do *local* factor prices need to fall at a particular location, relative to the Weber optimum location K^*, in order for the firm to move to that particular location?

If we take the case of location *H*, we can thus ask by how much do factor prices at *H* need to fall relative to the Weber optimum K^*, in order for the firm to move from K^* to *H*? As we see from Figure 3.6, *H* is on the $50 isodapane. If the costs of the labour and land factor inputs required to produce one unit of output m_3 at *H* are $40 less than at K^*, it will not be in the interest of the firm to move from K^* to *H*. The reason is that the fall in local factor input prices of $40 associated with a move from K^* to *H* would not be sufficient to compensate for the increased total transport costs of $50 if the firm's location were moved this far away from the Weber optimum K^*. As such, if the firm's location was moved from K^* to *H* under these circumstances it would experience profits which were $10 per unit of output m_3 less than at K^*. On the other hand, if the local labour and land prices per unit of output as *H* were $55 less than at K^*, it will be in the interest of the firm to move, because the reduction in the local input factor costs associated with a move from K^* to *H* will be greater than the total transportation costs incurred by the move. The lower local factor prices therefore more than compensate for the increased total transport costs. Under these circumstances the firm would experience profits which were $5 per unit of output m_3 greater than at K^*.

This type of argument can be equally applied to any alternative locations, such as *C, D, E, F G, I* or *J*, in order to determine whether or not a firm should move away from K^*, and which location it should relocate to.

For example, location *G* is on the $20 isodapane, *H* is on the $50 isodapane, *I* is on the $80 isodapane and *J* is on the $100 isodapane. Let us assume that the costs of the labour and land factor inputs required to produce one unit of output m_3 at *G*, *H*, *I*, and *J* are less than the factor costs at K^* by amounts of $22, $55, $84, and $106, respectively. We can determine that all these alternative locations from *G* to *J* are superior locations to K^*, in that they all will provide greater profits than K^*. However,

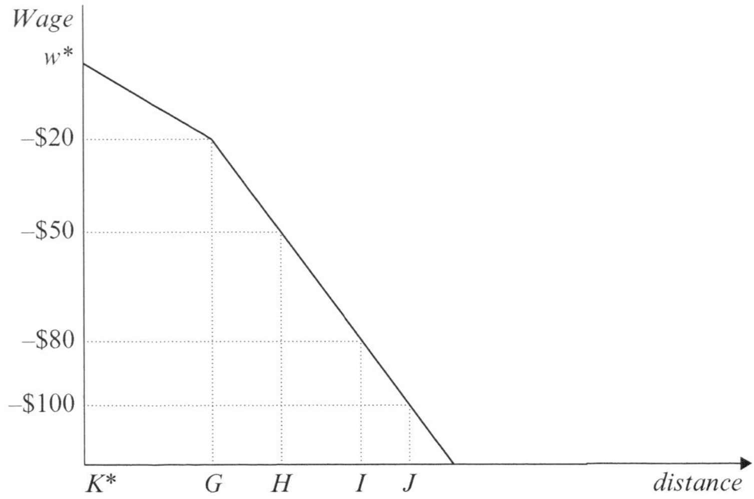

Figure 3.7 Equilibrium labour prices

of these superior alternatives, *J* is the best location because profits here are \$6 per unit of output greater than at *K**, whereas at *G* they are only \$2, at *H* only \$5, and at *I* only \$4 greater than at *K**. In preference ordering we can say that *J*>*H*>*I*>*G*>*K**, and that the firm will prefer to move to location *J* so as to maximize its profits. What this stylized example demonstrates is that as long as the variation in local land and labour factor prices is greater than the transport costs defined by the isodapanes, then the firm will be willing to move to alternative locations.

This type of approach now directly allows us to ask: by how much will local wages and land prices have to vary over space in order for the firm's profits to be the same for all locations? The answer implies to identify the critical change in local wage and land prices with respect to *K** just beyond which it will allow a firm to move away from *K**. This critical change in local factor prices can be understood with the help of Figure 3.7, in which the vertical axis represents the costs, and the horizontal axis is constructed by drawing a line from *K** eastwards passing through *G*, *H*, *I*, *J*, according to the geographical distance of each of these locations.

From the above example, we know that location *G* is on the \$20 isodapane, *H* on the \$50, *I* on the \$80 and *J* on the \$100 isodapane. We can therefore plot the positions where each of the \$20, \$50, \$80, and \$100 isodapanes intercepts with both the vertical axis and the horizontal axis.

The critical change in factor prices just beyond which a firm will be

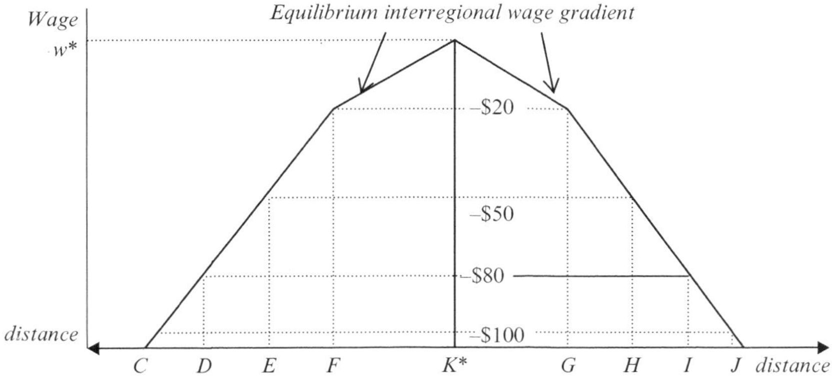

Figure 3.8 Interregional equilibrium wage gradient

willing to relocate away from K^* is exactly that change in factor prices which makes the firm indifferent between K^* and another location. The local factor price change relative to the prices at K^* is the change which ensures that the firm's profits will be exactly the same in all locations as that of K^*. In other words, the critical variation in local factor prices relative to the local factor prices at K^* is such that the local labour and land factor input prices at each location exactly compensate for the increased total transport costs associated with each location. Therefore in Figure 3.7 this allows us to plot the labour and land price gradient with respect to distance which ensures equal profits are made at all locations east of K^*, assuming the wage at K^* is w^*.

We can repeat the same exercise by drawing a line from K^* which passes west through *F*, *E*, *D* and *C*, and plotting the local factor prices which will ensure the firm makes profits equal to at K^* at all locations west of K^*. By combining all of this information we are now able to construct the interregional factor price gradient for a particular firm which ensures that the firm makes equal profits at all locations in the east-west direction as it would being located at K^*. This is shown in Figure 3.8.

This slope of the line in Figure 3.8 defines the interregional *equilibrium* factor price gradient for this particular firm along this particular axis, in which the firm is indifferent between all locations along the east-west line between *C* and *J* as the profits that the firm can earn are the same everywhere along this line. Moreover, although in Figure 3.8 here we only have a one-dimensional framework, in principle using exactly the same logic we can also construct similar factor price gradients for movements in any other direction away from K^*. This allows us to generate a two-dimensional equilibrium factor price map of the whole spatial economy.

Indeed, large firms often do exactly this with the help of GIS analysts when they are searching for new locations.

In Figure 3.8, if all local factor prices are interregional equilibrium prices, the firm will be indifferent between all locations. Under these circumstances, the firm will be equally likely to build its new production facility at any location. More generally, the fact that the firm is indifferent between each location means that the probability of a firm investing in a particular location will be exactly equal to the probability of investing in all the other locations. Therefore, over large numbers of firms with similar input requirements and similar output markets to our particular firm, the level of such investment in any single location should be approximately the same as the investment levels in all other locations.

With this type of analysis it becomes immediately obvious that geography and space confer different profitability advantages on different locations, which can only be compensated for by variations in local factor prices. The concept that locations can be perfect substitutes for each other from the point of view of a firm's profitability is therefore important in terms of our understanding of the spatial patterns of firm investment behaviour, as it defines the critical variation in local factor prices which ensures an even distribution of investment. From the isodapane analysis, we know that without any variations in factor prices across space, the Weber optimum will always automatically be the most profitable location for the firm's investment. Therefore, in order to make other alternative locations attractive for firm investment, local factor prices have to fall relative to the Weber optimum K^*.

If wages are not in equilibrium over space, in terms of profitability certain areas will automatically appear more attractive as locations for investment, and there are two possible general cases here. The first one is where the observed interregional wage gradient is less steep than would be required in order to generate the equilibrium wage gradient. In this case, all investment will tend to concentrate around the Weber optimum location. In contrast, if the observed wage gradient is steeper than the equilibrium wage gradient, then investment will tend to move away from the Weber optimum. More broadly, the attractiveness of any particular location as a new investment location for the firm will be positively related to the extent to which the local factor price falls can more than compensate for the increased transport costs associated with any sub-optimal geographical location.

So far, these generalizations have been made on the basis of a single firm with particular input and output coordinates, a particular production function, and facing particular transport costs and input and output prices. Yet, the fact that different firms exhibit both different production functions and also different spatial layouts of input suppliers and output markets, means that the Weber optimum location is uniquely defined for each individual firm. Therefore, different equilibrium interregional factor price gradients will exist for different types of firms. However, in order to find the ideal production location, this implies that firms must carefully consider where they purchase inputs from, where they sell outputs to, the costs that they face regarding input and output shipments, and how each of these features relates to their own particular production functions. Firm location decisions are major strategic investment decisions and a careful evaluation of the costs and benefits of each alternative location is required in order to ensure that a firm makes the locational decision which is the optimum for itself. Yet, allowing for the fact that the optimum location outcome is unique for every individual firm, it is still also true that similar optimum locational outcomes are to be expected for firms which are broadly similar in terms of the products they produce, their production functions, the input sources used and the markets they target (McCann 1995).

3.2.3 The Weber Location-Production Model: New Input Sources and Markets

One of the frequently observed aspects of MNE locational behaviour is that MNEs tend to search for an initial foreign investment location, and then over time they search for either new locations or additional investment locations. Yet, the Weber type of analysis which incorporates isodapanes and varying local factor prices is also very useful for analysing the conditions under which over time a firm will search for alternative locations. Until now, our Weber analysis has assumed initially that the firm's location decision is a 'one-shot' investment decision relating to a single location, which can then be extended to consider the conditions under which a firm will move to another location. However, this process of movement itself may engender changes in the input sources employed and the output markets served by a firm, the result of which is a dynamic process of firm spatial behaviour. Once again we can employ the isodapane type of analytical framework in order to examine these issues.

In Figure 3.9 we can consider the situation where the automobile manufacturing firm relocates from K^* to L in response to the lower factor prices at L. Following the previous arguments, the reason why the firm moves from K^* to L is not only that the lower local factor prices at L are more than sufficient to compensate for the additional input and output transport costs involved in consuming steel and plastic inputs from S_1 and S_2, and serving a market at M_3, but that the gain in profitability at L is greater

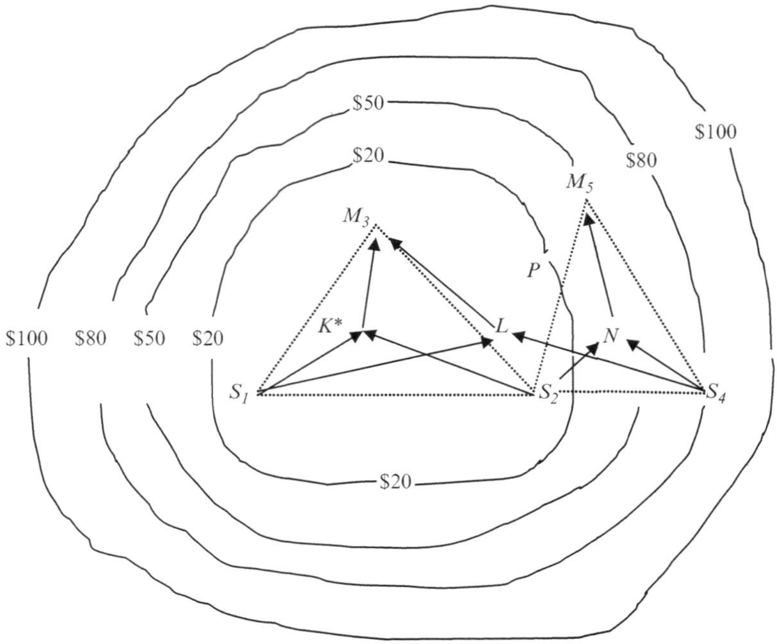

Figure 3.9 New suppliers and new markets

than for any other location. Location L has therefore been determined as the new optimum location with respect to S_1, S_2, and M_3. However, in moving from K^* to L, it may be that alternative suppliers of identical inputs now become available, which previously were not realistic supply options. For example, the input supply location S_4 may be able to supply exactly the same steel input as S_1 but from a location whose distance from S_4 to L, which we denote as d_4, is now much closer to the firm than the distance d_1 from S_1 to L. Assuming the delivered price at location L of the steel input produced at S_4 is less than that of the input produced at S_1, that is, $(p_4+t_4d_4)<(p_1+t_1d_1)$, the firm will be willing to substitute input supplier S_4 for S_1. This will produce a new Weber location-production problem, with the points S_4, S_2, and M_3 as the (triangular) spatial reference points.

This change in input suppliers itself changes the geography of the input-output linkages of the firm and obviously also implies that a new Weber optimum can be determined which must be different to both L and also K^*. For example, we can imagine that with the points S_4, S_2, and M_3 as the spatial reference points, the new Weber optimum for the automobile manufacturing firm is found to be at N. However, at point N, it also now becomes advantageous for the firm to serve market point M_5, rather than M_3. This is because M_5 is nearer to N than M_3, and $(p_5-t_5d_5)>(p_3-t_3d_3)$, so the firm will be able to make a greater profit from selling automobiles to market M_5 than to market M_3. One option is for the firm to simply switch markets completely from M_3 to M_5. Alternatively, it could decide to supply both markets M_3 to M_5 at the same time, under which conditions it may be that a new Weber optimum location of P arises, in which the firm at P buys from two supplier locations S_4 and S_2, and sells at two market locations, M_3 and M_5. More complex spatial arrangements are obviously also possible. For example, in order to guarantee sufficient supplies of steel inputs for the newly expanded automobile market of (M_3+M_5), the firm may decide to continue to purchase steel from both S_1 and S_4, as well as purchasing plastics from S_2. Now we have a Weber location-production problem with S_1, S_2, M_3, S_4 and M_5 as spatial reference points. Once again, this will move the Weber optimum away from point P.

The point here is that location changes in response to location-specific factor price changes also allow firms to take advantage of different input and output spatial arrangements which were previously not possible. As such, location behaviour is very rarely a one-shot phenomenon. Rather, this analysis suggests that location behaviour is generally a dynamic process by which firms make incremental adjustments over time to both their production locations and also their input supply and output market locations. Note that this type of dynamic process in which both the input sources employed and the output markets targeted are continually adjusted over time was triggered by a single initial relocation from K^* to L, that opened up possibilities for the firm to reconfigure its network of input supply and output market points, each change implying a new optimum location. Moreover, we also know that each new optimum locational arrangement will also be associated with a unique set of equilibrium local factor input prices which would render the firm indifferent between locations. If for whatever reason the differences in local factor prices adjust over time by a greater extent than the equilibrium wage gradient requires, then the firm relocation will once again be triggered, which itself will engender further induced changes in the input, output and location behaviour of firms. In the common situations where the firm is an intermediate goods producer such that both output market and input supply points are actually also other supplier and customer firms, then changes in local factor prices may cause these supplier and customer firms to relocate. What we see therefore is that the location process becomes a *dynamic process* in which firms are required to continuously re-assess the appropriateness of their location.

However, observation also tells us that firms in reality do not change their production locations very frequently and probably not nearly as frequently as they change their input suppliers or their output markets. Indeed, the relocation process itself usually incurs very significant costs, such as the dismantling, assembly or re-assembly of equipment, the relocation of some existing staff and the hiring and training of entirely new staff. Yet, the models described above assume that relocation costs are zero. However, this assumption is made simply for reasons of analytical simplicity and, importantly for our purposes, the conclusions from the models are not particularly affected by this assumption. In fact, these relocation costs can easily be incorporated within the above framework, by including the annualized (mortgage payment equivalent) cost of such one-off fixed relocation costs into our isodapane model. These extra costs obviously mean that firms will only move when the factor cost variations of alternative locations also compensate for the additional relocation costs, as well as the increased transport costs. In other words, in order to encourage relocation the negatively-sloped equilibrium interregional wage gradient will need to be increasingly steep the greater are such one-off relocation costs. In addition, other elements of the transactions costs connected with firm relocation are also related to information and uncertainty issues, to which we will now turn.

3.3 THE LOGISTICS-COSTS LOCATION PRODUCTION MODEL

Until now we have assumed that the prices of the input and output goods given as p_1, p_2 and p_3 have no effect on the optimum location of the firm. The reason is that the prices are assumed to be exogenously fixed, such that the Weber optimum location is seen to depend entirely on the quantities of goods m_1, m_2, m_3 shipped and the transport rates t_1, t_2, and t_3 involved in moving these goods. However, in reality, the prices of the goods do play a critical role in the determination of the optimum location, and this is because in practice firms make geographical shipment decisions with regard to the size and frequency of shipments. Such decisions allow for the fact that the firms face both the costs of transporting goods, but also the costs of *not* transporting goods, that is, the costs of holding goods as inventories. By taking these issues into account within a spatial framework, all firms involved in producing or handling merchandise or physical goods have to optimize on the combined costs of transporting versus the holding of goods, and this is the approach adopted by *logistics-costs* models (McCann 1993, 1996, 1998, 2001).

For firms producing and shipping physical commodities and goods, the *time* costs of distance are hidden in the costs of holding inventories. This is true in terms of holding input goods which have to be paid for on delivery, and whose delivered c.i.f. (cost, insurance, freight) inventory holding cost can be written in f.o.b. (free on board) terms as $(p_I + t_I d_I)$ or $(p_I + t_I d_I)$. As such, the cost of a unit of inventory is clearly a function of distance. The further an input has to travel to the plant, the higher is the cost of a unit of input inventory being held by the firm at K. The time costs of holding these input inventories depend on the average value of inventory held at any time, and assuming a stable flows of inputs into the factory, the average value of input inventory held by the plant can be written as $Q_1(p_1 + t_1d_1)$ or $Q_2(p_1 + t_1d_1)$, where Q_1 and Q_2 refer to the size of each individual input shipment. If the input quantities per time period being demanded by the firm are m_1 and m_2 respectively, then the number of shipments per time period, or in other words the shipment frequencies, can be written as $f_1 = m_1/Q_1$ and $f_2 = m_2/Q_2$, respectively. The logistics-costs framework demonstrates that the inventory holding costs for each of these inputs can therefore be written as $IQ_1(p_1 + t_1d_1)/2$ and $IQ_2(p_1 + t_1d_1)/2$, respectively. The inventory holding costs parameter I is comprised of the inventory capital financing interest rate i, the insurance rate r, and the land space-holding costs which are a combination of the rents R per square metre of inventory storage space and the labour prices w associated with managing inventories.

The same argument also holds for output inventories of finished goods which are waiting to be shipped to the market. In this case the output holding costs are the opportunity costs of time associated with a less-than-infinite shipment frequencies, whereby firms forego immediate inflows of funds which can reduce ongoing debt levels, in order to economize on shipment frequencies which increase the transport costs on outputs. In this case, using the same terminology as before, the inventory holding costs for finished goods can be written as $IQ_3(p_3 - t_3d_3)/2$. The total inventory holding costs faced by the firm are the sum of the input plus output inventory holding costs.

The point here is that the costs of economic geography are not just transportation costs as assumed in the simple Weber model and all of its subsequent variants, but also all of the distance-related costs tied up in inventory holding along with the location-specific land costs tied up in holding inventories. The total costs of time are embodied in the firm's total capital costs associated both with the capital tied up with transactions frequencies, and also with the stock of capital held in the form of inventory. This argument follows the Austrian school circulating capital concept of Böhm-Bawerk and closely mirrors the bathtub model of Dorfman (McCann 1995).

Finally, as well as these various capital costs, each time an input or output shipment is made the firm incurs fixed costs associated with switching or re-tooling the production line or reloading the delivery vehicles, and these costs are known as set-up costs S. The total set-up costs incurred per time period can be written as $S_1f_1 = S_1m_1/Q_1$, $S_2f_2 = S_2m_2/Q_2$ and $S_3f_3 = S_3m_3/Q_3$, respectively.

Broadly, the logistics-costs model demonstrates that the more frequently $[f_1, f_2, f_3 \ldots]$ input or output goods are shipped, the smaller is the individual shipment size $[Q_1, Q_2, Q_3 \ldots]$ and the lower are the inventory holding costs $[IQ_1(p_1 + t_1d_1)/2; IQ_2(p_1 + t_1d_1)/2; IQ_3(p_3 - t_3d_3)/2 \ldots]$. At the same time, the more frequently $[f_1, f_2, f_3 \ldots]$ input or output goods are shipped, the higher are both the total transport costs $[m_1t_1d_1; m_2t_2d_2; m_3t_3d_3 \ldots]$ and also the total set-up costs $[S_1m_1/Q_1; S_2m_2/Q_2; S_3m_3/Q_3,]$. As such, firms must balance these opposing trends and for each of the input and output linkages the firm must optimize its shipments so as to minimize the total {transport + inventory holding + set-up} costs incurred.

What becomes apparent from these logistics-costs models is that the combined logistics costs of distance are all square root functions of geographical distance; in other words they are concave with distance even for linear transport costs, and this is true for both input and output shipments. In other words, when time costs are incorporated into the true monetary calculation of distance costs, we see that total distance costs rise less than proportionately with geographical distance, with the distance cost function becoming shallower as distance increases. Moreover, these logistics costs model naturally incorporate the observed structure of transport costs (McCann 2001), without recourse to artificial and unrealistic transport specifications such as the 'iceberg' costs assumed in NEG models (McCann 2005).

The detailed mathematics of these models are beyond the scope of this book (McCann 1998), but in analytical terms the total input and output inventory holding costs can be treated as additive and separable; in other words the individual inventory costs components can be analysed individually. This is important because it allows us to consider the locational impacts of differences in the source prices of the inputs, the market prices of the outputs, the value-added at the plant, and the changes in value-bulk ratios of the products as the production process proceeds.

Importantly for our purposes, the logistics-costs model recasts the Weber optimization problem in terms of finding the location which minimizes the sum of the total logistics-costs of inputs and output shipments, whereby logistics costs equal the sum of the total transport + inventory holding + set-up costs incurred by the firm (McCann 1993, 1998). Apart from an additional layer of sophistication, empirically the justification for using the logistics costs model as an analytical tool is that for many facto-ries, the total inventory holding activities can account for a very high share of total productions costs (McCann 1998), while for distribution centres and retail outlets, these costs account for almost all of the establishments costs.

The logistics costs model provides some rich additional firm locational insights which are very powerful for the arguments in this book. If we allow for the much more complex interrelationships between input and output prices, inventory holding time costs, and transport costs captured by the logistics-costs model, it is possible to demonstrate that the optimum location of the firm K^* moves closer to the market the higher is the value-added by the firm (McCann 1993, 1998). The reason is that for any given set of input prices and input shipment distances, the greater is the value-added by the firm the greater are the financial opportunity costs associated with holding inventories of finished goods. As such, the firm should ship the finished goods frequently in small individual quantities in order to reduce the level of output inventory. However, these types of behaviour implies high transportation costs, which can only be reduced by reducing the distance between the firm and the market, such that the optimum location of the firm moves towards the market. Note that value-added in our Weber-model terminology is defined as $[m_3(p_3 - t_3) - m_1(p_1+t_1d_1) - m_2(p_2+t_2d_2)]$, and for fixed input output coefficients, can be proxied as $[p_3 - (p_1+ p_2)]$. As such, even with land prices and labour prices which do not vary over space, for a single establishment firm, high value-added activities will tend to be more market oriented in terms of location than lower value-added activities.

Moreover, for a single establishment firm situated within a value-chain production hierarchy, the observation that the optimum location of the firm moves towards the market is also found to be true for activities which are higher up the value-chain. This is because the average value of output inventories held will be high, whereby opportunity costs of output shipments will weight heavily on the total logistics costs calculation. Conversely, low value-added activities lower down the value-chain will tend to be located closer to the supply points, because the average value of the output inventories held will be low but those for the inputs will be relatively high in comparison to the output values. As such, a natural spatial ordering therefore tends to emerge, with establishment higher up the value-chain tending to be located closer to market locations than establishments lower down the value-chain.

Where land and labour prices are variable across space the logistic costs model suggest that the requisite equilibrium interregional wage gradients will be convex with distance, even for linear transport costs, whereby the downward slope of the gradient becomes shallower with distance.

The reason is that these convex wage gradients are required to offset the concave distance costs increases described above. However, the slopes of the requisite interregional wage gradients sufficient to engender relocation behaviour are themselves related to the square root of the product price, the value-added by the firm, and the value-bulk features of the products produced. In general plants producing more valuable products or products higher up the value chain are less responsive to wage and land price variations across space. Where establishment relocation is a possibility, lower price production or plants lower down the value chain are more responsive.

If we combine these logistics-costs insights with the basic Weber model insights we can conclude that for firms which are producing goods whose value and bulk are both increasing as inputs are transformed into outputs during the production process, the optimum location of the firm always moves towards the market. This would be represented by the case of furniture manufacturers who produce fully constructed furniture outputs, rather than flat-packed furniture. Meanwhile, for firms producing goods whose value increases while density decreases as inputs are transformed into outputs during the production process, whether the optimum location of the firm moves towards the market or towards the inputs depends on the relative change of the value-added and the transport rates, as these changes tend to 'pull' in opposite directions. This would be represented by the case of firms producing microprocessors. Here, the increasing density of the product during the production process of microprocessors (Arita and McCann 2002) will tend to pull the optimum location of the firm towards the inputs sources, whereas the high value added of the firm will tend to pull the optimum location towards the market (McCann 2008). Optimal location behaviour is therefore seen to be a result of a rich mixture of different 'pull' factors, including the transport rates, the quantities demanded, the values of the goods, and the bulkiness of the goods. In each case, the optimum location will need to be calculated.

In general, these logistics-costs conclusions both overlay and encompass the existing Weber solutions, but in addition provide a much richer and nuanced set of insights. In particular, the idea that the location of an establishment is related to the position of the establishment in the value-chain is a powerful insight for the behaviour of multinational firms. According to the logistics costs model MPDEs and MNEs will construct value-chains in space whereby higher order activities systematically operate closer to the market. This is as true for production activities as it is for knowledge-related service activities (McCann 2007).

Three final points are worth mentioning here as they pertain to the location behaviour of MNEs. Firstly, under this much broader logistics-costs framework, the simple Weber and Moses-type results can be shown to be no more than specific cases, which only hold for establishments where the value-added at the establishment is zero, such as in the case of a distribution warehouse or of a retail facility (McCann 1998). Indeed, it is somewhat ironic that most models of firm location are actually models of warehouse or store location. However, given that much of modern day globalization is about changes in the geography of distribution as well as production the models are still very applicable.

Secondly, the general insights of the logistics costs model can also be shown to hold for the movements of people (McCann 2007; McCann et al. 2010) across a range of knowledge-related transactions decisions. In particular, the overall schema mapped out by the logistics costs model (McCann 1998) is seen to largely hold for the movements of knowledge embodied in human capital, as is typical of service industries. Much of modern day globalization is also related to the interactions between key knowledge workers, and this is particularly critical for service industries or the knowledge-related activities within manufacturing, extraction or distribution sectors. As such, these broad logistics costs-type arguments are more or less equally applicable to service industries as well as to the manufacturing and distribution sectors.

Thirdly, as we will see in Chapters 6 and 7, one of the key features of modern globalization is an increase in consumer and producer preference for timeliness of delivery and speed of shipment, in response to a range product and service quality demands. These trends increase the optimum shipment frequencies, reduce the optimum shipment sizes, and consequently also increase the economic costs of distance, for any cartographical definition of distance (McCann and Fingleton 1996). These changes have profound implications for the emerging relationships between economic geography and contemporary globalization, implications which can be neatly captured on the basis of a logistics costs model.

3.4 MARKET AREA ANALYSIS: SPATIAL MONOPOLY POWER

As we mentioned at the beginning of the chapter, in terms of multinational firms the location behaviour characterized by these various Weber-type models is indicative of *resource seeking* strategies, and in particular of those MNEs aiming at host locations' physical resources and cheap labour (e.g., *general asset seekers*); though such models can be easily extended to represent those MNEs rationalizing across international differences in factor costs, endowments and levels of development (i.e. *efficiency seekers*, and in particular those named in Chapter 2 *global value chain efficiency seekers*).[2] However, MNEs also employ location strategies which are based primarily on *market seeking* behaviour and, in order to understand the nature of the latter, we need to make the geographical and spatial aspects of market areas faced by firms explicit in our models. This requires us to employ a different set of analytical location techniques. In what follows we employ two such models to illustrate the issues involved: the Hotelling and the Salop spatial competition models. They differ fundamentally from the Weber type models because of their treatment of the market area. This allows us to derive different insights into the nature of firm location strategies, and one of the key issues which becomes evident is that geography and space can confer *monopoly power* on firms. In particular, firms are seen to engage in spatial competition by adjusting their location behaviour so as to try to acquire monopoly power. This leads to questions regarding the *interdependence* between firms in terms of their strategic decision-making and, as we will see, in an oligopolistic setting the location behaviour of an individual firm becomes inextricably linked to the location decisions made by its competitors.

In order to understand the relationship between market areas and monopoly power we can adopt a diagrammatic approach based on the framework originally developed by Tord Palander (1935). In Figure 3.10 we have two firms *A* and *B* located at points *A* and *B* along a one-dimensional market area defined by *O-L*. For this model we assume that both firms are producing an identical product. In Figure 3.10 we examine the case where the production costs are different between the two firms while the transport rates faced by them are the same. In the diagram, the production costs p_a of firm *A* at location *A* can be represented by the vertical distance *a'*, and the production costs p_b of firm *B* at location *B* can be represented by the vertical distance *b'*. Meanwhile the transport costs faced by each firm as we move away in any direction from the location of the firm are represented by the slopes of the transport rate functions. As we see here the transport rates for the two firms in this case are identical, i.e. $t_a = t_b$. As is also very clear from the diagram, in terms of pure production firm *A* is much more efficient than firm *B*.

In a standard non-spatial model of competition between two firms, a duopoly, the firm which is more efficient will dominate all of the market. However, in the case of an explicitly spatial market such as *O-L*, this is not necessarily the case, as firm *B* is still able to survive.

For any location at a distance d_a away from firm *A*, the delivered price of the good is given as $(p_a+t_ad_a)$, and for any location at a distance d_b away

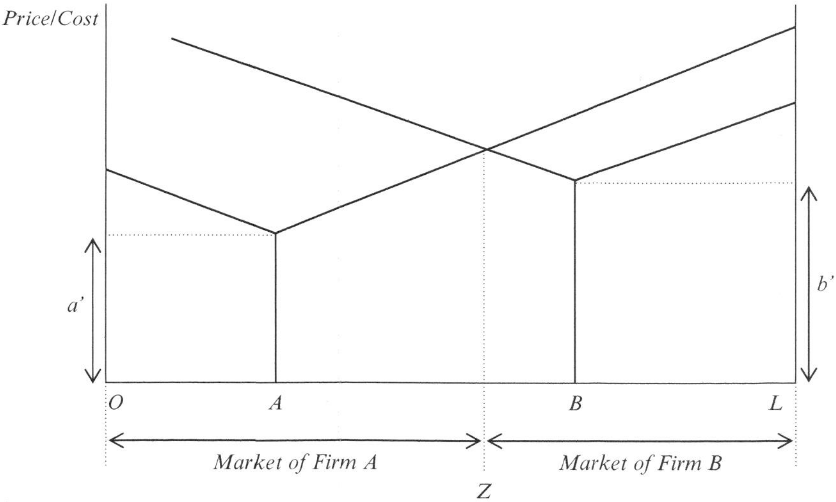

Figure 3.10 *Spatial market areas: a one-dimensional model with equal transport rates*

from firm *B*, the delivered price of the good is given as $(p_b+t_bd_b)$. In this model, for reasons of simplicity we can assume that consumers are evenly distributed along the line *OL*, and that rational consumers will buy from whichever firm is able to supply at the lowest delivered price at their particular location. In order to understand how the behaviour of the consumers determines the market areas under these assumptions, we first note that in Figure 3.10 the total market area will be divided into two areas *OZ* and *ZL*. The reason for this is that between *O* and *Z*, the delivered price of firm *A*, given as $(p_a+t_ad_a)$, is always lower than the delivered price of firm *B*, for all consumers located between *O* and *Z*. Similarly, at all locations between *Z* and *L* the delivered price of firm *B*, given as $(p_b+t_bd_b)$, is always lower than that of firm *A*, for all consumers located between *Z* and *L*.

Although firm *A* is more efficient than firm *B*, and although both firms produce an identical product, firm *A* does not gain all of the market, nor does firm *B* disappear. The reason is that geography, in terms of location and space, gives each firm some monopoly power over the area around itself. The result in Figure 3.10 is that even though firm *A* is more efficient that firm *B*, firm *A* is not able to capture all of firm *B*'s market, because the transport costs associated with shipping firm *A*'s goods to locations close to firm *B* increase the delivered price $(p_a+t_ad_a)$ of firm *A*'s goods to an uncompetitive level in the vicinity of firm *B*. As such, firm *A* is unable to compete for sales and market revenue in the vicinity of firm *B*, simply because it is too far away. In contrast, firm *B* is successful in selling close to its own location, even though it is less efficient in terms of production.

As we have seen, transport costs are thus a form of spatial transactions costs which provide a degree of protection for some inefficient firms against the more efficient firms. Other forms of spatial transactions costs are tariffs, customs duties and border taxes, as well as some knowledge and information costs which will be discussed in later chapters. The actual degree of spatial market protection which is provided to the firm by the transport-transactions costs associated with geography is manifested in terms of the scale of the local market monopoly power which an inefficient firm still exhibits. Only in the case where transport rates are zero is a lower production price sufficient to ensure a more efficient firm to capture all of the market.

The diagrammatic spatial monopoly analysis depicted in Figure 3.10 can now be extended to also allow for differences in transport rates between firms. Figure 3.11 provides three different hypothetical examples of how spatial market areas might be demarcated.

In the first hypothetical example depicted in Figure 3.11a, although producer *A* exhibits much lower production costs than producer *B*, firm *B* still has a monopoly power over its local market because its transportation costs are much lower than firm *A*. This is a more extreme case of the situation discussed in Figure 3.10 in that, as well as the pure locational advantage, the transport cost advantage of firm *B* serves to further limit the potential efficiency advantage of firm *A*. In contrast, in Figure 3.11b, firm *B* is able to maintain a degree of monopoly power over its local market, even though both its production costs and its transportation costs are higher than those of firm *A*. Of the cases which can often be observed empirically, in many ways this is the most extreme form of monopoly power protection which can be afforded by geography, because in an aspatial market firm *B* would automatically disappear. A common example of this phenomenon is the case where firms *A* and *B* are both bakeries, but firm *B* is a local bakery which manages to maintain a very small local market area in the face of national competition from firm *A*, a national bakery distributing through supermarkets. An even more extreme case of spatial monopoly power is indicated by Figure 3.11c, in which firm *B* does not dominate in its own immediate geographical vicinity, yet still manages to maintain monopoly power over some more remote region, beyond the reach of firm *A*.

In general, what we see is that the size of a firm's market area will be larger the lower are its production costs and its transportation rates, and the area over which a firm exhibits monopoly power is typically the area in which it is located. While the last case depicted in Figure 3.11c is an

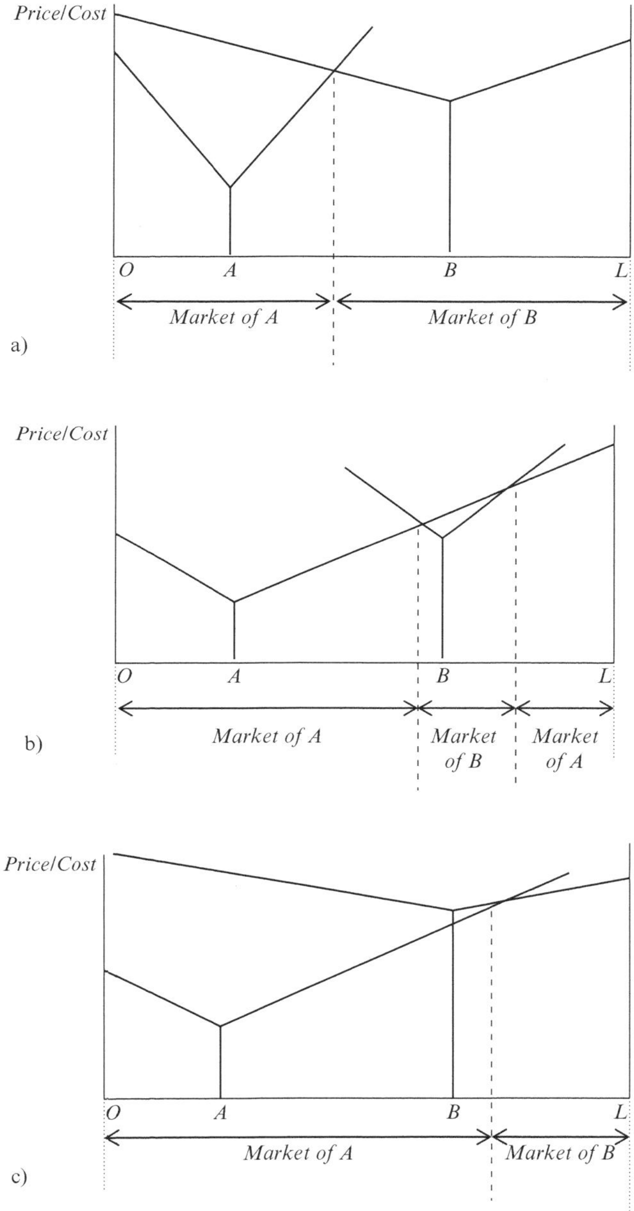

Figure 3.11 Spatial market areas: one-dimensional models with varying transport rates and production costs

exception and may only be a hypothetical possibility, it serves to highlight that fact that the distribution of spatial market areas is a highly complex phenomenon. Market areas are determined by the interplay of production costs, transportation costs and geographical location of consumers and competing producers. Obviously in the cases discussed here we have depicted a spatial market in its simplest form, namely a one-dimensional linear market *O-L*. If the market is defined in two-dimensional terms then the spatial areas become more complex. However, it is quite straightforward to use the one-dimensional model in order to provide some general rules governing the extent to which distance costs provide a firm with spatial monopoly power. These can be summarized as follows:

- The higher are the transport rate values t_a and t_b, the greater will be the monopoly power of the firm, and the smaller will be the reduction in the firm's own market area resulting from any marginal increase in the firm's own price, or from any marginal reduction in a competitor firm's price.
- The further apart are the locations of competing firms, the greater will be the monopoly power of the firm, and the smaller will be the fall in the firm's own market area resulting from any marginal increase in the firm's own price, or from any marginal reduction in a competitor firm's price.

Therefore, firms which are located at a great distance from each other, and which face significant transport costs, will consequently tend to exhibit significant local spatial monopoly power. That being the case, it might be expected that firms will therefore generally move away from each other in order to maintain their local monopoly power. However, observation suggests this is not always the case, and that while there are many situations in which firms locate away from each other, there are also many situations in which firms actually locate together. In order to distinguish between these different situations we need to consider the relationship between the spatial market area and the competition between the firms.

3.5 THE HOTELLING MODEL OF SPATIAL COMPETITION

The fact that geography and space confer local monopoly power on a firm provides an incentive for firms to use location behaviour as an explicit competitive weapon for acquiring greater monopoly power. In competitive environments characterized by oligopoly, the interdepend-

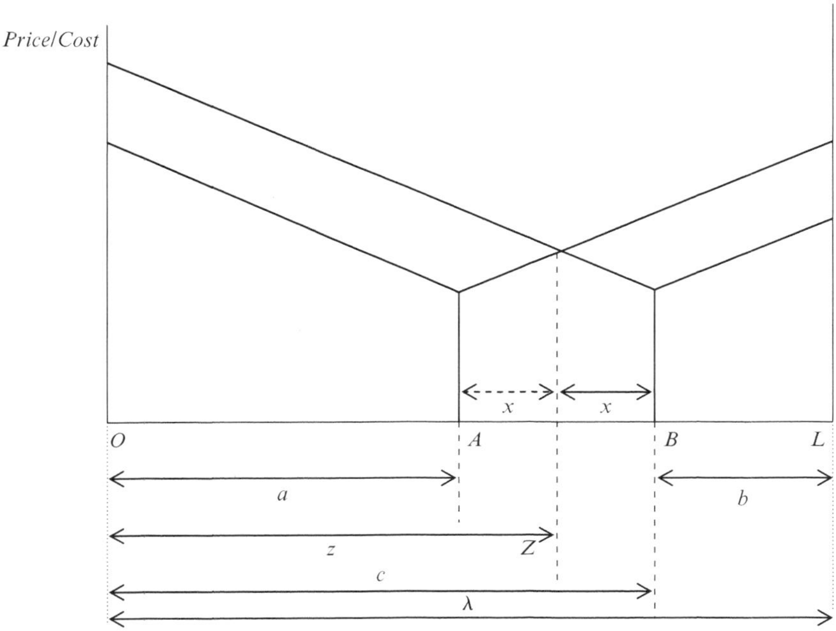

Figure 3.12 *The Hotelling spatial game setting*

ence between firms in the determination of output quantities and market share is also a result of locational considerations as well as pricing decisions. This is a particularly important strategy in oligopolistic industries where firms do not compete just in terms of price, but also engage to a large degree in non-price competition, such as a product or service quality competition. The simplest demonstration of this is the Hotelling (1921) model, which describes the spatial interdependence between competing firms within the context of a locational game.

In Figure 3.12 we adapt the model depicted in Figure 3.10 to the case where both production costs and transport rates exhibited by firm *A* and firm *B* are identical. In other words, we assume that $p_a = p_b$ and $t_a = t_b$. As before, we assume that consumers are evenly distributed along *OL* and we also introduce the Hotelling assumption that the demand of consumers is perfectly inelastic. As such, all consumers are assumed to consume a fixed quantity per time period irrespective of the price of the good. However, as consumers are also assumed to be rational, they will purchase from whichever producer can supply to them at their location for the cheapest delivered price.

In the initial Hotelling game, we will also assume that the firms do not compete in terms of output pricing, such that their only strategic decisions are with regard to their location. The reason is that changes in location can be used in order to change the overall spatial market area of the firm. In terms of the firm's strategy we assume that each firm makes a strategic decision regarding location on the basis of the assumption that its competitor will not change its existing location choice. As such, each firm is rather passive in as much as it takes the existing competitive environment as given and reacts accordingly. In the game theory literature this particular set of rules describing the nature of the competitive environment are known as 'Cournot conjectures'. If the firms react to each other over sequential time periods it is quite straightforward to determine the locational outcome.

In order to motivate our Hotelling game of spatial competition we once again assume a market space OL containing two firms A and B, with A located to the left of B. The distance from O to A is here denoted as a, and the distance from L to B is here denoted as b. The distance from each firm to the market boundary is x. The distance from O to the market boundary Z is denoted as z, and the total distance OL is denoted as λ. In order for this duopoly of two competing firms to exist there must be three conditions satisfied. The first condition is that a consumer located at point O must always buy from firm A. In other words, the delivered price of the output of A at O must always be less than the delivered price of the output of B at O. Following Richardson (1978) and McCann (2001) this can be written as:

$$p_A + t_A a < p_B + t_B(\lambda - b) \tag{3.5}$$

Similarly, a consumer located at point L must always buy from firm B. This can be written as:

$$p_B + t_B b < p_A + t_A(\lambda - a) \tag{3.6}$$

Thirdly, at the same time there must also be an indifferent consumer at a distance z somewhere between A and B. For this indifferent consumer the delivered prices must be the same. In other words:

$$p_A + t_A(z - a) = p_B + t_B(\lambda - b - z) \tag{3.7}$$

If we set $t_A = t_B$, then rearranging equation (3.7) gives:

$$p_A - p_B + 2tz - ta - t\lambda + tb = 0 \tag{3.8}$$

If the transport rates t_A and t_B for the two firms are the same, and the source prices, p_A and p_B, of the two firms are also the same, we have:

$$z = \frac{a + \lambda - b}{2} \tag{3.9}$$

The value of z given in equation (3.9) represents the size of the market captured by firm A, and the size of the market captured by firm B can thus be represented as:

$$\lambda - z = \lambda - \left(\frac{a + \lambda - b}{2}\right) = \frac{b + \lambda - a}{2} \tag{3.10}$$

Recalling from Figure 3.12 that $c = (\lambda - b)$, we can rewrite (3.10) as:

$$z = \frac{a + \lambda - (\lambda - c)}{2} = \frac{a + c}{2} \tag{3.11}$$

In other words, if the transport rates are the same and also the product source prices are the same, the market area boundary between the two firms is exactly half way between the two firms, and is independent of the transport rates. Therefore, for a given source production price p_A, known as a 'mill' price, the market revenue of firm A depends on maximizing the value of z. From Figure 3.12 we see that this is achieved by increasing the distance a and reducing the distance b as much as possible, while still ensuring that firm A is to the left of firm B. This is the 'one-shot' locational optimum strategy. However, the Hotelling model allows for the fact that the competition game can be played as a sequential location game under Cournot conjectures, and the location behaviour implied by equation (3.11) then triggers locational leapfrogging behaviour.

In Figure 3.13 we illustrate the outcome of this sequential Hotelling spatial competition game by examining the case where the firms A and B are initially located at one quarter and three quarters of the way along the market OL, respectively. As such, at the start of the game firm A will have monopoly power over OZ and firm B will have monopoly power over ZL, both of which are obviously identical market shares. In time period one, following the behaviour implied by equation (3.11), firm A will therefore move from its original location to a location at C, just to the left of B. In this way firm A will increase its market share from OZ to a new maximum value of OC. In contrast, although firm B will still retain market share over BL, its market share is now a minimum following the strategic location behaviour of firm A.

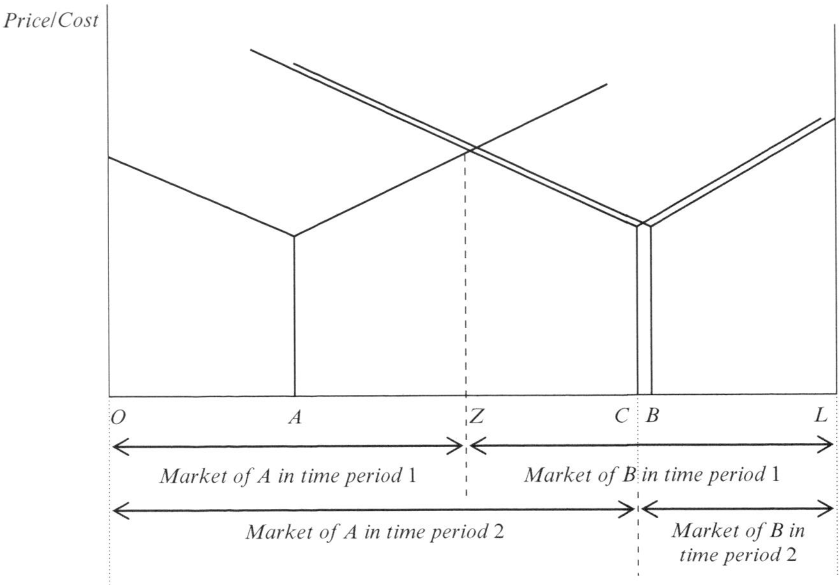

Figure 3.13 The Hotelling spatial framework

According to our assumptions or conjectures, firm *B* will now assume that firm *A* will maintain its location at *C*. Therefore, in time period two, in order to maximize its market, firm *B* will now move just to the left of *C*. Similarly, in the following time period, firm *A* will respond by moving to the left of firm *B*, and so on. This process will continue until both firms are located at *Z*, in the middle of the market. Once both firms are located at *Z*, neither firm has any incentive to change its location behaviour, because any location change will involve a reduction in market share of the firm that moves relative to the 50 per cent market share it gained when location is at *Z*. In game theory, any situation in which neither firm has any incentive to change their behaviour is known as a 'Nash equilibrium', and where the competition has been purely in output quantities determined by the market size rather than competition according to prices, then we have what is know as a Nash-in-quantities (Carlton and Perloff 2005). The locational result in which both firms are located at the centre of the market is therefore the Nash-in-quantities equilibrium for this particular locational game. Consequently, once the firms reach this point they no longer continue to move: this is the Hotelling result. From equation (3.11) we also see that this result holds irrespective of the initial distribution of the firms, so the case examined here in which both firms were originally located at

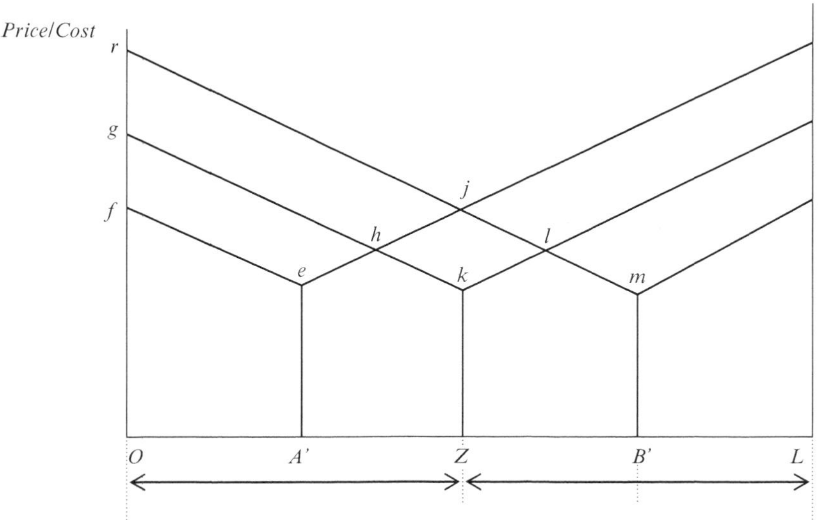

Figure 3.14 The welfare implications of the Hotelling spatial equilibrium

the quartiles is not a special case. The Nash equilibrium generated by the Hotelling spatial location game implies that under certain conditions regarding the nature of spatial market environments, there are situations in which firms will naturally cluster together in space, precisely as a result of competitive behaviour.

The idea that for competitive reasons firms might naturally cluster together in space does not necessarily imply that geographical clustering is efficient from a social perspective. In fact, the reality may be quite the opposite, and this is the case here. This can be demonstrated with the aid of Figure 3.14 where we compare the delivered prices paid by consumers in the original case at the start of the game where the firms were located at the quartiles, and the final case where they are located together at the centre. At both the start and the end of the Hotelling location game, the market share of each firm is 50 per cent. We have already assumed that all consumers at all locations have a perfectly inelastic demand curve. Therefore the change in delivered prices faced by these consumers between the start and the end of the Hotelling game will accurately reflect the change in welfare of the consumers at each location.

The net effect of these welfare gains and losses can be observed in Figure 3.14 in terms of the areas under the delivered price curves. In the original starting point where the firms were located at the quartiles *A*' and *B*', the delivered prices were traced by the lines going through the points *fes* for firm *A*, and *rmn* for firm *B*. At the end of the location game in the case of both firms the delivered prices are traced by the lines that go through the points *gko*. Now we can compare these delivered price curves in Figure 3.14 at the respective locations at the start and the end of the Hotelling location game. The consumers who are located in the centre of the market benefit by generally reduced delivered prices, represented by *jhkl*, whereas those located at the edges of the market lose by generally higher delivered prices, represented by (*efgh*)+(*lmno*). The gain in lower prices for the centrally-located consumers is outweighed by the increase in prices for the more peripheral consumers. The net effect is therefore a social welfare loss.

The Hotelling location conclusions discussed here also broadly hold in situations where firms charge the same delivered price for a given product at all locations. In such situations, the marginal profitability of each delivery will be different according to the distance between the firm and the location of the customer. This is because the transport costs of outputs must be absorbed by the firm. As such, the profits associated with deliveries to nearby customers will be much higher than those for deliveries to distant customers. The firm will therefore wish to locate so as to minimize the transport costs. In most cases, this will still also imply an equilibrium location at the centre of the market.

3.5.1 Price Competition, the Bertrand Problem and Non-price Competition

These stable Hotelling model's location conclusions start to break down if we also now allow for price competition. The problem can be explained with the help of Figure 3.15.

In Figure 3.15, we can consider what would happen if price competition is now allowed in the situation where both firms *A* and *B* are located at *Z* in the centre of the market, after having played the Hotelling location game. If firm *A* now lowers its sales price marginally in time period 1 when both firms are located at *Z*, in time period 2 firm *A* will gain all of the market. From our assumption about Cournot conjectures in which a firm assumes that its competitor keeps the same strategy it has already adopted, we know that firm *B* therefore now assumes that firm *A* will maintain both its new lower price and also its location at *Z*. Therefore in time period 3, firm *B* also lowers its market price just below that of firm *A*, in order to gain all of the market. This process will continue and the long-run Nash equilibrium of this price war is that both firms will end up selling at zero profit while still being located at point *Z*, because the market sale price will be driven down to the marginal costs of production. Within oligopolistic

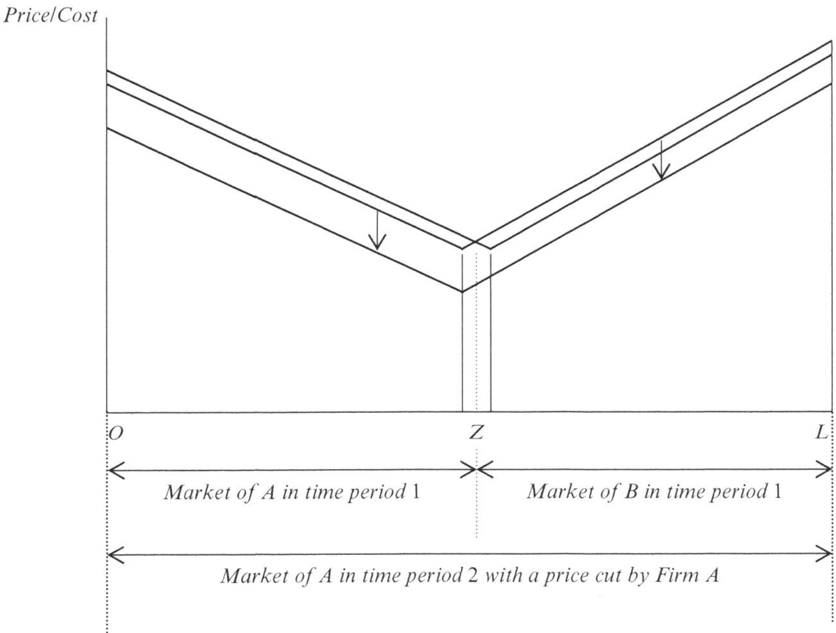

Figure 3.15 The effect of price competition on the Hotelling result

models this is known as the Bertrand problem and it is a problem which is faced by all oligopoly industries and firms.

The Bertrand problem can be depicted in price space as in Figure 3.16, in which the best price response of each firm is plotted against the price offered by the other competitor firm. The price p_a charged by firm *A* which is its best response $r_a(p_b)$ to the price offered p_b by firm *B*, is to just offer a price which just undercuts the price p_b. Similarly, the best price p_b offered by firm *B* is to offer a price response $r_b(p_a)$ which just undercuts the price p_a offered by firm *A*.

If price competition is engaged in when the firms are located at the centre of the market *Z* the Bertrand problem of an unending downwards price spiral becomes unavoidable. However, the objective of the firms in the Hotelling model is to acquire, preserve or gain monopoly power. Therefore, in order to generate localized monopoly power, as prices spiral downwards due to the Bertrand problem, each firm has an incentive to move away from its competitor in order to generate positive profits by maintaining a degree of localized monopoly power over some of the market area. Unfortunately, neither firm has an incentive to move away

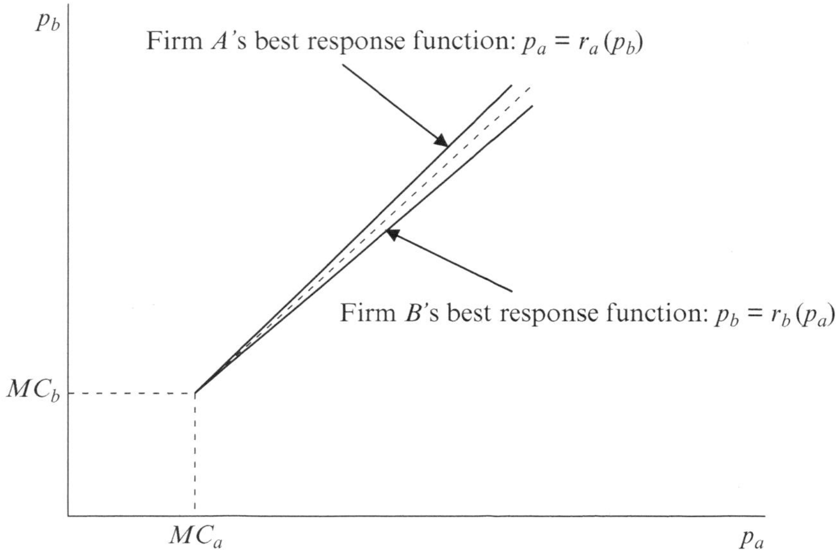

Figure 3.16 Bertrand best price responses

first, because in doing so, the other firm will then be able to maintain its current prices at the centre of the market and therefore will dominate a larger market area than the firm which moved away from the centre. In an environment of potential price as well as spatial competition this problem of instability in which all firms have an incentive to move away from each another, but no firm has any incentive to move first, can only be solved in one of two possible ways. One way to solve the problem is if firms are able to find a way to avoid moving too close together in the first place (d'Aspremont et al. 1979). However, this requires a level of implicit or explicit collusion. Therefore, unless there is some way in which the firms can cooperate and mutually agree to move away from each other, a price war always becomes inevitable and the Bertrand problem remains.

An alternative way of avoiding the Bertrand problem is to find a way to avoid price competition altogether. The most common means of doing this is to employ non-price competition, in which firms compete primarily on other issues, such as product or service quality, design and variation. In these non-price competition models, firms mutually agree not to cut prices, but to try to win consumers by the quality and variety of their products. This type of competition also involves issues such as branding, advertising, logos and customization. As such, when firms are competing for broadly the same market area, competitor firms will only be located next to each other in situations in which price competition is largely ruled out either by mutual agreement or by various forms of non-price competition.

A fundamental analytical issue now arises, in that the Hotelling location model which is developed above in the context of geographical space can be more or less equivalently applied by analogy to the case of product space.[3] If we reconsider the space defined by *OL* as now being transformed into a product characteristics space, the same Hotelling equilibrium locational result ensues as with the geographic space model. The way that this result can be interpreted in general is that if products exhibit a range of characteristics which can be ordered from *O* to *L*, the competition between the firms ensures that both competing firms produce products which are very close to each other in terms of product characteristics. As such, the locational proximity between the firms in the case of the geographical space is also mirrored in terms of the similarity of the competing products' features in terms of characteristics space.

Yet, it might appear that there is a slight inconsistency here in that, as we have already seen, firms located in the same place producing products with the same characteristics will be very vulnerable to the Bertrand problem. This apparent inconsistency gives rise to what is known as the 'Hotelling paradox', whereby in order to avoid the Bertrand problem, the mutual interdependence between firms often gives rise to firms producing basically the same product, but at the same time using non-price competition to make these products appear the most different to each other. As such, if we link the insights from both a geographical and a product space interpretation of the Hotelling model, the combined outcome is that where the firms are physically located in the same location, they will be selling products which are essentially the same products, although these will also made to appear as the most differentiated products in terms of non-price competition.[4] Yet this strict outcome may not always hold. Indeed, there are cases where firm clustering may give rise to some degree of price competition as well as product competition. In order to see this, we need to further investigate not only the parallels between the geographical concept of space and the product characteristics concept of space, but also the interaction between the two.

3.6 THE SALOP MODEL OF SPATIAL COMPETITION

The idea of the analogy between firm positioning strategies in geographical location and location strategies in terms of product characteristic space

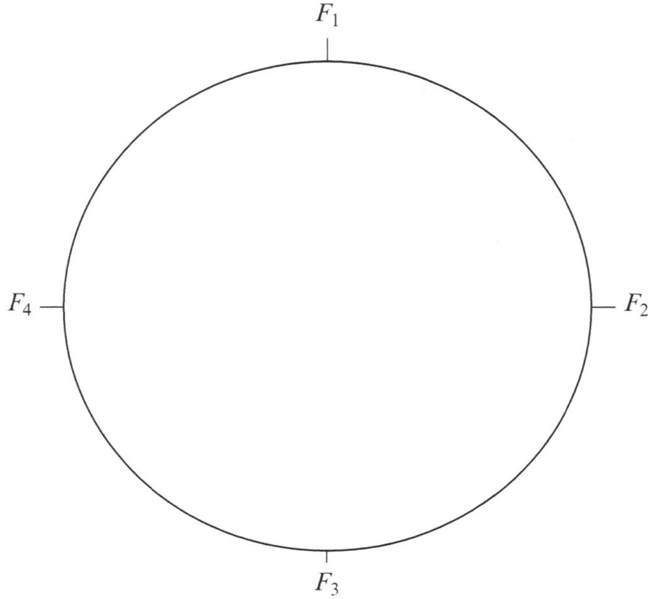

Figure 3.17 Salop circular model

can be further developed by analysing what is known as the Salop (1979) model of spatial competition. Whereas the Hotelling model essentially deals with the case of two firms in a finite geographical or product space, the Salop model applies to the case of three or more firms set in a continuous (non-finite) geographical space or in a continuous (non-finite) product characteristics space. Here we focus initially on the product space analogy, and then we will return to the insights that this approach provides for the relationships between geographical space and product space. The explanation here closely follows the excellent Carlton and Perloff (2005) simplified analysis of part of the broader Salop (1979) model, and the reader is referred to Carlton and Perloff (2006) for further details along with the original Salop (1979) model paper.

In order to understand the Salop model we begin by constructing a circular space with no finite borders, and allow four firms F_1, F_2, F_3, and F_4 to be located on this geographical and product characteristic space. The Salop model assumes a circular model space of distance L and this is depicted in Figure 3.17, in which there is an equal distribution of consumers along a circular market. As with the Hotelling model we assume that there is perfectly inelastic demand on the part of all consumers, and the products are differentiated here along a circular (non finite) a product space. The location of each firm F, $1 \ldots n$, implies a particular set of product characteristics as reflected in a particular brand. In terms of strategic behaviour we once again assume that the firms compete in an environment of Cournot conjectures, whereby each firm assumes that its competitor firm will maintain the same strategic behaviour as it has already adopted. In our case, this implies that the firm will maintain its same location behaviour along the product space, and the same principle can also be adopted if we were to treat the Salop model primarily in terms of geographical space.

As with the Hotelling model, in the Salop model each consumer is assumed to be rational. Yet, there is a subtle shift between the explicitly geographical space version of the Hotelling model described above and the product space Salop model described here. In this latter case, each location reflects a slightly different set of product characteristics, or put in another way, a different brand. As such, in this model a rational consumer therefore aims to relate the utility derived from consuming a particular set of product characteristics to the cost of purchasing that particular set of characteristics, in order to gain the 'best buy'. Once again, we can reinterpret this as saying that a rational consumer aims to relate the utility derived from consuming a particular brand to the cost of purchasing that brand, in order to gain the 'best buy'. The relationship between product characteristics, demand preferences and consumption in the Salop model can be represented with the aid of Figure 3.18.

In Figure 3.18 we denote U as the consumer's utility derived from consuming the favourite brand, l^* as the 'location' of consumer's favourite brand or taste defined in terms of characteristics space, and t as the rate of diminishing utility associated with consuming at a 'location' l which is not at l^*. In other words, in the Salop framework, the location l represents a particular sub-optimal set of product characteristics and t represents the strength of preferences associated with a movement along a unit 'distance' d away from l^* to l. As such, in product characteristics space, t here is analogous to the transport costs t in the explicitly geographical space of the Hotelling model above.

With this notation, we can now write a simple expression for the consumer's utility u given as:

$$u(l, l^*) = U - t|l - l^*| \tag{3.12}$$

in which the consumer's utility u is a function of the preference distance d between l and l^*. In other words the consumer's utility u depends on the distance between the brand characteristics of the product actually

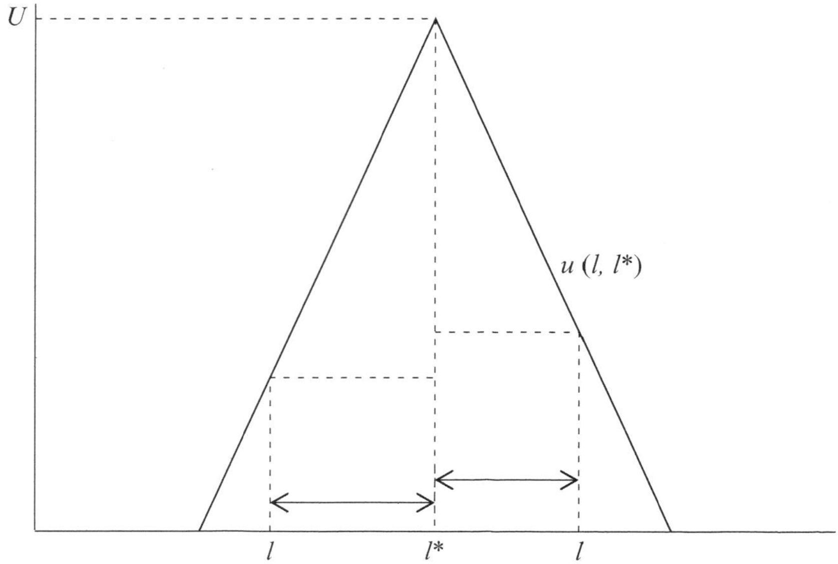

Figure 3.18 *Consumer utility function in a Salop circular model*

consumed and the consumer's ideal brand characteristics at l^*, as well as on the preference (transport) costs t per unit distance.

Now if we apply this Salop utility logic from that of a single product to multiple products, that is, from a single firm to multiple firms, by following the logic of the Hotelling model and applying it to the utility structure of Figure 3.18, it becomes immediately clear that in order for a firm to maximize its market area, it will want to locate as far away as possible from its competitor firms. As such, if there are n brands, and $L=1$, then the distance between each brand location is equal and is given as $1/n$, as depicted in the four firm example of Figure 3.17.

In the Salop model in which there are multiple firms we can imagine two broad alternative scenarios. The first scenario is where each firm has complete monopoly power over its own brand, while the second scenario is where the firms must compete, at least partially, between brands. The first case is depicted in Figure 3.19 and the second case in Figure 3.20.

In the case where all firms have complete monopoly power over their own brands then in spatial terms this implies that each firm has control over its own monopoly region, the length of which is $2d_m$ in Figure 3.19. In this case, all consumers whose preferences are located near to the brand characteristics of the firm will buy from that firm, and this will be true for

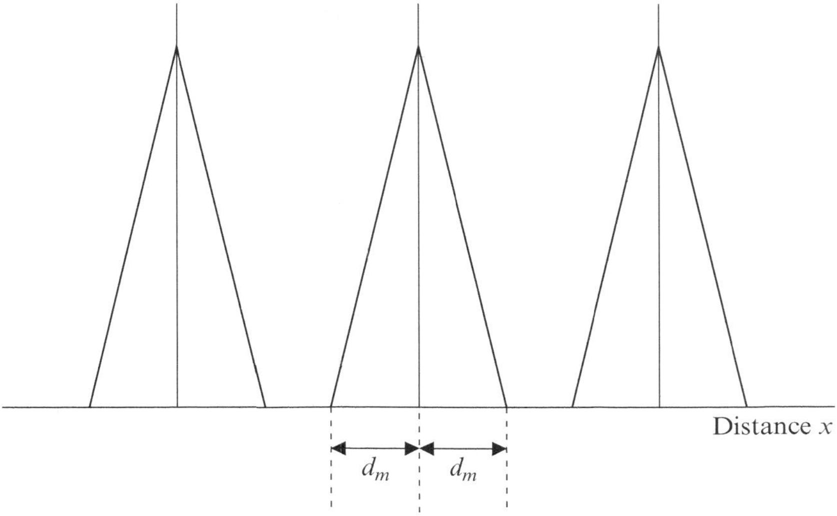

Figure 3.19 Monopoly region Salop model

all brands. As is clear from Figure 3.19, the market area of a firm will not impede on the market area of any other firm. The total market area for each firm will then depend on the rate of decline of utility with respect to differences in product characteristics.

In contrast, in the competitive region case depicted in Figure 3.20, competing firms producing competing brands must take account of their rivals. In this situation, there will be customers whose preferences are located in the potential market area of two competing brands, and rational consumers will buy from the firm whose brand offers the highest consumer surplus. In Figure 3.20 the monopoly area of the firm reflected in the distance of the market boundary between two competing brands is given as $2d_c$. In this quasi-competitive brand case, the market area is obviously less than the pure brand monopoly case depicted in Figure 3.19.

Part of the problem here is that if there are 'gaps' between the firms, as depicted in Figure 3.19, then this implies that there are market niches into which new competitor firms may enter. However, the new entrant firms will not want to be confined into selling purely in the market niches defined by the empty spaces between the existing firms as this will limit their market potential, unless the gaps are very large. If this is not the case, then entrant firms will also aim to encroach on the market regions of the incumbent firms, thereby transforming the monopoly region model of Figure 3.19 into something more akin to the competitive region model

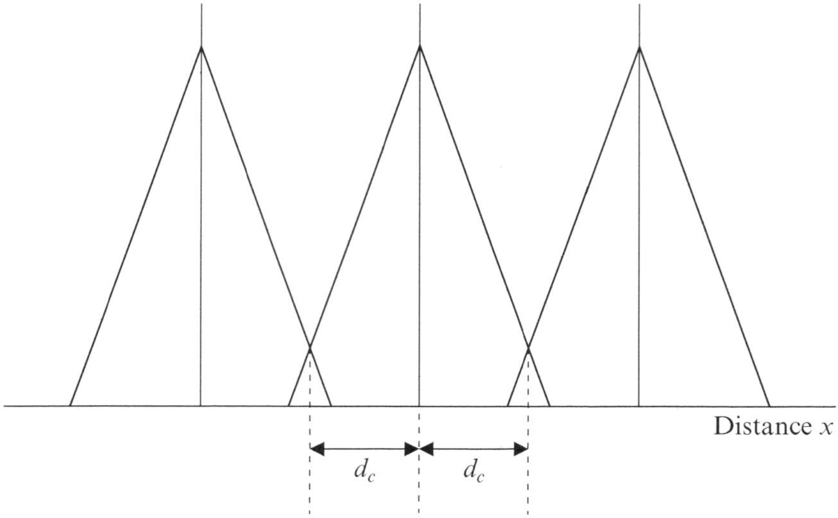

Figure 3.20 Competitive region Salop model

of Figure 3.20. However, given that the existing firms will also wish to maximize their market areas while new firms enter into the market niches, then the existing firms have two choices. Firstly, they can also employ price competition, but of course this will encourage the Bertrand problem. Alternatively, each firm can also relocate slightly so as to maximize the distance between itself and its two adjacent competitor firms. As explained above, if $L=1$ and there are n firms, this implies that the distance between each brand location is equal to $1/n$.

3.6.1 Price Competition in the Salop Model

In the Hotelling model, firms clustered together in geographical space which engage in price competition automatically face the Bertrand problem. At the same time, these firms have no power to increase the price of their own products, because unilateral price rises will mean that the firm's market area will immediately collapse to zero. As we have seen, this provides an incentive for strong non-price competition based on brand and design variety, giving rise to the Hotelling paradox. In the case of the Salop model, however, the characteristics space between each producer allows for slightly different pricing behaviour on the part of firms.

In order to help explain this, in a Salop model we first assume that there also exists what is known as an 'outside good'. This is another consumption alternative whose utility is fixed and which is also unaffected by the product or location strategies played by the firms in the Salop model. We denote the consumer surplus associated with consuming the outside good as w. If the consumer is 'located' at the location of their ideal brand tastes or preferences, in that the consumer is consuming their ideal good, the surplus gained by consuming this good can be written as $(U - p^*)$. However, we also know that the consumer will only consume this ideal good within the Salop model rather than the outside good if $(U - p^*) > w$, or alternatively if $(U - w) > p^*$. Therefore, the consumer's reservation price or reservation value v can be written as $v = (U - w)$, and this is the maximum utility gained by consuming the ideal good, allowing for the opportunity costs of consuming the outside good. As such, this also represents the minimum consumer surplus which a consumer will need to gain by consuming an alternative good which is not ideally located in product space, in order for the consumption of any alternative good in the Salop model to take place. Therefore, if we denote the price paid for the good located at a distance $d = l - l^*$ in the Salop model as p_l, then we know that the behaviour of the consumer in terms of consuming the good in the Salop model can be written as:

$$Max[u(l, l^*) - p_l] \geq w \tag{3.13}$$

which means that the consumer will maximize the net utility gain (consumer surplus) from consuming an alternative good within the Salop model, if the consumer surplus gained is greater than that which would be gained by consuming the ideal good instead of the outside good.

Assuming that L consumers are distributed evenly around the circle, and that a firm has a monopoly region as depicted in Figure 3.19, then we can write:

$$u(l, l^*) - p_l = v - td_m - p_l \geq 0 \tag{3.14}$$

i.e.:

$$\frac{v - p}{t} = d_m \tag{3.15}$$

This is the maximum distance between the firm and the edge of its market boundary in a Salop monopoly region model. From Figure 3.19 we know that each brand therefore captures:

$$2d_m = \frac{2(v - p)}{t} \tag{3.16}$$

and the number of consumers captured is:

$$q_m = \frac{2L}{t}(v - p) \tag{3.17}$$

Differentiating q_m with respect to p gives:

$$\frac{\partial q_m}{\partial p} = -\frac{2L}{t} \tag{3.18}$$

From equation (3.18) we see that the monopoly quantity demanded and sold falls by $-2L/t$ as the firm raises its sales price p by \$1.

We can now also apply a similar line of reasoning to the competitive Salop case depicted in Figure 3.20. If there are n competing brands, then following the logic above, if we also assume that both of a firm's immediately adjacent competitors are located at a distance of $1/n$ away and both charge a price P, we can investigate how much of a market a firm captures by selling at price of p.

In order to answer this question, we know already from Figure 3.20 that the firm captures all of the market within a distance d_c from itself, whereby the consumer surplus from each of the two competing brands is equal. Therefore we can write:

$$v - td_c - p = v - t\left[\frac{1}{n} - d_c\right] - P \tag{3.19}$$

in which the left hand side represents the consumer surplus from the firm's own brand, and the right hand side represents the consumer surplus from the competing brand. Given that:

$$q_c = 2d_cL \tag{3.20a}$$

such that:

$$d_c = \frac{q_c}{2L} \tag{3.20b}$$

We can therefore write:

$$v - \frac{tq_c}{2L} - p = v - \frac{t}{n} + \frac{tq_c}{2L} - P \tag{3.21}$$

and:

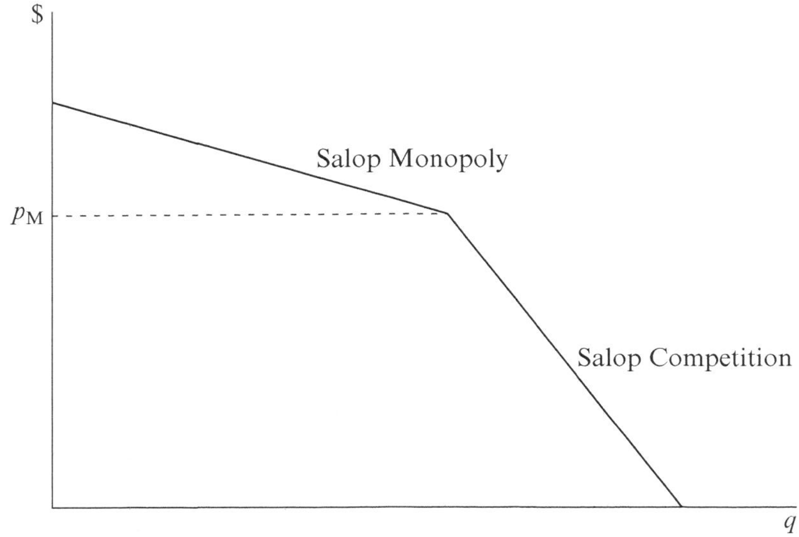

Figure 3.21 Demand in a Salop model

$$P - p + \frac{t}{n} = \frac{tq_c}{L} \tag{3.22}$$

Rearranging equation (3.22) gives:

$$q_c = \frac{L}{t}\left[\frac{t}{n} + P - p\right] \tag{3.23}$$

Once again, differentiating the quantity demanded and sold with respect to the price gives:

$$\frac{\partial q_c}{\partial p} = -\frac{L}{t} \tag{3.24}$$

As we see in equation (3.24) the competitive quantity demanded falls by – L/t as the price p rises by \$1.

If we compare equation (3.24) with equation (3.18) we see that the effect of a price rise of \$1 by one firm in the Salop monopoly model depicted in Figure 3.19 is twice as great as the effect of a \$1 price rise in the competitive Salop model depicted in Figure 3.20. This is represented in Figure 3.21.

The reason for this is that in a competitive model the consumers have the possibility to switch consumption between producers whose products are immediately adjacent to their preferred brand or combination of characteristics. As such, a more competitive market with more brands competing in the same product and geographical space leads to a lower overall elasticity of demand. Salop himself acknowledged that this observation is surprising (Salop 1979), although this interaction between the consumer surplus, the brand characteristics and the price mean that such competition only operates at lower price levels. At a very high monopoly price level, p_M, the individual firm still has a monopoly position, manifested in terms of elastic demand.

This major insight from the Salop model, namely that more competition implies a smaller demand effect of a unilateral price rise, also provides us with another fundamental insight into the nature of location behaviour. In particular, it encourages firms to both differentiate their various products and also to cluster their sales activities.

In order to see this we can reconsider the earlier insights of the Hotelling model. The general conclusion of the Hotelling model is that if firms are physically located together in geographical space, then the Bertrand problem can only be avoided if firms ensure that non-price competition operates. The problem with this is that the situation is often on a knife-edge if the competing products are still basically very close substitutes for each other, even if lots of branding and marketing is taking place in order to persuade consumers otherwise. Consequently, there is always the danger that consumers eventually come to realize the competing products are indeed very close substitutes for each other, such that the whole local equilibrium once again becomes unstable and subject to the Bertrand problem. This implies that a stable outcome is only really possible if the products are fundamentally different, in which case the firms may not actually be competing for the same consumers.

In contrast, the Salop model implies that a high degree of product differentiation acts as partial 'defence' against the Bertrand problem of instability, and encourages firms to both differentiate their products and cluster their sales outlets. In order to see this we recall that the distance in product space between each brand variety is equal to L/n, where L is the length of the Salop circle and n is the number of competing brands. As such, having closer substitutes implies that the number of competing firms n is larger, and the product space distance between the firms L/n is smaller. Under these conditions consumers have more possibilities to switch consumption between adjacent brands if the prices of particular brands rise, and in terms of Figure 3.21 this also implies that the monopoly price p_M rises. From the perspective of consumers, this implies that consumers have a high degree of 'protection' from a price rise by a particular individual producer, because the possibilities for switching producers are greater. At the same time, for a given number n of brand alternatives, a larger market choice space, defined as L, also implies that there is a greater characteristics distance between individual producers. This increases the monopoly brand power of each individual firm. Therefore, the greater is the distance L and the more brand alternatives that are available, the less will any particular firm be adversely affected by a downwards price adjustment by any other particular competing producer. This is because, on average, the characteristics distance between the two firms will be greater.

If we combine these two observations, then it becomes clear that a larger market characteristics space L combined with a greater number n of competing producers together increases the price stability and demand quantity of the whole market. However, while increasing the number n of producers has a clear meaning, what is not yet clear is how the size of the market space L can be increased. For this, we must once again turn to the issue of geography.

In terms of explicitly geographical space, if a greater number of competing firms producing alternative differentiated brands are located at the same geographical location, then the Salop arguments also imply that more local competition between product varieties will lead to a local market demand which is robust and stable. This is good for both producers and consumers, and geographical clustering thereby becomes a defence against the Bertrand problem, as long as it is also associated with high product variety. This conclusion itself provides a partial rationale as to why firms should cluster together in geographical space, because as a group, the clustered firms will benefit from these stable price and demand conditions.

In addition to the issue of price and demand stability, there is also a great deal of theoretical work as well as empirical evidence which suggests that consumers have an increasing preference for choice variety in general. If consumers enjoy greater choice variety for any given level of expenditure, this also implies that the more brand options are available at a particular location, the higher will be the consumers' utility. Therefore, if consumers know that a high degree of product variety is available in a particular locality this itself will encourage more consumers to consume at this particular geographical location. The geographical clustering of competing differentiated brand producers can therefore facilitate the growth of the overall local market space, as well as providing for greater overall market and price stability. These interrelationships between product variety, price stability and the growth of a local market all provide a clear, preliminary rationale as to why firms should cluster together in space. Indeed, modern multinationals frequently locate multiple stores, showrooms or retail outlets in the same location in a manner whereby the establishments are explicitly designed to appear as being from quite different producers, although in reality they are all part of the same corporate grouping. These practices are common in the case of clothing and fashion retailers, food and restaurant retailers, electronics retailers, media, musicals and shows, and also in the case of automobile showrooms. Indeed, multinationals use Salop strategies as a means of market entry into new markets, whereby *both* product space and geographical space clustering techniques are employed simultaneously in order to maintain price levels and avoid the Bertrand problem, while also extracting maximum consumer surplus rents from price discrimination. This is a particularly powerful strategy where market entry is a new phenomenon as it allows the firms to 'test' different combinations without facing the problems of collapsing prices.

3.7 LIMITED INFORMATION, BOUNDED RATIONALITY AND LOCATION BEHAVIOUR

As we saw in the Hotelling model, in order to achieve clustering as an outcome there must be conditions under which it is logical for firms to follow or copy each other in geographical terms. This following or copying type of behaviour suggests that many aspects of firm location choices typically exhibit dynamic features, an insight which is entirely consistent with the observed location behaviour of MNEs. The conditions under which such dynamic processes operate will here be seen to be ones of limited information and uncertainty.

3.7.1 Behavioural Theories of Firm Location

The Weber and Hotelling-type models discussed above are based on the assumption that rational firms use location behaviour as a strategy to maximize profits. However, in order for these models to be accurate we must also assume both that all of the necessary information is available to all firms, and that these latter have the capacity to make rational decisions. In reality, however, the information available to firms is often rather partial, limited or incomplete, with different firms even in the same industry often having access to different types and levels of information. Moreover, there may be other issues which militate against firms making purely profit-maximizing decisions. For these reasons, it could be argued that in many real-world situations it is not possible for firms to make decisions specifically in order to maximize their profits. At this stage it appears that the ability of our location models to provide clear insights into firm location strategies may therefore be somewhat limited.

An alternative perspective that many observers within the management field employ is to analyse how organizations make decisions based on insights from psychology. These are known as behavioural theories, and have also been applied to location behaviour. One of the key components of behavioural theories is the concept of 'bounded rationality', and this is originally most closely associated with the work of Herbert Simon (1952, 1959). Simon argues that in reality economic agents, whether individuals, firms, or organizations, all have limited information, much of which is of a partial nature and some of which is either conflicting or confusing. In such conditions, all economic agents will adopt fairly basic, routine and well-tried decision-making rules, which are often ad hoc or heuristic, in order to arrive at the best decision. However, there is no guarantee that the actual decisions made are indeed the best choices, because in the absence of complete information they also depend on the heuristics of the decision-making process itself. As such, Simon's work implies that in the real world the limited information available to a firm itself limits the firm's ability to be 'rational', at least in the narrowly-defined profit-maximizing sense adopted by traditional microeconomics models. In the terminology of behavioural theories, the ability of the agent to be rational is itself bounded by the limited information available and the decision-making process employed. This argument is particularly pertinent to the case of firm location or relocation behaviour, because the costs involved in such choices are often very high, and also because detailed cost and revenue information defined according to geographical market areas is usually very difficult to find.

A second key component of behavioural models follows the 'conflicting goals' critique of Cyert and March (1963) who focussed on the decision-making processes inherent in large corporate organizations. The key decision-making characteristic of these types of organizations is the separation of ownership from decision-making, which implies that different groups within the organization may have different goals. If this is indeed the case, then it is likely that the business objectives which are pursued may well be very different from simply profit maximization. This argument is based on the observation that different stakeholders within modern MNEs – thus multiplant, multidivisional and multiproduct firms – have different incentives and therefore may be pushed to pursue different objectives. The lack of alignment of many of these objectives militates against firms' decisions being based primarily on profit maximization. For example, only shareholders strictly have a desire for maximum profits, whereas directors' performance may be evaluated primarily in terms of a firm's market share, sales managers' performance may be considered in terms of sales growth, production managers' performance may be regarded primarily by inventory throughput efficiency, and human resources managers may be assessed on the basis of the number of days lost through industrial disputes. As such, while the shareholders desire maximum profits, the career performance of each of these various key decision-makers within the organization is evaluated on different criteria. Under these conditions, the conflicting goals critique suggests that firms will aim to make decisions in order to 'satisfice' rather than to optimize. In other words, the firms will aim to achieve a satisfactory level of performance across a range of measures. In particular, the firm will initially aim to achieve a level of profit sufficient to avoid both shareholder interference in directors' activities and the threat of a takeover. Once this objective is achieved, the other various goals of the firm can be satisfied.

However, in situations of limited information, it may be almost as difficult to satisfice as it is to optimize. In such circumstances Baumol (1962) argues that firms will focus on sales revenue maximization as the short-run objective of their decision-making, as it implies the maximum market share for the firm in the short term. Market share provides a very clear and understandable indicator of a firm's position in a market, and largely reflects its degree of monopoly power. This is important because the degree of a firm's monopoly power is also indicative of its ability to deter potential market entrants via techniques such as limit pricing and cross-subsidizing, whereby the firm will artificially reduce the prices of certain goods or services in order to deter potential competitors. The ability to undertake such predatory pricing measures is a good indicator of a firm's capacity to defend its position, and is therefore perceived by many observers to be the best indicator of a firm's long-run performance.

While these behavioural arguments reflect a more general critique of rationality within microeconomics as a whole, they have been applied particularly extensively in the field of economic geography, and especially regarding questions of firm location and re-location behaviour (Lever 1987). The reason is that information concerning space and location is very limited, due to the inherent heterogeneity of land, real estate, and local economic environments. Moreover, in terms of MNE strategies, this heterogeneity is magnified as we are considering the question of different locations across different countries and cultures. Therefore, when looking at location issues, not only would it appear that the ability of the firm to behave rationally is very much bounded by the limited information available to it, but also that the inherent complexity of modern multinational firms will very much militate against rational choices. The result of these behavioural critiques is therefore that MNEs will consequently make location decisions primarily in order to achieve satisfactory outcomes across a range of alternative goals, of which profit maximization is simply one objective. The ability of the profit-maximizing models outlined above to

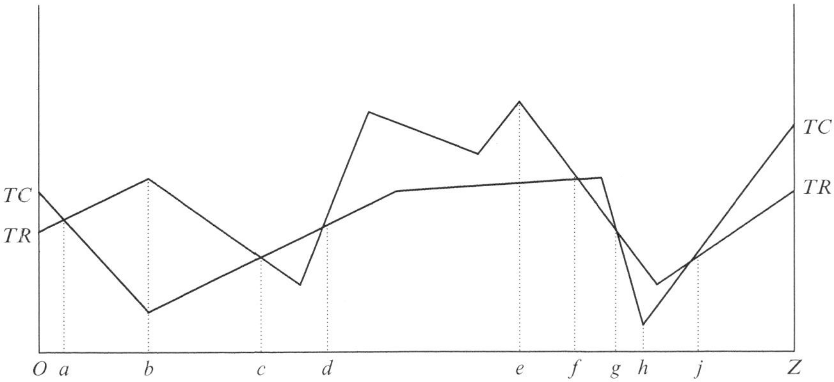

Figure 3.22 *Spatial margins of profitability and optimizing and satisficing location alternatives*

help us identify firm location strategies may thus be particularly limited in the case of multiplant and multinational firms.

3.7.2 Alternative Optimization and Location Behaviours

The kinds of implications of the behavioural arguments for firm location choices can be described with the help of Figure 3.22 which represents a one-dimensional geographical space *OZ* which varies significantly in terms of its location-specific factor costs (*TC*) and location specific market revenues (*TR*). In Figure 3.22, the areas in which positive profits are made, that is, where *TR* > *TC*, are known as 'spatial margins of profitability' (Rawstron 1958), and they represent the geographical areas in which the firm can locate and be profitable. As we see in Figure 3.22, the shapes of the spatial cost and revenue functions are neither co-linear nor symmetrical across the one-dimensional geographical space, and it is this lack of symmetry which gives rise to alternative location strategies according to the nature of the firm's decision-making processes and objectives.

In Figure 3.22, in an ideal world in which the firm had both perfect information regarding different spatial cost and revenue curves and no conflicting decision-making goals, our location theory models imply that the firm will always move to point *e*. However, in an environment of limited information and conflicting goals within the organization, the actual location behaviour of the firm will depend on which is the particular dominant objective of the firm. For example, if the firm is aiming to maximize its profits in the short run it will locate at point *b*, where the positive difference between the total revenue *TR* function and the total cost *TC* function is maximized. On the other hand, if the firm is aiming to maximize sales it will locate at *e*, and if it is aiming to minimize production costs and to maximize production efficiency it will locate at *h*.

These arguments regarding the lack of optimality in firm location behaviour also extend to the issue of firm *re*-location. The relationship between marginal location change and the profitability of the firm in these areas is given by the differences in the slopes of the spatial revenue and spatial cost functions as location changes. In the spatial margins of profitability in which the slopes of the spatial revenue and spatial cost functions are very shallow, the marginal benefit of relocation will be very low. Therefore, in the presence of high relocation costs the firm will not move to a superior location even if it knows which alternative is superior. In conditions of imperfect information, bounded rationality, conflicting goals, and significant relocation costs, the behavioural approach would argue that once a firm has chosen a location, it will tend to maintain it as long as profits are positive, and rarely use relocation as a competitive weapon. In particular, firms are unlikely to relocate in response to small variations in factor prices or market revenues. Nor are firms likely to relocate frequently, as has already been observed earlier. The basic reason for this is that in environments of limited information, the incurring of significant relocation costs will engender significant risks for the firm, which will therefore be unwilling to incur such costs unless it is very confident regarding the superiority of the alternative location. Rather the firm will attempt to undertake *locational adjustment* by reorganizing the resource allocations and activities between its current set of existing plants, as this type of reorganization within multiplant activities is often less costly to the firm than is the closing and relocation of a plant. This behaviour is very common amongst the multiplant organizational structures which are typical of MNEs. On the other hand, where relocation costs are very low, the firm will readily take advantage of spatial revenue and cost differences and will easily be able to adopt relocation as a competitive strategy.

At this point there are several issues we can raise regarding the relationship between rationality and optimality in terms of location behaviour. First, the logic of corporate multiplant locational adjustment still needs to be guided by the same cost and revenue considerations underpinning location theory. As such, our location theory arguments are still valid in terms of explaining where a firm should locate investment and activities in order to ensure high rates of return on the capital it invests. Firms whose locational adjustment processes follow the logic of location theory will achieve locational adjustment outcomes which are superior to those which do not. Second, the cautiousness which some firms exhibit in terms of relocation issues will depend on the level of sunk costs which would be incurred in association with any decision to relocate. Sunk costs are the costs incurred by a firm which cannot be recovered once a factor adjustment or an alternative investment has taken place. In the case of firm relocation behaviour, these sunk costs can often be very significant.

In contrast, if all or most of these costs can be easily recovered after relocating, the firm will readily undertake relocation behaviour. For example, if a firm's capital infrastructure and facilities can be dismantled and re-assembled rapidly, or if a firm's labour can either easily be relocated or new labour hired and trained quickly, then a firm will readily use re-location as a competitive weapon, even though the up-front costs associated with the relocation may still be very significant. Moreover, many modern developments in commercial building technology, such as 'turnkey' construction operations, pre-fabrication systems, modular construction, real estate leasing services, and serviced-office provisions, as well as the international harmonization and standardization of construction technologies, are all nowadays designed to facilitate the process of firm relocation. These developments would therefore imply that many of the potential sunk cost inhibitors to corporate relocation behaviour may actually be becoming relatively less significant over time.

3.7.3 Alchian's Competitive Selection and Leader-Follower Location Strategies

All of the behavioural location strategies discussed above suffer from the weakness that, unlike the location models described in the previous sections, they do not of themselves indicate why a firm chooses a particular location in the first place. In this sense the behavioural approach is not prescriptive. It is therefore necessary to reconsider the simple profit-maximizing location models in the light of the behavioural critique of bounded rationality, imperfect information, conflicting goals and relocation costs, as these are all features which are particularly characteristic of the spatial economy. In order to do this we can stand for the moment aside from locational issues and consider the original competitive selection argument of Alchian (1950) as applied to any form of rational optimization model. We can then reconsider the Alchian argument specifically in the context of location behaviour.

The Alchian approach is that the behaviour of firms in conditions of uncertainty can be understood by discussing the relationship between a firm and its environment. Here a firm's environment encompasses all the agents, information, and institutions which are competing and collaborating in the particular set of markets in which the firm operates. In the Alchian's framework, we can characterize the uncertain economy by two broad types of environments: the 'adoptive' and the 'adaptive' environment. These two types are not intended to be seen as mutually exclusive, but serve as the two extreme stylized ends of a continuum, somewhere along which the real economy will exist.

In Alchian's adoptive environment, all firms are more or less identical to each other in that no firm has any particular or systematic information advantage over any other. The results of the competitive process will imply *ex post* that some firms will be successful while others will not, although *ex ante* no individual firm had any a priori knowledge that their products or techniques would be superior to those of their competitors. This characterization of the economy is basically Darwinian, in that the environment 'adopts' the firms which are better suited to the needs of the economy, even though the firms have no particular knowledge beforehand that this was the case. In statistical terms, in any given time period in this adoptive environment, the probability of a particular single firm making a successful strategic decision is identical to that of all the other individual firms. In contrast, in Alchian's adaptive environment, some individual firms are able to gather and analyse market information more efficiently and effectively than others, simply by reason of their size. Large firms in general are able to utilize resources in order to acquire and process information relating to their market environment, with the purpose of subsequently using such information to their own advantage. In statistical terms, therefore, in any given time period in Alchian's adaptive environment, the probability of a particular firm making a profitable strategic decision is increased by reason of its size.

In the real world of heterogeneous firms and imperfect information, smaller firms will tend to perceive themselves to be at an information disadvantage relative to larger firms. Therefore, they will tend to make decisions which mimic or dovetail with those of the larger firms, in matters such as styles, protocols, formats and technology. In part this is because they perceive the market leaders to be the best barometers of market conditions, and also because the behaviour of the market leaders itself often contributes significantly to the overall economic environment simply because of their size. By copying the behaviour of the larger firms the small firms thus perceive that they will maximize the likelihood of their own success. The result is that large firms tend to overcome uncertainty by information gathering and analysis, and small firms tend to overcome uncertainty by imitation.

The type of leader-follower behaviour suggested by Alchian's framework is common in many models of oligopoly and uncertainty, but so far the Alchian logic has not been set in the context of geography, despite being particularly pertinent to questions of location. In environments of uncertainty, larger firms will generally have the information and financial resources to make more considered location decisions than small firms. In other words, large firms' location decisions will be more akin to those described by the Weber and Hotelling models, given that they will generally have sufficient resources to evaluate the cost and revenue implications of their potential location choices. These large firms will attempt to make rational and optimal decisions, and the results of their location choices can be largely analysed by the types of classical and neoclassical models described above. On the other hand, small firms will generally be located where their founders were initially resident. There will have been no explicit initial location decision as such, when the firm began operating. Yet, over time, competition between firms will be partly a result of spatial differences in costs and revenues, and the relationship between profitability and location will eventually become a decision-making issue. In subsequent location decisions, many small firms will tend to choose locations close to the major market leaders for the reasons outlined by Alchian. Imitation therefore also takes place in terms of spatial behaviour. For firms which are risk-averse, as we see from the Hotelling and Salop models, this is also a particularly good strategy for three reasons. Firstly, locating close to competitors ensures that an individual firm's market share is no lower than that of an equivalent firm; secondly, clustering allied with variety also limits the problems associated with price wars; and thirdly, such clustering behaviour can facilitate all sorts of additional benefits from externalities. As we will discuss further in Chapter 5, the clustering of many small firms around large ones is therefore a very common observation.

3.8 PATTERNS OF MULTIPLANT AND MNE SPATIAL AND ORGANIZATIONAL STRUCTURES

As we have already mentioned, the location behaviour of MNEs is more complex than individual single-plant firms because much of the geographical relocation of activities within MNEs consists of locational adjustment, in which the reallocation of activities and resources occurs within an existing spatial configuration of establishments, with little or no discernable external changes (Healey and Watts 1987). This is particularly complex to model because the largest component of geographical changes in foreign investment takes place via mergers and acquisitions rather than 'greenfield' investments, thus following the locational patterns of *efficiency*, *specific-* or *strategic-asset seeking* MNEs. This is particularly important, for example, in the case of financial service industries (Cohen 1998; Leyshon and Thrift 1997; Coleman 1996), whose firm and industry characteristics are largely unsuitable for modelling by orthodox microeconomic location theory. However, even allowing for these complexities we are still able to use the types of location theory principles outlined in the previous sections along with the organizational and information arguments raised by the behavioural critique, in order at least to begin to explore these issues.

As already said, multinational firms are by definition multiplant firms, in that multinationality means that such firms must have facilities in more than one country and thereby in more than one location. Indeed, the spatial arrangement whereby a firm has multiple input sources, multiple output market locations, and multiple production sites, is very much the norm for multinational firms. However, how we arrive at this observed outcome is itself important, in that firms rarely make a one-shot decision to site multiple facilities in different locations.

As we have already seen, some firm location decisions are indeed one-shot single establishment decisions and, as such, can be characterized very simply in term of the Weber model. In addition, there will also be cases where firms also make a one-shot decision to site several facilities in different locations. However, a key feature of the Weber, Hotelling and Salop models is also that they allow us to understand the factor price and market conditions under which other locations will become more attractive as locations for investment. These models also allow us to see location decisions in terms of a dynamic process, in which changes in factor cost or market competition conditions can engender changes in location behaviour, which themselves can change the supply linkages between suppliers, firms and markets. In other words, rather than simply assuming that all location decisions are one-shot single establishment location choices, if we assume that there is a certain element of 'stickiness' involved, they can be seen as being inherently incremental in nature as firms respond to new input and output conditions by changing their locations, and consequently the geography of their input and output linkages.

As we mentioned in Chapter 2, one of the frequently observed aspects of MNE locational behaviour is that MNEs tend to search for an initial overseas investment site, and then over time they search for either new or additional investment locations. This incremental process is often described in terms of the stage theory of the evolution of the firm (Johansson and Vahlne 1977), the stage theory of the evolution of the product (Vernon 1966) or issues related to market barriers to entry (Caves 1971, 1982). Each of these various approaches suggests that the process of internationalization tends to occur by firms setting up a new overseas investment 'platform', whereby an initial overseas facility is set up (Caves 1982a). This platform establishment then acts as a prototype overseas venture which, if successful, will then be replicated or adapted elsewhere. Much of the emphasis of these international business theories is focussed on the process of learning, by which the MNE establishes the initial prototype overseas facility and then monitors how successful the new venture is. Part of the role of the new establishment is therefore to facilitate a process of learning by trial and error, in which the firm uses the new establishment as a platform on which it can build capabilities related to the market it has entered.

The location decision is therefore in part also a reconnaissance exercise, a fact-finding mission, in which part of the role of the establishment is to act as a 'listening post', a radar designed to pick up local knowledge, and to develop local knowledge assets. However, none of these arguments help to explain *where* either the prototype establishment will be located, or where the subsequent additional establishments will be located, other than describing their locations as simply being in other countries. Under various circumstances, the Weber, Hotelling and Salop models and their various associated analytical developments (McCann and Mudambi 2004, 2005) can be used to deal with exactly these issues. These model frameworks predict that the internationalization process whereby firms establish an initial platform investment project will itself lead to incremental changes in both the optimum location of the investment and also the optimum spatial organization of its input and output linkages. As such, the Weber, Hotelling and Salop frameworks implicitly underlie a specific process of incremental learning, which focuses on the evolution of the optimum location of both the establishment and the spatial organization of its input supply and output market locations.

The process of incremental learning has implications in terms of the organizational and decision-making structure of the firm. As more complex geographical arrangements naturally evolve as a firm expands into new international markets, it may well also be the case that the decision-making logic of the firm may itself evolve so as to better control the emerging geographical structure. The relationship between firm organizational structures and decision-making processes is a highly complex issue in its own right, and largely beyond the realms of this book. However, using diagrams we can indicate how various alternative organizational and decision-making structures may be related to questions of multiplant and multinational geography. In these organizational diagrams, the thickness of the arrow lines represents the relative significance and dominance of the flows of knowledge and information, and the direction of the arrows represents the direction of decision-making authority and control. As such, the direction of the thick arrow lines indicates the relative importance of each location as a decision-making centre, and the relative ordering of subsidiarity between the establishments.

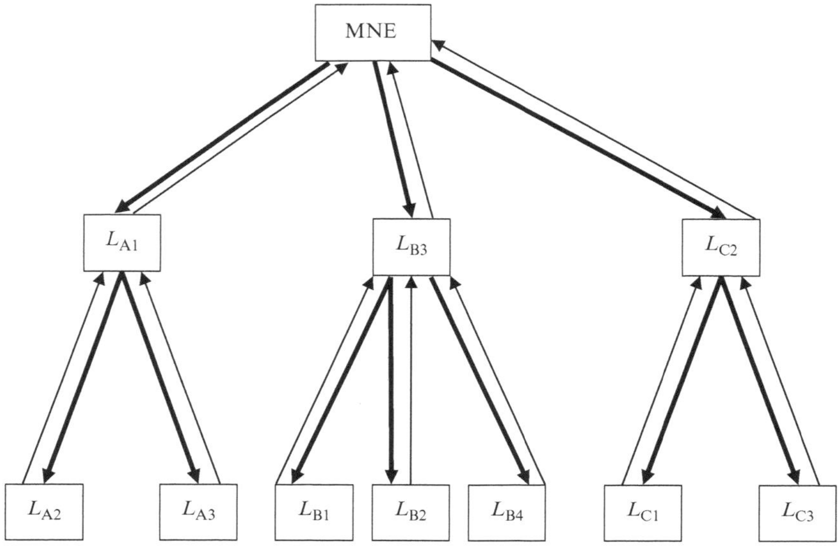

Figure 3.23 Hierarchical organizational decision-making structure

The simplest default case is where the multinational organizational structure of the firm exactly represents the decision-making logic of the MNE location decision represented by Figure 3.1 – in the introduction of the present chapter – and replicated in Figure 3.23. Here we also use the thickness and direction of the arrows to indicate both the strength and direction of the decision-making structure of the firm.

In Figure 3.23 we see that the MNE has developed seven different establishments in a variety of locations across three different countries A, B and C. If the multinational firm initially chose L_{A1} as its original overseas investment location, the decision-making structure will be represented by Figure 3.23 in that the firm will have first decided to invest in country A, and then secondly will decide in which particular location within country A to invest. The same argument holds for any alternative chosen location. Yet, while a strictly tiered hierarchical decision-making structure implied by Figure 3.23 is almost always applicable to an MNE's very first foreign direct movement overseas, in the long term as more international establishments are developed, it is not always the most appropriate organizational structure. The reason is that once the initial foreign investment project has become well-established, further overseas location choices will generally require more complex corporate decision-making and reporting structures than implied by Figure 3.23, as different establishments become

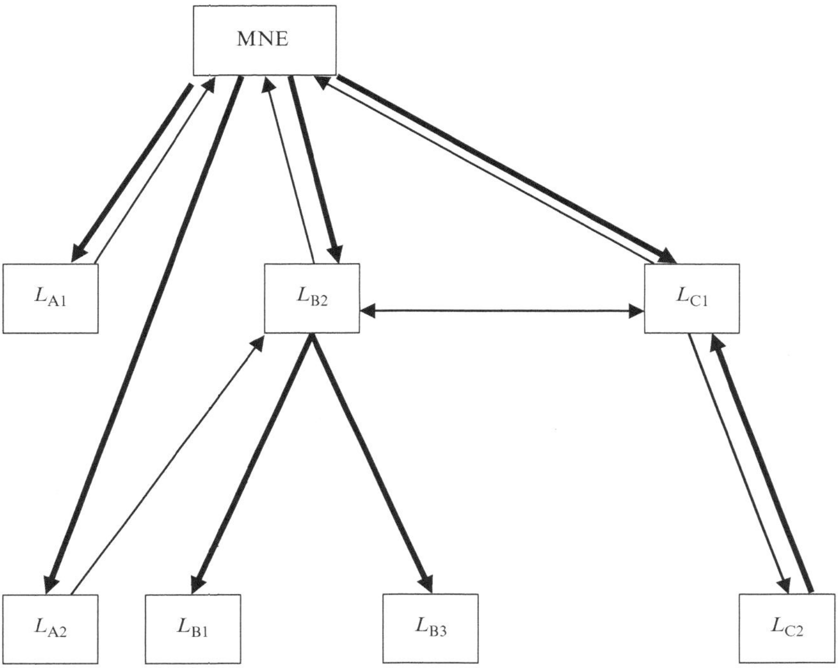

Figure 3.24 Quasi-hierarchical organizational decision-making structure

linked to each other in different ways due to technological and organizational issues which are also likely to be industry and geographic specific. The result is that over time, as the MNE makes subsequent investments in the alternative locations L_{Ai}, L_{Bi}, and L_{Ci}, it develops a more complex multiplant structure. This will usually require concomitant changes in the organizational and decision-making structure of the firm in order to manage the more complex knowledge, information and resource flows associated with this increased spatial complexity.

For example, in the case represented by Figure 3.24, the development over time of seven foreign establishments in seven different locations leads to the emergence of a quasi-hierarchical organizational and decision-making structure in which the establishments at locations L_{A1}, L_{B2} and L_{C1} represent a second tier of decision-making, each of which reports directly to the MNE headquarters in the country of origin, as well as mutually coordinating decisions in the case of L_{B2} and L_{C1}. Below this level of decision-making there are other third tier establishments L_{A2}, L_{B1}, L_{B3}, and L_{C2}, which report directly to the second tier establishments L_{B2} and L_{C1}, as

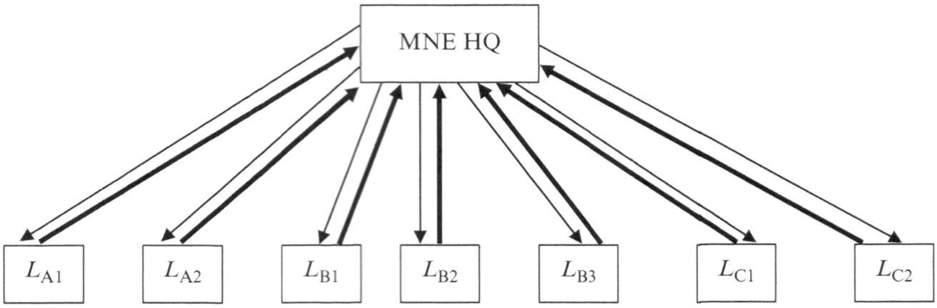

Figure 3.25 Heterarchical decision-making structure

well as directly to the parent headquarters in the case of L_{A2}. As we can see from both the strength and direction of the arrows in Figure 3.24, in this case the decision-making structure does not imply any specific hierarchical order between second and third tier affiliates.

As also highlighted in Chapter 2,[5] capabilities and competencies created in one location can be used in another, so that there may be a multi-directional flow of inputs, resources, knowledge and products between relatively autonomous subsidiaries. In order to realize these advantages, the 'heterarchical' MNE (Hedlund 1986) has to adopt more sophisticated means of coordination so as to continually maintain their local and global advantage, and this centres on allowing the subsidiaries not only high degrees of autonomy, but also a key role in reconnaissance, knowledge gathering and learning. The resulting flat organizational structure is depicted in Figure 3.25.

Figure 3.25 represents the case where all of the individual foreign investment establishments are largely autonomous both with respect to each other and also with respect to the parent establishment. This type of organizational structure would be representative of a very flat and 'democratic' decision-making structure in which the individual establishments are largely responsible for their own decision-making, and compete directly for resources emerging from the parent firm headquarters. We can consider the purely hierarchical and the basically flat organizational structures portrayed by Figures 3.23 and 3.25, respectively, as representing opposite extremes of a spectrum of multinational and multiplant organizational possibilities, with the quasi-hierarchical structured depicted in Figure 3.24 as being situated somewhere between these two polar cases. Moreover, linkages among subsidiaries and other network structures both within and outside the firm (e.g., outsourcing and offshoring agreements, alliances with competitors, and so on) can enrich these stylized representations, showing at the same time more hierarchical or networked structured according to the different operations and functions carried out by the firm. This will allow the depiction of the locational patterns of other types of MNEs described in Chapter 2: those that are simultaneously investing to acquire specific technological capabilities, management or organizational skills (i.e. *specific-asset seekers*), to rationalize previous resource- or market-led investments (i.e. *efficiency seekers*, and in particular those termed in Chapter 2 as *scale and scope economies seekers*, or MNEs looking for rationalization across different institutional settings, market and industrial structures, and policies, often as a consequence of economic integration processes), and to enhance long-term strategic objectives relative to their competitors (i.e. *strategic-asset seekers*).

3.9 CONCLUSIONS

The four types of location analytical frameworks discussed in this chapter, namely the Weber, the logistics-costs, the Hotelling, and the Salop models, each provide us with different insights into the nature of firm location strategies. The Weber model demonstrated how firm location behaviour can be an evolutionary and iterative process, as firms balance the costs of spatial transactions against location-specific factor prices in their search over time for optimal, or at least superior, locations. The insights of the logistics costs model go further, arguing that the key transactions costs which firms face are related to time, and in particular the costs of capital and knowledge-acquisition. This framework naturally leads to a spatial ordering in which activities which are higher up the value-chain tend to be located closer to the market whereas those lower down the value chain are located closer to the supply points. These locational orderings are a result of the ongoing relationships between the costs of time and the costs of space, both of which are reflected in the cost of inventory working capital. For MNEs which operate establishments in numerous locations, these capital funding costs are critical. In particular, as we have already seen, in modern globalization a huge share of international transactions are actually movements of goods and people within individual multinational corporate structures.

The market area location models of Hotelling and Salop provided us with spatial competition insights which in some ways are similar to those generated by the Weber and logistics-costs framework in that firm location processes are perceived to be dynamic, whereby firms adjust their spatial behaviour over time in response to other factors. In the case of the Bertrand-Hotelling model firms adjust their pricing strategies in a manner which is optimal for their location, but spatial clustering leads to a collapse of the market. As a response to this, the Salop model demonstrates that a combination of geographic clustering and product variety clustering enhances both the level and growth of demand as well as price stability. Indeed, this type of behaviour is typical of MNEs entering new markets. Yet, the dynamic nature of locational behaviour is not so obvious in the basic Salop model, unless we also allow for the fact that all firms in the industry will want to increase the overall size of the market. This is achieved by the additional expenditure behaviour on the part of the consumers, which is a particular type of externality associated with geography. As we will see in Chapter 5, externalities not only provide the main rationale for industrial clustering in geographical space, but also underlie the dynamic mechanisms by which such clustering evolves.

Finally, discussions of spatial profit and cost variations along with the role played by uncertainty in driving evolutionary processes, allow us to use the models of geographical- and product-space competition in order to understand how concentrations of activities arise in space.

Surprisingly, in some respects the simple location theory models discussed in this chapter work particularly well for MPDEs and MNEs. While MPDEs and MNEs are very different from the representative firm of microeconomic location models, both MPDEs and MNEs are typically firms characterized by relatively good information and knowledge resources and also low relocation costs, in comparison to other firms. As such, theoretical models of location behaviour and market competition based on profit maximizing principles, allied with models also allowing for uncertainty and evolution, are actually very relevant for understanding the investment behaviour of modern multiplant and multinational firms. This is particularly the case for MNEs representing either *general-asset seeking* or *market seeking* types of behaviour.

As already anticipated in Chapter 2, both MNEs and multiplant domestic enterprises (MPDEs) carry out their operations in multiple geographical locations and, from the perspective of the types of location theory discussed in this chapter, they can be treated in a largely analogous manner. However, a critical distinction between the two types of firms has also been highlighted here. Beyond the most apparent fact that MNEs operate across contexts which differ substantially in terms of economic, social, cultural and institutional characteristics, the crucial difference between MNEs and MPDEs lies precisely in the degree of access that MNEs have to a much broader variety of sources of information and new knowledge. These knowledge sources, which can be either intra-firm or inter-firm knowledge in origin, provide the MNE with huge opportunities for acquiring competitive advantages through the combining, blending and integration of complementary kinds of technical and organizational knowledge. Most notably, our discussions in this chapter of the advantages of heterarchical versus hierarchical organizational structures in multiplant MPDE and multinational MNE firms, allows us to take on board the importance of local knowledge accessibility in driving location decisions, and for the first time to begin to re-shape the lines of argument implicit in the OLI paradigm. In particular, this type of analysis highlights the links between the location L aspects, the knowledge internalization I aspects, and the ownership-organizational O aspects embodied in any multiplant location decision. If we move beyond general asset-seeking or resource seeking behaviour and instead follow Alchian's (1950) knowledge evolutionary arguments, we discover an analysis which is particularly pertinent to MNEs. These issues relating to the firm's internal and external sources of knowledge and the ways in which these shape the firm's strategy, have profound implications for the geography of MNEs. These are the issues we now discuss in more detail in Chapter 4.

NOTES

1. An examination of the insights of these three models is critical in order to allow us to develop these broader themes. Much of the material in the following sections of this chapter draws on elements of the analysis contained in McCann (2001), where a more formal mathematical treatment of these issues can be found than is necessary for our purposes here.
2. See Section 2.2 in Chapter 2.
3. The idea that a product or service is basically a bundle of characteristics which a consumer will pay according to their preferences and tastes was originally associated with the ground-breaking consumer theory work of Lancaster (1966, 1971, 1979) and Becker (1965).
4. Common examples here are automobile showrooms and dealerships, which are often clustered together on the edge of cities.
5. See Section 2.3.5 in Chapter 2, and also Chapter 4, on the changing role of MNE subsidiaries.

4. The sources of innovation: the firm and the local system

4.1 INTRODUCTION

Over recent years, the interrelationships between innovation, technology and industrial location behaviour have come to be seen as essential features of regional development. Technical progress has been a key element of the major transformations brought about by globalization. In the case of the MNE, these consist of increasing cross-border interdependence and the integration of all kinds of operations and functions, including those aimed at creating new knowledge and technology. On the other hand, the importance of contextual factors and systemic interactions in the process of the generation and diffusion of innovation has also been recognized in the literature as a key determinant of the technological and economic performance of firms, countries and regions. The significance of the geographical dimension of innovation systems has naturally emerged as the logical consequence of the emphasis put on the relations with sources of knowledge and innovation external to the firm.

The multiple linkages which exist between internal and external knowledge sources and conditions will be discussed here by following a micro to meso logic. Thus, in this chapter we will start from the firm as the main actor at the centre of the analysis of innovation processes; then we will move on to the industrial and geographical system, eventually integrating the two levels of analysis with particular reference to the technological creation and accumulation processes of MNEs.

The first two sections of the chapter are devoted to briefly recall some of the main theoretical perspectives on the firm as the principal economic agent and social institution generating and governing innovation and technological change. As also explained in Chapter 1, the need for rehearsing here the major economic views on the growth of firms is purely functional and is intended to highlight their treatment of innovation sources and the relevance of technological competence, capabilities and learning to growth arguments. These latter concepts find application not only at the micro level of the firm, but also, as we will see in the following Sections 4.4 and 4.5, in the meso analysis of sectoral, technological and geographical innovation systems, where firms – in particular MNEs – and other economic and social actors locate and establish linkages and networks of mutual interdependence for the creation of new technical and non-technical knowledge.

As we have already observed, primarily in the previous Chapters 2 and 3, one of the main reasons why the case of MNEs differs from that of multiplant domestic enterprises (MPDEs) – and even more so from that of uni-located firms – lies precisely in the much greater variety of knowledge sources MNEs tap into and integrate in their own innovation capabilities and strategies. On the other hand, differentiated technological competencies and capabilities are increasingly observed to be concentrated in particular sub-national locations. This suggests that further clarifications need to be made as to why, how, and to what extent, locational advantages of regions and clusters emerge and vary greatly across and within national boundaries. The geography of knowledge sources is in fact crucial for explaining the behaviour of the other types of MNEs reported in the classification of Chapter 2, that is multinational firms aimed at acquiring specific capabilities, expertise or skills in particular locality or rival firm (i.e. *specific-* and *strategic-asset seeking* MNEs), or at restructuring across different institutional, industrial, and policy settings in areas often characterized by processes of economic integration (i.e. *scale and scope economies seeking* MNEs).[1] The two levels of analysis of innovation and knowledge phenomena adopted here – namely the micro and the meso – will help us provide the reader with the basic building blocks for understanding the phenomenon of the globalization of innovation and MNE technological networks.

4.2 VIEWS OF INNOVATION AND THE GROWTH OF THE FIRM

4.2.1 Traditional Economic Theory

Technological change has long been understood as a driving force of economic growth. In its very simplest form it is treated in the neoclassical economic theory, where firms translate inputs into outputs according to the principle of profit maximization, and choose between productive inputs on the basis of their relative marginal productivity. Inputs or factors of production can be combined along a production possibility frontier, which is assumed to be smooth, continuous, and allows for an infinite number of possible production techniques or input combinations. The knowledge of how to reach all points on the production possibility frontier is available to every firm freely and instantaneously, that is, knowledge is considered as a fully public good.

The aim of the neoclassical theory of the firm, as part of the wider theory of value, is to explain how both prices and the allocation of resources are determined. The notion of the firm corresponds to a model of 'price-output decision maker', where growth is conceived as an increase in the output of given products carried out by a movement along the production function. The *optimum size* of the firm is determined by the output level at which the marginal cost is equal to marginal revenue, and for a price-taking firm this corresponds to the lowest point on the firm's average cost curve. In models of perfect competition and constant returns to scale the limits to the firm's output expansion are found to be cost increases, whereas under monopolistic competition and increasing returns to scale the limits to output growth are in terms of falling marginal revenues. The implications are that firm 'growth', that is, the observation of a larger size firm at a later point in time, must be a consequence of the changes in supply or demand conditions which determine changes in the optimum size of the firm in equilibrium. As such, by implication, in a particular industry serving a particular market, all firms must exhibit essentially the same size, thereby making it possible for us to discuss a 'representative' firm for each given market. In this fundamentally static type of theory, the market limits the size of the firm and determines its equilibrium size. Moreover, the introduction of uncertainty and risk does not substantially alter this theoretical framework.

Neoclassical economics has obviously acknowledged the existence of technological change. When it became apparent that growth could not be explained simply by changes in factor endowments, the unexplained growth was attributed to technical advance (Swan 1956; Solow 1956). However, in this basic neoclassical framework technical progress is treated as an exogenous phenomenon, as 'manna from heaven', which leads to shifts in the production frontier. In other words, growth is seen primarily as a *state*, reflected in terms of an increase in quantity of output, rather than a *process* involving changes in either the quality of production or the nature of production (Koutsoyiannis 1982). As such, in this schema, the sources of innovation are thus treated as being external to the firm and also indeterminate.

The economists of the New Growth Theory (NGT) of the late 1980s and early 1990s started to overcome some of the major limitations of the orthodox neoclassical view of economic growth, and in particular the exogeneity of innovation and technology. In the newer generation of models, technological progress began to be treated as an input factor in its own right, as a kind of 'intangible capital' which results from the knowledge derived and generated by either spillovers, human capital or 'learning by doing' (Romer 1986; Lucas 1988) or alternatively by investment in R&D (Romer 1990; Grossman and Helpman 1991). The NGT appreciation of the importance of knowledge and technology in growth processes is widely considered to be a crucial contribution to modern economic theory, allowing for better account of the empirical evidence on the subject which has been accumulating since at least the 1950s (Abramovitz 1956; Solow 1956). Nevertheless, the endogenous and, at least to some extent, appropriable characters of technical knowledge emphasized by the New Growth Theory still leave unresolved the problem that technology and innovation are far from being simple inputs into the production process. The emphasis on undefined 'spillovers' within the New Growth Theory analytical framework still does not help explain the fundamental questions of how knowledge and technology affect growth and why growth rates differ among firms with the same input endowments. Indeed, what emerges from the discussions of spillovers is that some forms of *external* effects give rise to *endogenous* growth.

Within mainstream economic and industrial organization theory, the transactions costs approach has had the merit of expanding out the 'limits of the market' in constraining firms' growth. According to its proponents, the market is imperfect because most activities, such as discovering the relevant prices, negotiating transactions for both inputs and outputs, investing in monitoring and enforcing contracts, forecasting long-term exchanges, coping with different regulations and institutional settings, all involve costs. The firm thus emerges as an organizational form for coordinating of economic activity by direction rather than by the price mechanism (Coase 1937). Potential limitations of the market therefore stem from the behavioural assumptions of transactions costs economics, such as bounded rationality and opportunistic behaviour. In this theoretical approach, the organization of economic activity is carried out within and between *markets* and *hierarchies*. Hierarchies are the main governance structures which vary with the nature and dimensions of the transaction, or with other economic environment factors such as the investment characteristics (*asset specificity*), frequency of transactions, and uncertainty and complexity (Williamson 1975, 1985, 1987). For example, non-specific, occasional or recurrent transactions, such as the purchase of standard machinery, are effectively organized by the market. In contrast, uncertain, idiosyncratic and non-recurrent transactions, such as the production of a customized intermediate input, are often removed from the market and integrated into the firm, where they can be more efficiently undertaken, because the internalization process better militates against opportunism.

As is also highlighted in Chapter 2, in the transactions costs framework both the nature and the growth of the firm differ substantially from the neoclassical approach. Indeed, transactions costs theory is very appropriate for analysing the existence of MNEs, which by definition are excluded in the neoclassical notion of a 'representative firm'. The firm's size is determined by the efficiency of its own operations relative to that of the market (Coase 1937). In the words of Williamson, 'the modern corporation is mainly to be understood as the product of a series of organizational innovations that have had the purpose and effect of economising on transactions costs' (Williamson 1987, p. 273). Therefore, the tension between markets and hierarchies may create different types of competencies or different levels of transactions costs for certain activities, requiring firms to match their competitors' distinct advantages by developing their own.

Technological competence and organizational innovation are brought into the transactions costs picture, thereby underpinning the growth and heterogeneity of firms. On the basis of its behavioural assumptions and the importance attributed to complexity, uncertainty, and asset specificity, transactions costs economics allows for the inclusion of technological change as an explanation of firm behaviour. In addition, the dichotomy between markets and hierarchies theoretically allows for the possibility of an interaction between external and internal sources of innovation. From this perspective, transactions costs theory substantially advances our understanding of innovation processes. This is because the behavioural and environmental assumptions embodied within the transactions costs framework introduce an interface between the internal and the external dimensions of the firm, and the impact of the sources of innovation depends on the balance between these two dimensions. Nevertheless, as in neoclassical economics, explaining a firm's growth is not the main aim of the transactions costs approach. Rather, its objective is to explain the existence and nature of the firm as an alternative to the market as a form of governance.[2] A more complex task, however, is to understand *how* technological change occurs and in particular *where* and *why* it tends to thrive. In order to do this, the analytical perspective has to shift from the investigation regarding the existence of the firm to one which focuses on the firm's dynamics. In other words, we must move the attention from the firm's features as a representative agent within, or because of, the market, to the firm's internal attributes and resources and the changes in these as a response to changing environments. The two other major approaches which have tried to shed light on some aspects of this complex debate are the Penrosian resource based view of the firm, and evolutionary economics of technological change.

4.2.2 Penrose and the Resource-based View of the Firm

A dissatisfaction with traditional economic theory in setting the principles underlying the growth of the firm motivated Edith Penrose's seminal contribution in *The Theory of the Growth of the Firm* (1959). In the neoclassical framework the firm is the abstract microeconomic unit of production, actually suitable for the purpose of a theory whose focus is on the relationship between output and prices. Meanwhile, transactions costs economics had shifted the attention to the transaction as the main target of observation. In neither theoretical approach was the firm regarded as an economic institution. In Penrose words, however, the firm 'is a complex institution, impinging on economic and social life in many directions, comprising numerous and diverse activities, making a variety of significant decisions, influenced by miscellaneous and unpredictable human whims, yet generally directed in the light of human reason' (1959, p. 9). The firm is seen both as an administrative planning unit and also as a 'bundle of resources'. In order to understand the motives for 'the expansion of the innovating, multiproduct, 'flesh-and-blood' organizations that businessmen call firms' (p. 13), it is necessary to consider the organization as a 'whole'. The firm is therefore treated as a *collection of productive resources*, that is, physical and human resources, the latter being the most important. Resources per se are never seen as 'inputs', but only in terms of the services that they can render, which are regarded as being unique to each individual firm. Entrepreneurial and managerial services are crucial, as they determine the direction of change and the 'single-mindedness' of the firm.

The motivations of the firm therefore also become much wider compared to the orthodox economic analysis as both profits and growth are equivalent criteria for the selection of new investment programmes. The limits to a firm's growth are found not only in the external market, but rather in its failure to provide specific resources. Growth requires management with firm-specific knowledge, which is not available on a simple demand and supply basis. Indeed, the prominence attributed by Penrose to the firm's management is clear-cut in relation to growth, and both the market and the demand are treated as perceptions ('images': see Penrose 1959, p. 42) in the manager's mind. The dynamic interaction between the internal resources of the firm and the perceived external environment gives rise to the *productive opportunity* of the firm, that is, all the productive possibilities that the entrepreneur envisages (Pitelis 2007a). Opportunities to produce new products therefore arise from changes in the productive services and knowledge available in the firm, or from changes in the external supply and market conditions as perceived by the firm. Most importantly, diversification in response to specific opportunities or to solve demand problems is seen as a general policy for growth, concomitant with the development of the technological base of the firm.

Penrose distinguishes between two kinds of knowledge: 'objective' knowledge, which is codifiable and transmittable, and 'experience', which results from learning and can only be transferred to a limited extent. All activities or functions carried out within the firm generate new knowledge through intra-firm learning, resulting in 'excess' resources that can motivate growth and further innovation. Resources and their services are thus the engines of the firm's growth, whilst the mechanism of transmission is the learning process. Changes in the productive opportunities of firms are intrinsically associated with knowledge increases "built into" the very nature of firms (Penrose 1959, p. 78). The importance of increased knowledge in the external world should not be ignored (see, for example, p. 79), but the focus is indeed on the experience and knowledge of the management, as the latter also determines what the firm "sees" in the external world' (p. 80).

Edith Penrose can be viewed as the forerunner of modern 'managerial' theories, pursuing economics in a manner that was common before the neoclassical paradigm became dominant. Subsequent managerial developments of Penrose's ideas are grouped under the heading of the Resource-Based View (RBV) of the firm. Drawing on Penrose, although not always explicitly, in the RBV the firm is seen as a bundle of organized resources, which are idiosyncratic, relatively immobile and built cumulatively over time, and the task of its management is to adjust and renew the firm's resources as time, competition and change erode their value (Wernerfelt 1984; Rumelt 1984; Foss 1997a, b, 1998).

In the last few decades, Penrose's contribution and the RBV developments have acquired significant recognition in many circles, arguably challenging transactions costs as the leading alternative economic theory of the firm. Curiously, however, both Penrose's work and the RBV extensive literature has found comparatively limited application in relation to MNEs (Pitelis 2007a). This is even more surprising because, as mentioned in Chapter 2, Penrose herself carried out substantial work on MNEs (e.g., Penrose 1956, 1968, 1990), and had obviously the multinational corporation in her mind when writing about the growth of the firm (see her own foreword to the third edition of *The Theory of the Growth of the Firm*, Penrose 1995). As pointed out by Pitelis (2007a, b), the major reason for such an apparent anomaly is that Penrose saw multinationality simply as a natural result of the growth process. Therefore, in her seminal contribution of 1959, she did not devote particular attention to distinguishing the features of the phenomenon per se, though her applications of the firm as an 'evolving collection of resources' were made with particular reference to MNEs.[3] In contrast, Penrose's theory of the firm's growth, and its suitability for explaining the evolution of multinational corporations, has received much more attention in the literature on the economics of technological change.

Nevertheless, in Penrose's approach and in most of the RBV, the multiplicity and heterogeneity of knowledge sources, as well as the complementarity and interdependence of their internal and external natures, are largely overlooked. In contrast to traditional economic analysis, the emphasis is put on understanding a firm's internal functions and resources, while the investigation of the external environment, including the market, has been relegated to a secondary level, that is, that of the management's perception.[4] However, as acknowledged by Penrose herself, the concepts of entrepreneurship and management in this schema are rather slippery because they are only related to the personal qualities of individuals. This also implies that the influence of the factors characterizing the firm's external environment are strongly underestimated in explaining growth. Technological change, which is exogenously determined in the traditional economic theory, in the RBV becomes endogenously governed by unique and heterogeneous resources which are specific to the individual firm.

Innovation processes and sources, and their situated and contingent nature, are therefore either unexplained, that is, treated as 'manna from heaven' as in traditional neoclassical economics, or are made essentially idiosyncratic, that is, built into the exclusive nature of each business organization, as in most of the RBV literature. In the case of neoclassical economic theory, so far the power of its formalized theoretical apparatus has been unable to make up for its critical limitations when it comes to the study of technological change. In particular, innovation cannot be rationally planned due to uncertainty; firms cannot have perfect knowledge of production techniques when knowledge is gained through learning and experience, and innovation is often continuous and disequilibriating. Transactions costs economics does bring some 'life' into the economic model by taking into account human as well as environmental factors in explaining the nature and existence of the firm. However, a degree of determinism and only a minor interest in the firm's growth has so far limited the applicability of these approaches to the analysis of innovation sources. On the other hand, the emphasis of Penrose and the resource based view on firms' internal mechanisms and processes, and also on new conceptions of corporate strategy, are of the utmost importance for understanding technical change as an active component of firms' behaviour, and particularly in the case of multinational firms. It is, however, largely ineffective in analysing the contextual nature and the interdependence of the many sources of innovation.

4.2.3 Economics of Technological Change and Evolutionary Views

As in the case of neoclassical economics, the economics of technical change, and its Schumpeterian and evolutionary foundations, is not a theory of the firm. Its central issue is rather technological progress and its relationship with growth. This approach focuses directly on the questions of how, where and why technological change occurs, and how it affects economic processes such as industrial dynamics and growth. The basic tenet here is that such processes take place in historical time, therefore bringing into the picture many ostensibly non-economic factors, such as science, society, culture, institutions and policy. Two concepts are particularly important here in considering the microeconomic level of the firm. *Heterogeneity*, implying that the representative firm of the neoclassic theory no longer makes sense, and *selection*, which suggests that firms with 'good' strategies will tend to grow, while those with 'bad' strategies will tend to lose. As a consequence, the other axiom of the orthodox theory that any feasible pattern of productive activity can be faultlessly replicated becomes incongruous.

The evolutionary framework, which was first articulated as a modern theoretical framework in Nelson and Winter's seminal book *An Evolutionary Theory of Economic Change* (1982), draws significantly on the work of other scholars including Simon (1947), Alchian (1950), Cyert and March (1963) – seen also in Chapter 3 – and most notably Schumpeter (1934, 1950). As with Penrose, in the evolutionary approach attention is paid to the nature and the sources of continuity in the behavioural patterns of the individual organization. In the evolutionary approach, the resource base of the firm is seen as being subject to improvement through learning, adaptation, imitation and innovation. These changes may or may not be transmitted to other firms, thereby accounting for differences in competence and capabilities between firms. Such variety actually enhances the evolutionary fitness of the population of firms as a whole, and this is the actual focus of Nelson and Winter's contribution.

Within the firm, change involves the disruption of *routines* which provide the basic structure for everyday operations. According to Nelson and Winter, routines are 'regular and predictable behavioural patterns of firms' (1982, p. 14). They are the persistent and organization-specific characteristics that shape the firm's potential behaviour, though, importantly for our purposes here, the actual behaviour of the firm is explicitly determined *also* by the external context. Routines are inheritable, path-dependent, and selectable, as variety among firms will render some routines 'better' than others. Skills and routines are not deliberately chosen as against alternatives, but rather they are developed through trial and error and include various stochastic elements. Over time routines change because of innovation, which reassemble old routines into new combinations, or substitute them with entirely new ones (Zollo and Winter 2002). Thus, firms grow over time through the search and selection of the 'best' routines, shaping the condition for the whole industry. Routines are therefore seen as largely analogous to 'genes' in biological evolutionary theories. Every firm thus has distinct options and strategies for responding to technical change which are drawn from its own history, the personalities of its management, the variety, experience and insight of its personnel, the contributions of the culture and locale in which the firm is sited, and the network of suppliers and customers related to the firm, all of which might be seen as 'identity creating' elements of the firm. The evolutionary approach emphasizes the value of diversity in producing 'evolutionary fitness' and the role of routines as stable constructs by which the firm is governed in everyday activities, and from which innovation is a departure. Routines act as coordinating mechanisms for individual knowledge, skills and activities, and they reflect the skills of the organization as a whole. Routines therefore embed organizational knowledge and are at the basis of the firm's organizational capabilities (Nelson and Winter 1982).

In discussing the limits of traditional economics for explaining technical progress, three critical questions are addressed by Nelson and Winter (1982, pp. 62–64). The first one is *where does knowledge reside?* The answer given is in the organization's memory, as knowledge is an attribute of the firm as a whole, and is not reducible to any of its single components. The second question is *what are the conditions that separate what is technically possible from what is impossible?* The answer given is 'the boundary is the boundary of knowledge' (p. 63). For the firm, the purpose of carrying out innovative and R&D activities is to widen up technological possibilities and create its own technology. Yet, technology is constrained by a number of factors and technological possibilities are not an infinite set. Problem-solving activities can search for sources that provide an answer, thus improving the state of knowledge, but solutions to some problems may not exist at all. The third question, which is arguably the most relevant for us here, is *what is the relationship between the knowledge of a firm and that of other firms or of the environment in general?* The indefinite boundaries of a firm's knowledge point to different ways in which 'firms can augment their own knowledge by reaching out into the environment – into their industry or into society more broadly' (p. 64). As skills, organization and technology are all seen as being intimately intertwined in routines, knowledge sources are therefore entangled in the internal and external environments of the firm, therefore making it rather difficult to separate neatly the importance of the two dimensions.

From what has briefly been discussed so far here it emerges that different understandings of knowledge, innovation and technology underlie different theories of the firm's growth. Yet, in contrast to other theoretical frameworks, evolutionary economics has found extensive applications both in the theory of the MNE and in the economic geography arena. This underlies our choice to favour – although not in an exclusive sense – such a perspective in considering the interdependence between the micro level of the firm – both MNEs and domestic firms – and the meso level of the industry and the spatial system, particularly when looking at the changes imposed by the globalization of technology and innovation sources.

4.3 TECHNOLOGICAL COMPETENCE, CAPABILITIES AND LEARNING

4.3.1 The Nature of Technological Progress

Innovation is a crucial source of profits and growth. Profits derive from continual advances in production, as wage increases tend to lag behind productivity increases. This is usually achieved by adapting the resources and expertise of the firm to the changing economic circumstances, often also drawing on the complementary skills and resources of other firms, organizations and institutions. The degree of firms' technological competence therefore affects both the unit production costs and also the output product range or quality. As such, advanced technological competence increases the profit margins of advantaged firms relative to other firms in the same industry, and thereby affects their market shares. It also facilitates the entry of firms producing at low unit costs or with high quality outputs into new or related fields of activity.

The conceptualization of technological competence suggests that its dissimilarities and uneven distribution are a consequence of the nature of technology and the way in which it develops. The central proposition on this issue is first clearly articulated in Nelson and Winter's theoretical framework, although it was already implicit in Penrose's thinking of the firm's growth (1959) and in Rosenberg's history of technology (Rosenberg 1969). The basic principles are that technological knowledge is partially tacit, is specific to the context in which it has been created, used or adapted, and is dependent upon the learning and skills of those that have developed and operate it. Technology is a product of a steady search for improvements and a learning process in any particular firm.

As pointed out by Penrose, two elements of technology should be distinguished: (1) potentially public knowledge, including information and patents; this is a latent public good which incorporates generic technological knowledge, and which is tradable, though it may still be difficult to transfer; (2) tacit knowledge; this is private, is locally specific and non-codifiable, as it depends on common codes and organizational routines which are non-tradable, and it may be imitated but never exactly copied (Nelson and Winter 1982). As mentioned earlier, the skilful behaviour which is required for the generation and application of technology normally becomes embodied in a set of routines which characterize the individual firm, and change only gradually. In their treatment of skills and tacit knowledge, Nelson and Winter acknowledge the observation 'We know more than we can tell' made by Polanyi with reference to that kind of human knowledge that is difficult to articulate (Polanyi 1967, p. 4). Collective skills are even harder to express to outsiders than individual skills, posing serious problems for the teaching and learning processes. It is important to note, however, that knowledge tacitness is a matter of degree: the same knowledge can appear *more tacit* to some individuals or organizations than to others, implying also an influence of contextual and institutional circumstances in the potential communication of knowledge.

In this framework, different attributes can be put forward with respect to the nature of technological change (Nelson and Winter 1982). The first one is that technological innovation is a *cumulative* process, partly as a consequence of the gradual learning and establishment of specifically refined skills and routines upon which it depends. Routines only change slowly through careful experimentation, learning by doing and learning by using. Cumulativeness also results from the need for critical revision caused by the emergence of imbalances, or when one advance gives rise to others in the same field of activity. Second, innovation progresses *incrementally*, so that firms tend to move gradually between related types of technological activity. The solutions found by a problem-solving activity raise other difficulties, and an advance may require others in related fields for its effective implementation. Incremental changes also often follow from a shift in technological opportunities or in the availability of inputs. Third, technology is *differentiated* between firms and locations, given that it is specific to the context in which it is created. Due to its tacit component, technology has unique and characteristic features in each new learning environment. Fourth, technological change is partially irreversible: once technology and the accompanying skills and routines have moved on, previous or simpler technologies are 'forgotten', and to reintroduce them would require a new learning process and the modification of individual and collective skills, organizational practices and institutions. Technological development is frequently non-ergodic in the sense that more than one outcome may have been possible, but beyond some threshold the path or trajectory becomes established and earlier alternatives are excluded; in other words it is a *path-dependent* process (see also Arthur 1989). Fifth, the specific path of innovation in each firm and location is constrained by a system of technological *interrelatedness* between firms, types of activity and knowledge sources (Nelson and Winter 1982).

All these attributes have crucial implications for a spatial analysis of the sources of innovation and for explaining the dynamics of industrial agglomeration. The importance of space in lowering the barriers and costs of knowledge sharing and transmission is related to the basic properties of knowledge and learning processes, most of all their degree of complexity and tacitness (Breschi and Malerba 2005). Cumulative and incremental innovative processes operate through non-linear and self-reinforcing feedbacks between technological and structural change. There is an array of constraints and conditions that are basically idiosyncratic to economic and social structures and the historical paths of locations affect learning and knowledge accumulation (Antonelli 2008; Patrucco 2009). Differentiation means heterogeneity, which is found at the micro-level of the firm as well as at the meso-level of the industrial, regional or local environment (Boschma and Lambooy 1999; Garofoli 2003). The latter is heterogeneous due to a combination of chance or stochastic events, and path dependent as a result of historical contingency.[5] Geographical space therefore acts as a selection mechanism that may, or may not provide conditions favourable to meeting the requirements of technical change. Interrelatedness has an important influence on the locational specificity of innovation processes, as it may be costly to change the methods prevailing in an individual firm or industry without complementary changes elsewhere in the same context (thus, outside the control of any particular firm). Hence, the main reason for knowledge to be confined to certain geographical contexts is assumed to be its inherent complexity, which may make it difficult to share among different interacting actors or organizations. Such complexity may prevent knowledge from being codified and made explicit and mobile, and thereby stored and transmitted by way of information (Steinmueller 2000; Breschi and Lissoni 2001a, b). It is these arguments which underlie the knowledge 'filter' hypotheses (Acs 2002). Moreover, this wider notion of knowledge may also embrace cultural and institutional differences, which shape the spatial patterns of knowledge production, absorption and diffusion.

By viewing knowledge-creation processes as complex, cumulative, partially tacit and path dependent phenomena, there are strong grounds for arguing that innovation is very likely to stay highly concentrated across space, organizations and hierarchies, thereby giving rise to rather distinctive growth patterns. Technological innovation is in fact generally more 'sticky' than production, and this may be explained through the distinction first introduced by Henderson and Clark (1990) and developed within spatial systems by Phene and Tallman (2002), between *component* and *architectural* knowledge. The first encompasses specific resources, assets, skills and technical systems that refer to particular constituents of the organizational system rather than to the whole. Architectural knowledge, in contrast, is related to an organizational system as a whole, which through its institutional structure and routines arranges the components for productive use (Phene and Tallman 2002). Such architectural knowledge is region-specific or cluster-specific, has a path-dependent and evolutionary nature, and is developed through relationships at both inter-firm and intra-community level. It provides a common framework through which the creation, adaptation, and diffusion of component knowledge is greatly facilitated, thereby raising intra-regional absorptive capacity and extra-regional linkages. As we will see at the end of this chapter, to global actors such as MNEs, it is these knowledge features that will therefore lead to the attractiveness of a particular region, vis-à-vis of other locations.

4.3.2 Dynamic Capabilities and Learning

Since Nelson and Winter's seminal book, contributions in the area of firm-specific capabilities have proliferated within the resource based view, within evolutionary economics, within the economics and history of technical change, and within strategic management. The dynamic capabilities' framework has also been widely used to investigate how firms (Iansiti and Clark 1994; Teece and Pisano 1994; Iansiti 1995a, b; Teece et al. 1997; Eisenhardt and Martin 2000; Teece 2000) and countries (Porter 1990; Hobday 1996) create and sustain competitive advantage over time while facing ongoing changes in the economic environment. The assumption of this particular conceptualization of a firm's dynamic capabilities is that firms act in a Schumpeterian competition context which is characterized by winners and losers, and which is in a continuous disequilibrium. The constant learning processes are the mechanisms by which capabilities are accumulated and these determine the selection of winners against losers.

The dynamic capabilities idea takes into account the firm's specific assets, which are the assets that are generally difficult to imitate or protect by strong intellectual property rights. Likewise, it encompasses the tacit elements of innovation such as skills acquisition, knowledge and learning, all of which are represented by routines. Dynamic capabilities are thus defined as the firm's ability to integrate, build and reconfigure internal and external competencies to address rapidly changing environments (Teece et al. 1997). As such, they reflect an organization's ability to achieve new and innovative forms of competitive advantage (Leonard-Barton 1995).[6]

Dynamic capabilities consist of several dimensions which are either technological or organizational (Leonard-Barton 1995; Tsekouras 1998). Technological capabilities imply that knowledge, artefacts, operations and processes are industry or sector specific, while organizational capabilities refer to the knowledge which is related to the specific arrangements in a particular order or system in which a business unit or any other type of organization operates. Hence, despite their highly situated (within the firm) character, the processes of capabilities' accumulation allow for the identification of commonalities within and across industries (technological dimension) and systems (organizational dimension). That is, firms attempt to imitate and transfer the particular processes of capability accumulation that they observe in other firms, locations or industries. The search for firms' 'best practices' in strategic and organizational processes are therefore instances of such generalized efforts, even though, on close examination, these processes will often appear to be idiosyncratic in their details (Eisenhardt and Martin 2000). From this perspective, the creation of competitive advantages for any firm therefore relies on its ability to reconfigure its resources and to acquire and use the resources that come from other actors in a specific production system.

In spite of the somewhat limited application of the resource based view to the specific study of the multinational firm, the concept of dynamic capabilities is especially suitable in exploring MNEs' behaviour. In fact, as Teece and Pisano (1994) point out, the replicability of dynamic capabilities based on complex routines is particularly difficult when carried out in a context which is different from that of origin of the firm. This would imply that there is a high level of effectiveness on the part of MNEs in mastering the learning processes across multiple locations and environments. On the one hand, capabilities and learning can be spurred by multinational experience. On the other hand, they can also be moulded by the different contexts which are external to the multi-located international firm. As discussed below, capabilities become an important bridge between the microeconomic actors, such as firms, and sectoral and geographical innovation systems.

Along with an acknowledgement of the analytical advantages of the literature on capabilities, in terms of the identification of new ways of conceiving technology and corporate strategy, several criticisms have also been raised. For example 'the term "capabilities" floats in the literature like an iceberg in a foggy arctic sea, one iceberg among many, not easily recognised as different from several icebergs nearby' (Dosi et al. 1999, pp. 5–6). In response to this, von Tunzelmann and Wang (2003) have helped clarify the distinction between competencies and capabilities. Competencies are understood as stemming from inputs to produce goods and services; in this sense, they are pre-set attributes of individuals and firms, with the enhancements typically produced either internally to the firm or by a different organization. For example, one may think of the firm's endowment of adequate skills as the necessary internal competencies to obtain value from R&D and innovation investments (Piva and Vivarelli 2009). Meanwhile, capabilities involve both learning and the accumulation of new knowledge, and also the integration of behavioural, social and economic factors. Consequently, capabilities are to be taken as outputs of learning processes that in their collective dimension can be highly localized, giving rise to 'system' capabilities, that is, referring to a specific spatial and industrial setting (von Tunzelmann, 2009b). In other words, while technological competencies are prerequisites or resources for innovation activity, technological capabilities correspond to knowledge that, through learning and absorption, is ready to be incorporated into new products and processes. For instance, a pharmaceutical firm endowed with an adequate R&D lab and performing research on a new vaccine is a firm with technological competencies, while a competitor already testing a new vaccine on patients is a firm with technological capabilities (Iammarino et al. 2012).

A further criticism of the literature on capabilities has been the limited emphasis put on the increasing multi-technological nature of products brought about by technical advances. The main implication here is that, in multi-technology settings, the number of technologies is simply too high to be managed entirely within any single firm, even if the firm is an MNE. This observation emphasizes two strictly connected issues, namely the relevance of sources of technology which are external to the firm, and in particular linkages with science; and also the role played by relational and interactive capabilities for creating new knowledge.

The first issue is the specific focus of the systems of innovation approach discussed in detail below. Briefly, other actors of the production system (universities and public research institutes, government, finance sector, and so on) also develop dynamic capabilities in their interactions with firms (Coriat and Weinstein 2002). Thus, the system as a whole builds on dynamic capabilities because its agents are able to deploy effectively inter-organizational routines (Fujimoto 2001; Zollo et al. 2002). These links may promote the system's capability for action through effective execution of problem-solving processes in the face of an uncertain business environment (Eisenhardt and Martin 2000; Griffith and Harvey 2001; Leoncini et al. 2006). Thus inter-organizational and collective learning processes occur at the system level (Cowan et al. 2000). Nevertheless, the possibility also exists that systems, like firms, may neither accumulate knowledge efficiently, nor be able to learn and develop capabilities (Winter 2003).

The second issue, which is the role of relational and interactive capabilities, has been addressed in some recent developments within the literature on dynamic capabilities. In particular, von Tunzelmann and Wang (2007) observe that firms act simultaneously as producers of intermediate goods that are inputs for other industrial firms, thus playing the role of suppliers, and also as consumers of different kinds of inputs themselves. They suggest that the dichotomy of 'demand-supply' should therefore be replaced by the triad 'supplier-producer-consumer', so as to take into account the multiple roles concurrently played by producing firms. Consumers and suppliers are in turn producers which are related to other consumers and suppliers, thereby creating a complex network of transactions and knowledge flows. Within this framework, dynamic capabilities involve the producers managing timely changes on the suppliers' side (reflecting technological change) and also on the consumers' side (reflecting market change), in a context of dynamic competition characterized by 'ever changing products, market structures, technologies, resources, forms of industrial organization and so forth' (von Tunzelmann and Wang 2007, p. 202). Drawing on Malerba (1992), von Tunzelmann and Wang (2007) illustrate firms' internal and external sources of knowledge, highlighting the fact that it is the successful interaction between the capabilities of consumers, producers and suppliers, which is at the origin of the *dynamic interactive capabilities* that allows a network to compete in the long run. As we will see, such concepts are crucial in explaining the behaviour and strategies of MNEs in generating and diffusing new technologies across borders and locations of technological excellence.[7]

Over recent years, the appreciation of the nature and characteristics of the innovation process has therefore put learning processes at the centre of attention. Understanding technological innovation as a collective and interactive learning process helps to explain many of the issues related to innovation and to the most internationally competitive MNEs. One such issue is the importance of R&D, which is usually the major source of new knowledge for most MNEs. The significance of R&D varies across industries, just as the relationship between public or potentially generic knowledge and tacit capability differs across industries. R&D contributes to technological innovation by providing a supply of potentially public knowledge, which is then used in combination with the continuous development of tacit capabilities through the evolution of collaborative skills and organizational routines. Therefore, it is misleading to think of R&D as the general driver of innovation, even though it plays a crucial role in science-based sectors, since innovation also reflects what is achieved in the broader learning process. Appreciating that innovation is a collective learning process also helps to resolve the longstanding debate as to whether the essential source of technological progress is research-push or demand-pull. Innovative learning instead gathers together cumulative and incremental processes that interact with, but are not driven by, the development of either science or market demand. Technology affects science and demand as much as the other way round.

The dynamic capabilities framework has evolved from initially considering that the competitive advantage of firms is explained by the accumulation of specific assets, to the inclusion of organizational aspects and managerial processes which also promote learning. From the typical internal (to the firm) learning processes such as learning by doing, learning by using, learning by carrying out R&D, the emphasis has shifted to external learning processes, such as learning from spillovers, learning from interacting, and learning from science (Malerba 1992). Therefore, in the next section we will briefly consider the body of literature which proposes that the generation and diffusion of economically useful knowledge involves the systemic arrangement of a series of actors and institutions beyond the individual firm (Nelson 1992, 1993; Lundvall 1992). Industry structures, linkages between firms and other organizations, and also institutions and specific government policies, all have an impact on the sources of innovation available to firms, thereby giving rise to the notion that there are indeed systems of innovation.

4.4 SECTORAL PATTERNS OF INNOVATION

4.4.1 Industry Dynamics and Diversity

The emphasis on firm heterogeneity which is central to both the evolutionary economics of technological change and the resource based view literatures, has challenged both orthodox industrial economics and also strategy studies in stressing that firms are indeed different *beyond* industry boundaries. In both orthodox industrial economics and strategy studies, firms are regarded as being different *because of* industries, whilst variety in the patterns of firm behaviour within the same industry are basically not emphasized, neither in a traditional industrial organization setting (Bain 1956; Sutton 1991) nor in the competitiveness framework proposed by Porter (1980). Yet, the observation that much of the cross-sectional and temporal variation in the characteristics, behaviour and performance of firms is within the same industries, is not at odds with the recognition of sector-specific properties. These properties actually influence the intra-industry heterogeneity and its persistence over time. The fact that high/low innovators at time t have, *ceteris paribus*, a higher probability of being high/low innovators at time $t+1$, is explained on the basis of the co-evolution of knowledge sources which are internal and external to the firm, with the external sources being at least partially internal to the industry (Griliches 1986; Pavitt and Patel 1991; Cefis 1996). Indeed, the empirical evidence in industrial economics shows that there are significant differences across sectors in the patterns of industrial competition (Dosi et al. 1995, 1997; Marsili 2001). These differences exist across various dimensions in terms of both static and dynamic measures of competition, such as the degree of market concentration, firm size distributions, rates of entry and exit, growth rates, the persistence in firms' attributes and performance, the volatility of market shares of incumbent firms, and the degree of diversification. For example, entry rates are supposed to depend on the economic incentives for entry into an industry relating to the trade-off between profits and entry barriers. However, when looking at the industry-specific factors that might explain such differences, the empirical evidence suggests that those factors which are related to the nature of economic incentives, such as profitability, capital intensity, demand size and growth, have little explanatory power with regard to the observed variety of sectoral patterns. In order to explain this variety, it is necessary to take into account the nature of technology in different industrial sectors (Marsili 2001).

Some stylized facts on the evolution of industrial structures are pointed to by evolutionary economists within the so-called Schumpeterian scenario of 'institutionalizing innovation' (Nelson and Winter 1982; Dosi et al. 1988, 1995; Audretsch 1997, 1998). Firstly, industrial demographics in a particular sector are shaped by the rate and direction of technological change in that sector. These in turn depend on the nature of technology, such as whether it favours entrants over incumbents or the reverse. Secondly, the sources and nature of knowledge influence the decisions to enter the industry in the first place, while technological learning and market selection determine the growth and survival of firms.[8] Hence, if cumulative learning occurs mostly through sources inside the sector, such as via in-house R&D and knowledge which is specific to industrial applications, a pattern of industrial dynamics characterized by low entry and high concentration is likely to prevail. If instead new opportunities come from sources outside the sector, such as academic research and generic and non-systemic knowledge, high entry and low concentration are likely to be dominant. Thirdly, the overall process of industrial evolution is characterized by structural change, with the emergence of new industries and the decline of old ones. Radical innovation will often lead to the creation of new economic activities, while conversely, incremental innovation acts as the growth engine of existing activities in a specific institutional context (Verspagen 2000). Consequently, the evolution of firms in an industry is fundamentally shaped by the underlying conditions affecting the creation and reproduction of knowledge. These 'technological regimes' (Winter 1984), as we will see in Chapter 5, are argued to play an important role in determining the interdependencies between industry characteristics and spatial agglomeration.

4.4.2 Technological Regimes, Paradigms and Sectoral Taxonomies

The characteristics of knowledge and technological learning processes are known as a 'technological regime' (Winter 1984). A technological regime is defined (Malerba and Orsenigo 1990, 1992, 1993) as being a particular combination of: appropriability conditions (i.e. the returns to innovation); technological opportunities (i.e. the likelihood of innovation); degree of cumulativeness of technological knowledge (i.e. the extent to which the amount of innovation produced in the past raises the probability of current innovation); and the characteristics of the knowledge base (i.e. the type of knowledge upon which the firm's activities are based). Technologies and sectors exhibit substantial differences in the extent to which innovation can be protected from imitation by competitors. In addition, the size and scope of technological opportunities vary greatly across different sectors. Technological opportunities, and thereby a firm's growth opportunities, are governed by the prevailing technological paradigm. This is defined as 'a system of scientific and production activities based on a widespread cluster of innovations, representing a response to a related set of technological problems, and relying on a common set of scientific principles and on similar organizational methods' (Cantwell 2001, p.447; Dosi 1984; Perez 1985; Freeman 1987). The characteristics of the technological paradigm are that the degree of tacitness and specificity, specialization or pervasiveness of knowledge, and the extent of cumulativeness of innovation (or its correlation over time) show particular industry characteristics (Dosi et al. 1995). The technological paradigm is therefore a consequence of the interrelatedness in both the knowledge conditions and the problem solving activities between firms and industries.

Technological regimes identify common properties of innovative processes in different sets of production activities, and this helps to explain the asymmetries in the dynamics of industrial competition at a sectoral level, and as we will see in Chapter 5, also at a geographical level. Changes in knowledge bases, technological procedures and trajectories, determined by changes in the prevailing technological paradigm,[9] affect the combination of the properties and conditions which make up the technological regime in each industrial sector. Changes in paradigms ultimately explain industrial structures and their innovation dynamics (Nelson and Winter

1977, 1982; Dosi 1988; Malerba and Orsenigo 1993, 1995, 1996; Dosi et al. 1995; Marsili 2001). In the words of Nelson and Winter, 'industries differ significantly in the extent to which they can exploit the prevailing general natural trajectories, and these differences influence the rise and fall of different industries and technologies' (1982, pp. 59–60). The same concept of technological paradigm also implies that there is a close link between technical progress, organization and also socio-economic institutions. By definition, any radical innovation to some extent brings about transformations in the organization of markets, production and communities. Therefore, organizational and institutional changes are inextricably associated with technological change and these interrelationships characterize the evolutionary development of socio-economic systems (Freeman and Perez 1988; Foray and Freeman 1993).

The two archetypical types of technological regimes can be traced back to the different interpretations of innovative processes given by Schumpeter (Freeman 1982a). The first type of technological regime, which is labelled as *Schumpeter Mark I* (*SMI*), is that of 'creative destruction' and this is first discussed in the *Theory of Economic Development* (Schumpeter 1934). This is also known also as the science-based regime (Nelson and Winter 1982; Kamien and Schwartz 1982), in which the evolutionary dynamics are driven by entrepreneurs who launch new firms that embody new business conceptions and ideas and consequently challenge the incumbent firms, thereby continuously disrupting the established procedures of production, organization and distribution. In contrast, the second type of technological regime, labelled as *Schumpeter Mark II* (*SMII*), is that of 'creative accumulation', the nature of which was first discussed in *Capitalism, Socialism and Democracy* (Schumpeter 1942). This technological regime is also known as the cumulative technology regime (Nelson and Winter 1982), in which learning is highly cumulative in established large firms which rely on large in-house stocks of knowledge, strong institutionalized R&D capacity, tightly controlled production and distribution facilities, and sophisticated managerial and financial resources, in order to create high barriers to entry into the industry.

These two profoundly different insights have been investigated by different strands of empirical research. By adopting an industry life cycle approach, some scholars have pointed out the regularities which exist regarding how firm entry and exit, market structure and technological change all vary from the birth of technologically progressive industries through to maturity (e.g., Utterback and Abernathy, 1975; Klepper, 1996). In the early stages of an industry, when technology and demand are not established, industries are typically associated with a high degree of uncertainty and barriers to entry which are generally very low and the patterns of innovation are likely to be represented by the SMI type of technological regime. This is consistent with the findings of Winter (1984), whose *entrepreneurial regime* is characterized by intense R&D activity, much of which is externally undertaken, and often in the public as well as the private sector, thereby contributing to a broad knowledge base. Here, innovation is typically undertaken mostly outside of established firm routines, and is therefore very favourable to the entry of new and often small firms. In contrast, at later stages of industry maturity, when scale economies, entry barriers and capital accumulation become highly relevant, a *routinized regime* (SMII type) is likely to prevail, in which the knowledge base is narrow, targeted and cumulatively built through the learning processes within the firm. In other words, innovation takes place typically inside its established routines of the firm (Winter 1984).

Other research which investigates firms' innovative activities and sectoral technological trajectories typically makes reference to Pavitt's (1984) classic taxonomy, which is widely used to explain the firm and sectoral patterns of technical change and innovation among innovative manufacturing sectors. The Pavitt taxonomy originally splits firms and sectors into three broad categories, namely supplier dominated, production intensive (distinguished in terms of specialized suppliers and scale economies), and science based firms and industries. The logic of these categories is related to the sources of knowledge, the production and use of innovations, and the characteristics of innovative firms. Subsequent work based on Pavitt's seminal contribution (e.g., Pavitt et al. 1989; Archibugi et al. 1991; Veugelers and Cassiman 1999; Evangelista 1999, 2000; Archibugi 2001; Marsili 2001; Miozzo and Soete 2001; Castellacci 2008a; Capasso et al. 2011a) have adapted the framework to take into account structural changes such as, for example, the emergence of information-intensive and life science-based firms, the increasingly blurred boundary between manufacturing and services activities, and even the shifting patterns of these Pavitt classification sectors over time.[10] The Pavitt-related literature has emphasized the crucial role played by vertical linkages and interactions among innovative firms both within and across sectors. The intensity and effectiveness of such networks rely on firms' dynamic capabilities, as well as on the social and institutional conditions (Perez 1983; Freeman and Perez 1988). The perception that techno-economic and socio-institutional factors together explain economic change and growth thereby opens up the perspective of thinking of innovation in terms of innovation systems, which can be considered across multiple dimensions.

A complementary line of research has focussed on the sector specific features of innovative activities and industrial dynamics, proposing categories that group sectors on the basis of the properties of the processes through which firms innovate (Malerba and Orsenigo 1993, 1995, 1996; Breschi et al. 2000; Marsili and Verspagen 2002; Malerba 2002). On the one hand, widening technological patterns (SMI) are determined by low degree of cumulativeness, low appropriability and high opportunities, and also a high importance of applied science and sources of innovation which are outside the firm and the sector, thereby favouring new firms entering the industry. Conversely, deepening technological patterns (SMII) are shaped by a high degree of cumulativeness, and high appropriability and opportunities, with a high importance of basic science and sources of innovation which are inside the firm and the sector, thereby favouring the incumbent firms. Obviously, these two patterns are just the extremes of a range of intermediate cases that reflect the different combinations of knowledge conditions in different industries. However, these are seen to evolve and shift from one regime to another depending on their life-cycle (Malerba and Orsenigo 1997; Marsili 2001).

Interestingly, and importantly for our purposes, if the appropriability and cumulativeness conditions are seen as being relatively invariant properties across advanced economies (Breschi and Malerba 2005), technological opportunities and knowledge bases are instead more likely to differ according to their geographical locations. As we have already highlighted in Chapter 2, this geographical dimension, namely the L in the OLI paradigm, has been heavily under-researched when it comes to MNEs in comparison to the O and I, and this is very problematic because all three of these OLI components are related to knowledge and knowledge flows. Following this argument, the reason why technological opportunities and knowledge bases are instead more likely to differ according to geography is because opportunities and knowledge bases depend both on the balance between internal and external knowledge sources and on the interactions among firms and other localized actors. It has been argued that also the strength of cumulative processes within any particular industry may vary substantially across space, because learning processes are strongly shaped by contextual features (Patrucco 2009; Rodríguez-Pose and Bilbao-Osorio 2004; Antonelli 2005). Thus, at any given historical time, those regions and industrial clusters whose knowledge base and capabilities are correlated with new emerging radical technologies will be the locations which provide the greatest opportunities and possibilities for dynamic trajectories. In contrast, those locations whose knowledge base is locked-in to previous technological paradigms will find it very difficult to accumulate capabilities based on new general purpose technologies (Carlaw and Lipsey 2006). Not surprisingly, they will therefore also be the regions which face fewer opportunities to move along dynamic paths (Breschi 2000; Castellacci 2008a, b), and this is a critical issue which merits much deeper investigation.

4.5 DIMENSIONS AND BOUNDARIES OF INNOVATION SYSTEMS (ISs)

4.5.1 Geographical Innovation Systems: from National to Regional ISs

In evolutionary economics the determinants of technical change are a synthesis of supply and demand factors,[11] and technical change is the result of complex interactions among scientific, technical, economic and social changes. Instead of being seen exclusively as a linear process, innovation is also seen as a chain-linked process that is achieved in an interactive development which is both internal and external to the firm (Kline and Rosenberg 1986). Thus, various economic and non-economic factors combine to cause change, and these factors range from firm competence to the clustering of resources according to either geography or technology, and from organizational features such as hierarchies, markets or networks, to 'development blocks'[12] and institutional infrastructures.

In this vein, the prolific literature on national innovation systems (NIS) introduced by evolutionary economists in the late 1980s (Freeman 1987; Lundvall 1992; Nelson 1993; Nelson and Rosenberg 1993; Edquist 1997), argues that the performance of national economies cannot be analysed only in terms of the strategies and performance of firms. There are other factors and actors that play vital roles in favouring the generation and diffusion of knowledge, including: inter-organizational networks, financial and legal institutions, technical agencies and science research infrastructures, education and training systems, governance structures, and innovation policies. Institutions broadly defined which include laws, rules, social norms and routines, taken together shape the interactions among different organizations, and thereby both provide incentives for innovation and also create obstacles to innovation. Such complex sets of relationships, feedbacks and networks between firms, between firms and innovation-oriented organizations, and between producers and users, are strongly influenced by spatial proximity mechanisms that favour a process of polarization and cumulativeness. The highly uneven pattern and spread of innovation in space suggests that innovation can be better understood by considering sub-national units of analysis, rather than thinking of national systems of innovation as homogeneous entities (Morgan 2004). As Carlsson and Stankiewicz point out 'high technological density and diversity are properties of regions rather than countries' (1991, p. 115). Hence, the notion of a regional innovation system (RIS) has emerged as a territorially-focussed perspective of analysis which is derived from the broader concept of NIS.[13] A regional innovation system RIS may be defined as 'the *localised* network of actors and institutions in the public and private sectors whose activities and interactions generate, import, modify and diffuse new technologies *within and outside the region*' (Cooke et al. 1997; Howells 1999; Cooke 2001; Evangelista et al. 2002). Importantly, the RIS framework embraces different territorial scales and utilizes various geographical units of analysis (Doloreux and Parto 2004). Its broad scope allows for a sufficient generalization and it is therefore considered a suitable background for devising innovation and industrial policy.

In the RIS approach it is implicitly assumed that the elements characterizing the national system can be largely transposed to a smaller territorial scale to help define the local system. Following Dopfer et al. (2004), micro-structures refer to individual components or actors, that is, individuals, firms and other organizations that through their relationships (market and non-market) and their attributes (competencies and functions) build up and maintain systems of rules (Carlsson 2003). Macro-structures, on the other hand, are comprised of the population of meso-systems and their interdependence. The top-down (*macro-to-micro*) view of the RIS thus implies identification of characteristics which are already recognized for the NIS (Howells 1999; Iammarino 2005). These characteristics include the internal organizations of firms, inter-organizational relationships and, more specifically, the type and intensity of interactions between the business sector and the rest of the economic system. The RIS relationships and interactions relate variously to: the role of the public sector and innovation policy; the institutional framework (administrative, political, financial, legal, fiscal, educational, and so on); the industrial structure, including the organization of both public and private R&D activities; the spatial structure, including the relative geographical position, the degree of urbanization, the scale and scope of agglomeration; the degree of openness, and the capacity to attract, absorb or tap into external resources including the integration into global innovation networks, often flag-shipped by multinational corporations.

Such a list is clearly not exhaustive and underlies a broad definition of territorial systems of innovation. The elements listed encompass the types of specific institutions that help to promote innovation and knowledge, as well as the wider socio-economic system in which political and cultural influences and specific modes of governance foster innovation structures and performances (Lundvall 1992; Rodríguez-Pose 1999; Cooke et al. 2000; Freeman 2002; Garofoli 2003). However, the identification of components, relationships and attributes, is in many ways more complex at the regional level than at other levels, such as the national, sectoral or technological levels (Carlsson 2003). The reason is that regional systemic advantages (or disadvantages) are also argued to depend on attributes known as 'untraded interdependencies', such as informal flows of knowledge, interactive learning, and the degree of embeddedness,[14] which generate the bulk of territorial externalities at the meso level (Saxenian 1994; Storper 1998). Features such as dense social networks appear to be critical channels for knowledge diffusion and learning, allowing for the recombination of old and new pieces of knowledge. Indeed, it has been shown that geographical proximity per se does not produce knowledge spillovers: other forms of proximity, interacting with spatial propinquity, allow for knowledge spillovers (e.g., Breschi and Lissoni 2001a, b; Faggian and McCann 2006; Boschma 2005; Breschi et al. 2005, 2007; Boschma and Ter Wal 2007; Ponds et al. 2007; D'Este et al. 2012). Thus, the top-down perspective only accounts for the necessary but not the sufficient conditions needed to identify a RIS. Such a perspective also needs to be integrated with a bottom-up (*micro-to-meso*) perspective, which takes account of the internal and dynamic regularities of territorially embedded socio-economic structures (Asheim 1995; Asheim and Gertler 2003; Dopfer et al. 2004; Laursen et al. 2007; Pike et al. 2007, 2010; Crescenzi and Rodríguez-Pose 2011a, b). Among the idiosyncratic characteristics of a RIS which emerge from bottom-up views (Howells 1996, 1999; Vilanova and Leydesdorff 2001; Tödtling and Trippl 2005; Iammarino 2005; von Tunzelmann 2009a, b; Crescenzi and Rodríguez-Pose 2009; Uyarra and Flanagan 2010a) there are: communication patterns relating to innovation processes, both at individual and corporate levels; invention and learning patterns; localized knowledge sharing between and within individuals and organizations; search and scanning procedures relating to innovation and technology; network integration (within and between networks), operating both intra- and extra-region; structure and direction of regional governance; and the historical path dependency of localized innovation processes.

Following the insights of the technological gap approach (Abramovitz 1986; Fagerberg 1987, 1994; Fagerberg et al. 1994; Fagerberg and Verspagen 2002, 2007), concepts such as 'social capability' and 'technological congruence' are relevant when considering a meso level of analysis such as the RIS, in so far as both appear to be highly variable across space, even within the same national system. Social capability refers to the overall ability of the region to engage in innovative and organizational processes and to undertake institutional changes. Technological congruence relates instead to the distance of the region from the technological frontier, with a smaller distance improving its capacity to implement the technical properties embedded in new technologies (Fagerberg et al. 1994). In addition, recent applications of the capabilities framework to geographical innovation systems have also emphasized that regions and clusters can be considered as spatial configurations of suppliers, producers and consumers, each with their own unique level of capabilities (von Tunzelmann, 2009a, b; Iammarino et al. 2012). Shifting the logic from mere co-location to co-evolution, it is argued that, for a region to be progressive, the capabilities of all its supplier and customer firms need to be in tune and sufficiently dynamic to be able to cope with the continuous change of actors' needs and abilities (von Tunzelmann 2009b).

Therefore, regional systems with stronger capabilities and a progressive knowledge base will also tend to be better equipped to exploit new technological opportunities, to adapt existing activities to emerging business environments, and to learn more rapidly about how to build new capabilities and advantages. Conversely, and again in analogy with the firms, several reasons may account for a system failure, including local actors' poor absorptive capacity (Cohen and Levinthal 1989), the underdevelopment of system social capabilities (Abramovitz 1999), or the lack of any dynamic interactive capabilities and system integration (von Tunzelmann 2009a, b). As a result, a particular regional innovation system may lag far behind others (even within the same national borders), while the weakest regions may not even show systemic features at all (Doloreux and Parto 2004; Iammarino 2005; Rodríguez-Pose and Crescenzi 2008; Uyarra and Flanagan 2010a).

This explicitly geographical RIS view therefore emphasizes evolutionary mechanisms such as routines, technological trajectories, selection environments, heterogeneity and path dependency. New opportunities are perceived to be shaped and constrained by variety and path dependency at the meso level, or, in other words, by the inheritance of local structural regularities from past knowledge accumulation and learning. In such a perspective, the interdependence between structures and actors can be regarded as a feedback mechanism in that not only do the characteristics of the selection environment and their changes influence the actors, but the actors also change the environment (Boschma and Lambooy 1999; Lambooy and Boschma 2001). However, in spite of the central theoretical role played by dynamics and co-evolution in the innovation system conceptualization, only a small percentage of geographical innovation systems studies 'can be considered 'dynamic' in the sense that they focus on a historical process or development over time rather than on a snapshot of a system in a particular time period' (Carlsson 2003, p. 11).[15] In the main, the prevailing approach for studying the processes of co-evolution within sub-national systems of innovation still largely adheres to a static view of the world (von Tunzelmann 2003; Doloreux and Parto 2004; Uyarra 2010; Uyarra and Flanagan 2010a).

4.5.2 Sectoral and Technological ISs and Interactive Learning

The approach to innovation known as the sectoral innovation system (SIS) is described as 'a set of new and established products for specific uses and the set of agents carrying out market and non-market interactions for the creation, production and sale of those products' (Malerba 2002 p. 248). In line with the NIS and RIS structures, the main components of a sectoral system are: products; agents such as individuals, firms, universities, financial institutions, central and local government, and firm consortia; knowledge and learning processes; basic technologies, input demand conditions and their related links, evolution and complementarities; mechanisms of interactions both within and between firms; processes of competition and selection; institutions, including industry standards, regulations and labour relations. Major differences have been highlighted between the type and structure of SIS in terms of links and relationships between firms, and among firms and other organizations in generating and diffusing knowledge; in terms of the innovative process, and particularly with regard to the objectives and sources of innovation; and in terms of the organization of production and distribution. Some of these differences, as seen above, are clearly related to the type of technological regime which governs innovation processes in each sector.

Along the same lines, a technological innovation system (TIS) is defined as 'a network (or set of networks) of interacting actors that functions in a specific techno-economic area and that is involved in the generation, diffusion and utilisation of technology' (Carlsson and Stankiewicz 1991, p. 111). The central features of technological systems are economic competencies and the ability to exploit new business opportunities, and the clustering of resources and institutional infrastructure (Carlsson and Stankiewicz 1991; Metcalfe 1995). The TIS notion is very similar to that of a SIS, although the TIS can include more than one industry under different technological regimes. A TIS, as the other innovation systems, is a multi-dimensional concept and in many cases, its constituent elements such as its knowledge or competence networks, its industrial networks and development blocks, and its institutional infrastructure, are spatially correlated. However, the SIS and TIS analytical frameworks are both basically agnostic with respect to the role of space, with the primary focus being on explaining the coherence of both horizontal and vertical industrial 'filières' and of identifying new technological opportunities. The SIS and TIS concepts are largely motivated by the limited explanatory power of the national systems of innovation approach when dealing with institutional settings regarding technological and sectoral exchanges of knowledge across national boundaries. The basic problem here is that nation states are often considered to be the natural 'containers' of many sectoral and technological systems, whereas sectoral and technological systems may also be essentially regional or local in nature, with local/regional systems being both multi-technological and/or multisectoral. In some other cases sectoral and technological systems can be international or global. As such, where the boundaries are drawn depends on various circumstances and issues such as the technological or market requirements, the capabilities of the various actors, and the economic interdependencies between agents and institutions. In summary, with respect to geographical innovation systems (NISs and RISs), sectoral innovation systems and technological innovation systems differ primarily in the following ways:

1. the SIS and TIS approaches refer to specific production and techno-industrial areas, whilst the NIS and RIS approaches refer to all or many areas as a whole;
2. the geographical boundaries of an SIS or TIS do not necessarily coincide with those of the NIS or RIS;
3. the SIS or TIS approaches stress the relevance of microeconomic aspects rather than that of institutional infrastructures: therefore, they focus far more on the adoption and utilization of technology;
4. in contrast, NIS and RIS approaches focus much more on the processes of generating and distributing knowledge, through linkages and networks.

The innovation systems literature as a whole emphasizes the importance of the relations between organizations and institutions for the innovation processes and for the functionality of any system. Organizations are strongly influenced by institutions which are also called 'social technologies', or organizational capabilities, as opposed to 'physical technologies', or technological capabilities (Nelson and Nelson 2002; Nelson 2003; Schienstock 2009).[16] Firms are embedded in institutional set ups and environments, which include the legal system, social norms, routines and standards. But institutions are also both embedded in and develop within organizations and among them. Therefore, complex relationships of mutual embedding between organizations and institutions underlie the diversity in all types of innovation systems (Edquist 2005, p. 197). Organizations and individuals perform the activities, many of them carrying out more than one activity and eventually developing more than one capability. Institutions can foster both the incentives and also the obstacles influencing innovation activities and capability accumulation.

Learning processes are at the centre of the innovation systems' approach and interactive learning is regarded as the crucial factor for innovation inside and across organizations (Edquist 2005), in both advanced and emerging economic systems (e.g., Chaminade and Vang 2008; Lundvall et al. 2009; Belussi and Sammarra 2010; Pietrobelli and Rabellotti, 2010). The mechanisms for developing dynamic capabilities are those which foster interactive learning. Therefore, understanding the interactive learning process as a combination of geographical, sectoral and technological systems should lead to a more effective synthesis between the theoretical and empirical work aimed at explaining the dynamics of innovation systems (Lundvall et al. 2002; Chaminade and Edquist 2006). Indeed, the notion of sectoral systems of production and innovation may actually complement both geographical and technological systems of innovation (Breschi and Malerba 1997), as interdependence and complementarity define the real boundaries of a system at all levels of analysis. One can observe a strong overlap between geographical, technological and sectoral systems of innovation, particularly when considering industry boundaries as being multidimensional, integrated and dynamic. Most importantly, it is precisely such interdependence between systems of innovation that has been incorporated within modern multinational firms, thereby changing their structure and strategies and, at the same time, being influenced by their behaviour.

4.6 TECHNOLOGICAL ACCUMULATION, MNEs AND KNOWLEDGE NETWORKS

4.6.1 Research-intensity and International Expansion

As discussed above, firm heterogeneity is a crucial variable which affects economic growth and performance. Indeed, the observation of heterogeneity is all the more important when analysing MNEs organization and behaviour. This is because, following also the Alchian argument discussed in Chapter 3, the size of large MNEs means that their behaviour often has some bearing even on industry overall structures. In the words of Nelson 'One cannot study the computer industry sensitively without paying attention to the peculiarities of IBM. The recent history of the automobile industry cannot be understood without understanding Toyota and GM' (Nelson 1994, p. 254). A major cause of firm heterogeneity lies in the differences between firms in terms of their knowledge, innovation, technology and R&D. As exemplified in the innovation patterns illustrated by the Schumpeter Mark II type of regime, historical evidence shows that the institutionalization of R&D, particularly in large multinational firms, has been one of the most important drivers in the contemporary patterns of industrial competition (Freeman 1982a; Mowery 1985).

As has long been acknowledged in the international business literature, a tight link exists between multinational expansion and the innovativeness of firms, and authors from different theoretical backgrounds have pointed to such interactions. Yet, different interpretations have been put forward regarding the links between multinationality and innovation-intensity. Firstly, the degree of multinational expansion has been seen as a function of the firm's R&D-intensity, which is typically used as a proxy for the level and complexity of the firm's accumulated technological competence. Along these lines, two approaches can be distinguished. On the one hand, there is the industrial organization view, where, on the basis of the linear model of technological processes, investment in R&D leads directly to cost reductions and higher quality products, increased corporate competitiveness and larger market shares and, ultimately to a stronger international expansion (Dunning 1970; Markusen 1984). On the other hand, there are the transactions cost theories which propose that R&D activities generate more intensive knowledge flows and a greater complexity in transactions which, in turn, lead to a greater degree of vertical integration, industrial concentration, and the expansion of firms across national boundaries (Buckley and Casson 1976; Hennart 1977; Rugman 1981).

A second interpretation has instead emphasized the two-way relationship between multinational expansion and innovation. According to the resource- and competence-based theories of firms' growth, it is misleading to treat high R&D-intensity as the cause of greater multinational expansion or increasing international market shares. Instead, high R&D-intensity and internationalization are viewed as complements in the accumulation of more advanced levels of technological competence and capabilities, and these account for strong firm heterogeneity (e.g., Cantwell 1989, 1992; Patel and Pavitt, 1991; Kuemmerle 1999; Petit and Sanna-Randaccio 2000; Castellani and Zanfei 2006; Frenz and Ietto-Gillies 2007a, b). Technological competence is seen to provide firms with inherent (dynamic) capabilities for expansion through continued learning in production. In turn, faster and more effective learning generates greater competence, and hence rising market shares. Learning processes display various modalities, such as R&D, but at the same time they also involve design, redesign, product and organizational improvements, and cultural and institutional learning which derive from multiple international locational structures. The MNE therefore becomes a repository of technological competence and a device for the accumulation of capabilities (Cantwell 1991).

It should be noted, however, that beyond the different interpretations of the relationship between multinationality and innovativeness, it still remains true that R&D functions, which are seen as part of wider innovation processes, gain in importance as technological progress becomes more complex. A higher level of skills and organizational expertise is associated with the use and generation of larger stocks of knowledge. Therefore

MNEs, which on average have relatively high levels of accumulated competencies, tend to be more research-intensive than other domestic firms in the same industry. They are thereby more capable of overcoming innovation bottlenecks and barriers (see Box 4.1 for an illustrative case study), and of taking advantage from the opportunities offered by the most promising technological fields (e.g., Cantwell 1991; Iammarino et al. 2009). Furthermore, a growing complementarity between technological diversification and internationalization has emerged over time, thereby merging two of the crucial firm strategies for growth originally envisaged by Penrose (e.g., Pearce 1983; Pavitt and Patel 1991; Granstrand and Sjölander 1992; Zander 1997, 1999a, b; Blanc and Sierra 1999; Santangelo 2001; Cantwell and Piscitello 2000; Patel and Le Bas 2005). In order not just to exploit effectively but also to consolidate existing productive capabilities, it is generally necessary for a firm to grow by extending its capabilities into new related technological fields, and across a variety of geographical sites. The firm is therefore able to benefit from complementarities both between related fields of production and between related paths of innovation, and from corporate learning in different institutional settings (e.g., Cantwell 1992; Chesnais 1992; Cantwell and Iammarino 2005; Castellani and Zanfei 2006; Cantwell and Piscitello 2010).

4.6.2 The Globalization of Technology and Global-Local Knowledge Interactions

Over time, the core of the MNE analysis has therefore shifted from a unilateral link between innovation and internationalization, to a multi-way relationship between these phenomena. Such an approach seems more suitable for explaining and interpreting the current organization of technological activities across space in a context of increasing globalization. Indeed, the concept of the globalization of innovation has been depicted as the 'zip' between the two fundamental phenomena of current economic systems, namely the increased international integration of economic activities carried out especially by MNEs, and the raising importance of knowledge in socio-economic processes (Archibugi and Iammarino 1999; Narula and Zanfei 2005). Attempts have been made to critically assess the concept of globalization as applied to technology (Archibugi and Michie 1995), and in this respect three complementary categories have been identified. First, there is the international exploitation of technology which is produced on a national basis, in which the innovators aim to gain economic advantages from their own technological competencies in international markets. Second, there is the global generation of innovations, which are those innovations which were originally conceived by MNEs on

BOX 4.1 THE PERCEPTION OF OBSTACLES TO INNOVATION: A CASE STUDY ON FOREIGN MNEs AND DOMESTIC FIRMS IN ITALY

In this case study we analyse how the perception of different types of obstacles to innovation in the case of Italian firms is affected by both the type of firm (by organizational structure and ownership) and the regional location. The ultimate aim is to shed light on the role of foreign MNEs and on the regional context.

Obstacles to innovation – of a different nature, that is, economic/financial, organizational, institutional, and so on – are largely context-specific. They play a key role in shaping the characteristics of the external technological environment, and thus in determining the performance of existing producers and also the attractiveness of a region for foreign MNEs and domestic firms. The decision of (both nationally-owned and foreign-owned) firms to operate in particular regions and to engage in innovative activities may be affected by their evaluation of the difficulties that will be encountered in the process of innovation.

This is likely to be the case in a country such as Italy, which historically has been characterized by strong territorial imbalances that are among the sharpest in the European Union. The territorial distribution of innovation in the country is highly concentrated in a few northern regions, a phenomenon known as the Italian 'innovative divide'. Regional innovation patterns differ not only with respect to the specific technological strategies and performances of firms, but also in terms of the relevance of systemic interactions and contextual factors favourable (or unfavourable) to innovation.

The micro-data used in the empirical analysis come from the Italian third Community Innovation Survey (CIS3) and refer to the period 1998–2000; the sample is composed of 15,512 firms. The analysis considers at the same time all nine individual obstacles to innovation listed in the Italian CIS3; the estimation method used a Multivariate Probit Model to control for the potential presence of any unobserved error correlation structure amongst obstacles assessment.

The study shows that important differences in firms' perception of obstacles to innovation occur both across regional locations and types of firms. A visible 'innovation divide' pattern emerges in terms of the perception of obstacles, in which firms in the North and the Centre of Italy tend to perceive the obstacles to innovation as less significant than those located in the South. In particular, firms located in the North and the Centre generally tend to evaluate a lack of financial resources as an impediment to innovative activity significantly less than firms located in the South; the result is the same in relation to information on technology and markets, especially for firms located in the North-West of Italy. Furthermore, when the estimation is carried out on subsamples of firms by type, geographical specificities in the perception of the obstacles to innovation are shown to characterize only single domestic firms. In other words, the perception of obstacles to innovation does not significantly differ across regions, unless the firm is a single domestic firm.

In terms of firm types, the major difference in the perception of obstacles occurs between firms belonging to a group (foreign and Italian) and single domestic firms, rather than between firms with different nationality ownership. We find that firms belonging to a multiplant group structure tend to perceive obstacles to innovation as being relevant significantly less than single domestic firms, with the exception of a lack of organizational flexibility. Thus, firms belonging to a group – both foreign and Italian – appear to be less sensitive to the socio-economic and institutional context than single domestic firms. This pattern is however more pronounced in the case of foreign MNEs, which also emerge as the most innovative firms, regardless of their geographical location. Interestingly, foreign MNEs show coefficients significantly lower than those of Italian groups for most obstacles. This is the case for obstacles of an economic/financial nature, for those related to the lack of information of technology and also in the case of regulation rigidities, which one might have expected to be more of a hindrance for foreign-owned than for Italian-owned firms. Italian groups, instead, tend to be less sensitive than foreign MNEs to barriers related to the lack of information on markets – which is not entirely surprising as they have stronger ties with the local economy – and to the lack of organizational flexibility.

Source: Adapted from Iammarino et al. 2009.

a global scale. Third, there are global technological collaborations, facilitated by national and international agreements, and which involve both private and public actors for the joint development of specific technological discoveries. These categories are strongly interdependent and imply both degrees of competition and collaboration among the main agents (Archibugi and Iammarino 1999, 2002).

Although firms, and especially large ones, innovate and compete globally in all three different ways described by these categories, MNEs are undoubtedly the most important actors in the worldwide cross-border creation of new technical knowledge. The location of their innovative activities is often linked to the location of their productive activity but, however strong the correspondence between production and R&D activities may be, it is far from being unambiguous.

The question of how globalization affects the (macro-) geography of technology and innovation and whether this fosters or constrains local technological capabilities has been extensively addressed in the international business literature and, more generally, in the OLI framework (e.g., Dunning 1993, Dunning and Lundan 2008), as well as in innovation theory and the theory of the firm (Chandler et al. 1998; Chesbrough 2006). As already noted, traditional analyses of multinational activity emphasized the centralized nature of the research and development efforts of MNEs. The main advantages of the centralization of R&D services in the location of origin of the parent company are basically connected to economies of scale and scope in R&D, control over innovation, and linkages with national business and non-business sectors (Pavitt and Patel 1991; Patel and Vega 1999). In this centralized structure the headquarters of the MNE provides knowledge to the rest of the firm, which is to all affiliates elsewhere located. However, as already highlighted in Chapter 2, many scholars have now argued the obsolescence of this thesis (e.g., Cantwell 1995; Birkinshaw and Hood 1998; Blanc and Sierra 1999; Frost 2001; Rugman and Verbeke 2003; Birkinshaw et al. 2005; Cantwell and Mudambi 2005; Sullivan and Daniels 2008; Piscitello 2011), because such advantages seem to be increasingly counterbalanced by the advantages associated with decentralization or multiple geographical location of MNE innovative and technological activities. The latter can be summarized in terms of the linkages between innovative activity and local production, markets, suppliers and clients, and the acquisition of foreign knowledge and capabilities in technological fields of excellence of host locations (Pearce and Singh 1992; Howells and Wood 1993; Miller 1994). All these factors acquire a greater or lesser importance depending on innovation systems' features, firm organization and strategy, products and technologies involved both in the origin and host locations.

From a geographical standpoint, even within the same national borders different regions exhibit highly differentiated characteristics and these differences determine their degree of attractiveness as investment locations for MNEs in terms of the nature and scale of the knowledge linkages and spillovers that they are able to generate. The choice of location in which an MNE might invest in R&D and innovation activities is thus driven by several factors, among which, as seen above, are social capabilities and the technological congruence of the local systems. Yet, being able to build new competencies rapidly involves the ability to establish links at all levels, from the global down to the local and vice versa. As such, the extent to which a region attracts innovative resources from outside, thereby spurring its external integration, depends first and foremost upon its absorptive capacity (e.g., Asheim and Isaksen 2002; Cantwell and Iammarino 1998, 2003; Garofoli 2003; Simmie 2003; Criscuolo et al. 2005). As learning advantages are mainly embodied in people and institutions, human capital and the upgrading of skills increase the ability of a region to absorb, diffuse and generate new knowledge. Absorptive capacity therefore depends significantly on diversity, in that innovation occurs where there is a diverse technological, social, and economic culture, and the most dynamic capabilities lie in the combination of both the exploration and exploitation of new and existing assets (Rantisi 2002).[17]

Where cumulative causation mechanisms occur they give rise to vicious and virtuous circles, thereby contributing to the rise or fall of national or regional innovation systems (Niosi and Bellon 1996; Malmberg et al. 1996; Zanfei 2000; Castellani and Zanfei 2002). These mechanisms are sometimes explained in terms of path-dependent processes in which localized learning and dynamic capabilities reinforce differentiation amongst national and regional innovation systems and lock them into particular patterns of growth. In this sense, as we will see in the following chapters, the relationship between institutions and business behaviour of MNEs across geography and time evolves along paths in which history plays a critical role (Jones and Khanna 2006; Morck and Yeung 2007).

As we mentioned in Chapter 2, an increasing awareness of the role of MNEs in the generation of new technology has led to research which focuses on both internal and external networks for innovation (e.g., Bartlett and Ghoshal 1989; Ghoshal and Bartlett 1990; Andersson et al. 2002). MNEs may adopt strategies of technological differentiation and related product diversification, so as to obtain access to locally based technological expertise (Cantwell and Janne 1999; Cantwell and Santangelo 1999; 2006), thereby promoting complex networks of intra-firm cross-country specialization in innovative activities. When entering new related fields of technology across a variety of locations across national boundaries, the firm is thus able to benefit from the dynamic economies of scope that derive from the technological complementarities between related paths of innovation in spatially distinct institutional settings (Cantwell and Iammarino 1998, 2003; Cantwell and Noonan 2002; Cantwell and Piscitello 2007; Iammarino et al. 2001; Kramer et al. 2011). Inter-firm networks established between an MNE's foreign affiliates and local firms and actors mean that the MNE's sources of knowledge for innovation increasingly extend beyond their internal resources, and encompass the relations of affiliates with their external environment (Almeida 1996; Almeida and Kogut 1999). For their part, indigenous firms and local innovation systems may benefit from a variety of linkages and externalities from MNEs, given the access of the latter to complementary streams of knowledge being developed in other regions or clusters. A detailed case study illustrating such interactions in the regions of the UK and Germany is reported in Box 4.2.

In this way, MNEs spread the competence base of the firm and acquire new technological assets, or sources of technological advantage. As also discussed above, resources and capabilities that are critical for firms' competitive success 'can often be found inside a region, rather than within any single firm' (Enright 1998, p. 315). Hence, the MNE assumes a critical role in bridging capabilities at both the micro and the system level. From the point of view of the latter, MNEs can therefore be seen at the same time as being both 'internal' actors, contributing to the creation and diffusion of new technical knowledge within the region, and also as 'external' players, channelling knowledge created elsewhere within the firm into the local system. As such, the MNEs can play the role of technological 'gatekeepers', influencing the regional knowledge base without necessarily being located *in situ*, as in the case of global buyers or distributors.

Scale and scope economies in research and technological activities, which allow for the solving of problems at a very high degree of technological complexity, are primarily a prerogative of large firms, and in particular of MNEs. On the other hand, the local knowledge-related external economies available to an MNE also depend on the incremental innovative processes within small and medium sized enterprises. Indeed, the nature and extent of local knowledge externalities within a region cannot be separated from the patterns of collaboration and competition between small and medium-sized local firms, as well as those with the large indigenous firms and MNEs. As discussed previously, all these factors are largely characterized in the technological regime prevailing in the regional system or industrial cluster.

From the point of view of MNEs' locational choices, although it is difficult to define the 'optimal' cluster or region as the cluster boundaries

BOX 4.2 TECHNOLOGICAL CAPABILITIES AND REGIONAL EMBEDDEDNESS OF MNEs IN ADVANCED REGIONAL SYSTEMS: A CASE STUDY ON GERMANY AND THE UK

Looking at the process of MNEs' spatial embeddedness through the lenses of technological capabilities means to understand how MNEs carry out their innovative activity within the firm and the region and how, in so doing, they rely upon and contribute to regional knowledge creation. We study six MNEs belonging to the ICT, automotive and life science industries (two for each sector) operating both in Germany – specifically in the regions of Baden-Württemberg, Munich, North-Rhine Westphalia and Lower Saxonia – and in the UK – in Cambridge, Thames Valley, West Midlands and North West. For each industry, we consider two MNE headquarters, one UK-owned and one German-owned, as well as one respective subsidiary in both Germany and the UK. The analysis is based on 44 semi-structured interviews with senior managerial and technical personnel from different business units. We focus on three main aspects: human resources, organization of R&D, and intra-firm and external knowledge linkages. All such mechanisms become territorially embedded focussing on the two-way relationship between the region and the MNEs and vice versa.

Human Resources: how MNEs support and upgrade their employees' skills

MNEs strategies broadly fall into three categories: (1) *Hiring methods and talent management*. As for the former, traditional strategies (specialist press, head hunters, and so on) are being complemented by social network software systems, to attract especially the youngest highly qualified segments of the labour market. As for the management of talent, firms are investing increasingly in leadership programs for high profile candidates, developing systems to define clearer career paths with individually tailored training; (2) *Strategic skills forecast*, that is, awareness of future skills demand and supply. In the UK, for example, one of the firms in the automotive sector produced a report to estimate the entity and cost of the future skill-gap, advocating urgent action from the government. The human resources unit of one German automotive firm instead reported to being actively engaged with a regional chamber of commerce, particularly by initiating specialized educational programmes in fields where demand was expected to grow (such as electro-mechanical engineering). Additionally, this firm established a corporate university at its headquarter location to secure the life-long learning of its employees; (3) *Engagement with higher education.* All of the interviewees highlighted the importance of structured graduate programmes. For instance, those from HR departments valued graduate programmes as a means for talent identification, which can then be developed through talent management and talent retention initiatives. On the other hand, interviewees from R&D units pointed out that these programmes are crucial to integrate graduates, and particularly PhDs, into industrial research, for instance by offering the opportunity to conduct small research projects in site.

From the MNE to the region: There are several ways in which firms' capabilities merge with regional ones. For instance, MNEs often have strong collaborative ties with regional partners from higher education, such as investments in polytechnics and support their scientific and technical education. In some cases, especially in German-owned MNEs operating in Germany, local skills are improved through collaborations with regional policy makers, by supporting regional talent management and talent retention initiatives. For instance, the automotive MNE we investigated operating in Lower Saxony, collaborates actively with state and city government officials in a 'Study and Stay' programme, a place branding and talent management initiative designed to retain university graduates in the regional labour market.

From the region to the MNE: The regional availability of scientists and of innovative industries has important consequences for MNEs' human resources and, therefore, technological capabilities. Indeed, although international recruiting is a key priority for all of the analysed MNEs, the sourcing of regional scientific personnel and the recruitment of graduates remains a central objective in the development of strategic partnerships with local universities. In some cases, the local availability of scientists has in fact directly driven the R&D investment decisions of MNEs. Moreover, in a regional labour market with a high density of innovative actors, MNEs' human capital also benefits from high inter-firm interregional mobility. According to our interviews in the UK, MNEs located in the Cambridge Cluster and in the Thames Valley have high degrees of inter-firm mobility in which outflows generally balance inflows.

Organization of R&D: how MNEs organize their R&D activity
Also in this case we can distinguish three different organizational modes adopted by MNEs: (1) *Infrastructure assisting R&D decision-making*. Two rather distinctive models were identified. In some firms, highly centralized basic research facilities are used in order to catalyse economic resources and capabilities. This strategy is followed for instance by one German-owned MNE in the automotive sector, in which the central research facilities serve the network of globally spread subsidiaries, carrying out strategic R&D for the whole group. A similar approach is taken by a German pharmaceutical company, which consolidated its pre-clinical research in its headquarters, whilst it kept carrying out clinical research in several major markets. In these cases, whilst strategic R&D decisions are taken in the headquarters, development activities are highly decentralized and spread across various locations. On the other hand, some MNEs are developing highly decentralized and entrepreneurial R&D units, with large degrees of budget and planning autonomy. This model is chosen as it enables the firm to manage the research portfolio more efficiently by accessing external knowledge at earlier stages of the innovation process. Moreover, they are often complemented by venture capital units, which support the identification of new growth opportunities. At the same time, intense interactions of such R&D centres with local business and service delivery units ensures inflows of relevant market information into the corporate R&D network. For instance, one UK-owned pharmaceutical firm has introduced new managerial positions to coordinate globally independent research units and their collaborative strategies; other MNEs have established specific departments that manage internal knowledge transfer from the R&D units to the rest of the company and vice versa. Similarly, research portfolio offices have been set up by one ICT firm to evaluate and match internal and external research findings. Supported by intense and structured communication processes, these units secure the transfer

and utilization of research through intra-firm channels; (2) *Internal knowledge transfer processes*. First, face-to-face contact has emerged as the most important mechanism of internal and external communication for all interviewed MNEs. This is especially the case at the beginning of technologically complex projects, where knowledge gaps with collaborators (internal or external) or clients are larger. Second, IT-based support structures, such as workshare-platforms, directories, intranet-based document systems or wikis are broadly used across the interviewed MNEs. Third, as mentioned above, some companies have developed specialized business units or positions focusing on intra-firm knowledge transfer, collecting and evaluating all internal research results and comparing them against external sources; (3) *Management practices for interdisciplinary projects*. Innovative activity relies on a diversified mix of skills; therefore adequate organizational approaches to manage knowledge exchange across different disciplines are critical. An interesting example occurs where MNEs, particularly from the ICT sector, pilot new research products with lead users in a real-life environment – an approach adapted from clinical research in the pharmaceutical industry.

From the MNE to the region: Most obviously, MNEs' impact on the regional organization of innovation occurs through their R&D network infrastructure, especially when firms follow a decentralized approach with the aim to interact with the local knowledge base. We focus here on three additional channels. Firstly, inter-industry initiatives sponsored by MNEs are used to stimulate local and cross industry knowledge spillovers: a relevant example is provided by a regional research and transfer platform developed by an ICT firm, which funded inter-industry spin-outs and invested in new interdisciplinary research lines in Germany. A second type of influence occurs through MNEs institutional participation as active stakeholders in regional cluster initiatives, chambers of commerce or science advisory boards. Through such participation MNEs can affect coordinated investments in the quality of place, regional talent management projects or schemes to financially support star scientists in regional universities. For instance, in the UK, the Thames Valley Economic Partnership, to which one of the interviewed MNEs (a German ICT subsidiary) belongs, has established the TV Investment Network, a *business angel* *network* that funds high-tech start ups in the region. Thirdly, regional organizational capacity benefits from corporate venture capital units, which are often found in knowledge centres. Such units present in Germany across the three sectors, foster the commercialization of regionally created knowledge and IP by supplying local innovative partners (universities, SMEs, and so on) not only with venture capital, but also with access to the MNE knowledge network.

From the region to the MNE: The regional organization of innovation can have important effects on the degree of territorial embeddedness of MNEs' operations. In particular, we identify three key aspects that strengthen MNEs' regional integration: the presence of networking platforms and cluster initiatives; the presence of regional policy incentives for innovation; and, most importantly, a critical mass of innovative actors. In other words, for MNEs to be attracted and effectively embedded in a location, the presence of scientific excellence is a necessary condition. This, however, is not sufficient, as multinationals can easily access global scientific knowledge.

Internal-external knowledge linkages: how MNEs establish knowledge network within and outside the firm

Three different stages are taken into account here, and for each of them, the analysis takes into account both industrial and scientific networks: (1) *Identification of research partners.* Based on our interviews we identify two types of sources for the selection of potential industry research collaborators: those which are internal to the MNE and those which are external. The MNE itself provides the first internal avenue to select potential partners: not only R&D centres in the headquarters, or in other branches and subsidiaries which are obvious and common options for collaborations, but, in certain cases, the R&D facilities of the parent company can act as a 'repository' of partners by guiding subsidiaries in the choice of collaborators within and beyond the corporate boundaries. As for external sources of partnerships, these can be found with competitors, customers and suppliers or contract research organizations. The identification of these collaborators relies on active and constant participation to industry activities (conferences, publications, innovation competitions) and organizations (industry boards). Although partnering options are scanned on a national and, even more, global level, a regional environment rich with industrial partners is critical for supporting networking activity. Sectoral differences are extremely relevant. In the ICT business-services sector, R&D networks with clients and suppliers are very common. On the other hand, collaborations in the pharmaceutical industry tend to be mostly with biotechnology companies and seldom with direct competitors. Finally, MNEs in the automotive sector actively source external knowledge from development partners/suppliers, for instance by establishing supplier parks. R&D collaborations with academia are highly valued and efforts are made, across industries, to increase them at the global as well as regional scale. Scientific partners are normally met first at conferences and industry related events or – as especially stressed in the pharmaceutical industry – by active participation in the scientific community (e.g., producing and/or publishing high quality research results). One of the major obstacles to network formation is the inability of academic institutions to adequately advertise their skills (so far universities rely mostly on peer reviewed publications or conference participation). As a consequence, several MNEs are trying to devise new mechanisms to identify the strengths of universities and design a more efficient way to interact with them; (2) *Creation and exchange of knowledge*. Two types of industrial collaboration can be identified: firstly technology partnerships, which focus primarily on early-stage R&D at the pre-competitive level; secondly product specific collaborations, which focus on later stages of the product/service development processes. These collaborations can either be bilateral or multilateral (as, for instance, in joint EU projects involving both academic and industrial members) and are generally supported and coordinated through MNEs' dedicated departments. Collaborations with universities and research institutes are, in most cases, at the very early stage of the innovation stream (blue sky). Universities often act as a filter mechanism for the MNE to identify new targets, as well as to validate and evaluate whether an idea has commercial potential. For instance, one of the UK-owned pharmaceutical firms operating in Germany established a research advisory board composed by selected senior academics from leading universities, to ensure high level scientific-management decisions (such as clinical study protocols). Other major motivations for MNE collaboration with academic partners are early access to human capital, including the possibility to support talent creation, and publicly funded joint research projects, in which most of the interviewed MNEs were involved. In such projects, the inflow of complementary knowledge from a range of industrial and academic partners was reported to be beneficial over and beyond the actual financial R&D incentive; (3) *Support to continuous collaboration.* All our respondents reported difficulties in engaging with the academic community, mostly related to its partial understanding of business and industrial R&D processes (including its legal requirements), and to the differences in professional incentive structures: for instance academic publication needs might interfere with confidentiality protocols, moreover universities' specific funding requirements (needed for example to accommodate a PhD degree) could not always be met by MNEs. To overcome the differences in incentive structures, collaboration programmes with universities are often led by managers with a strong academic background (usually a PhD graduate), and try to incorporate academic incentives by, for instance, facilitating peer review publishing. IPR management is also critical in academic collaborations, especially concerning joint IP and the temporary exclusion of competitors from the newly created knowledge. Finally, it is acknowledged that continuity in the relationship ensures successful collaborations and, although partnerships are firstly developed ad hoc, this is done with the long-term objective of establishing a durable contact.

From the MNE to the region: Although knowledge flows are facilitated by geographical proximity, MNEs are not spatially constrained in their choice of partners and, indeed, in the majority of cases, they do not explicitly differentiate among regional and global collaborators. Nevertheless, the need of spatial proximity in specific types of collaborations has strongly influenced, if not driven, the R&D investment strategy of some MNEs. Those firms with a decentralized R&D infrastructure have in fact devised their strategy on the basis of the local availability of a critical mass of other innovative public and private actors. As such, they have built research facilities in different areas across the world, in close proximity to local universities and research centres. For instance, one of the ICT firms based its R&D centres in outstanding engineering campuses worldwide and developed joint PhD programs with the university. Other firms, across sectors and countries, have invested in science parks (e.g., supplier parks, partner ports) close to their major R&D location, to benefit from local knowledge and/or established strategic research partnerships with top-universities. As reported by one ICT firm, future opportunities for regional network creation are also seen in fostering spin-outs or start-ups. Furthermore, MNEs can act as global antennas opening up international networks and knowledge inflows for SMEs in the region. For instance, an ICT firm reported that involving regional SMEs (and potential customers) in innovative product developments, such as technology or service delivery platforms, would particularly enable the local system to access global markets.

From the region to the MNE: As MNEs are not necessarily engaged in regional networks in order to exploit spatial knowledge flows, other institutions with a specific regional mandate need be in place to support such processes. Many of the interviewed MNEs have taken part in regional network initiatives, which broadly fall in two categories: (1) activities financing R&D collaborations among regional actors, which include funding streams from national governments or the EU, and (2) activities aimed at creating networks across regional actors with the specific target of exchanging knowledge. In the UK these include, for example, sectoral *Knowledge Transfer Networks* (where Regional Development Agencies have a strong involvement). Similarly, some of the firms located in Germany were, or are, actively involved in the Clusters of Excellence initiative, funded by the German Federal Ministry of Education and Research.

Source: Adapted from Iammarino et al. 2011.

are subject to change such that the system never reaches a 'steady state', it is possible to identify a clear logic which is driving global knowledge and technology flows. It has been argued that MNE networks for innovation turns out to be a function of a *geographical hierarchy* of clusters and regional centres (Cantwell and Iammarino 1998, 2001, 2003). In particular, these networks depend on the strategic importance of the host location and of the innovative dynamism of local competitors, suppliers and institutions (see Box 4.3). The sectoral composition of technological strengths is seen to differ across regions, while the technological specialization of foreign affiliates depends upon the rank of the location in the

BOX 4.3 MNEs TECHNOLOGICAL PROFILES AND THE GEOGRAPHICAL HIERARCHY: A CASE STUDY ON FOUR EUROPEAN UNION ECONOMIES

The existence of a *geographical hierarchy* of European research centres has been explored by examining the technological profile of both nationally- and foreign-owned MNEs located within regional centres of excellence. Patent data (USPTO) are used to analyse the location of research activity across space in Italy, France, Germany and the UK over the 1969–95 period. In terms of the distribution of foreign MNE technological activity, the UK was the main host in Europe of foreign-owned patenting activity in the early years, whilst since the 1970s Germany has been the most attractive location. Italy is further behind, although recording an increase in the proportion of foreign activity carried out in the country since the mid-1980s, whilst France displays an intermediate position, with a rather stable share of foreign-owned patenting through time, increasing particularly since the beginning of the 1990s. In terms of penetration – that is, the percentage of foreign patenting on total national patenting by large MNEs – the overall proportion of foreign-owned research located in Europe has on average increased over this time period with a most noticeable rise in recent years (almost 29 per cent in 1991–95). Whilst in the UK and Italy foreign-owned firms constitute a substantial and rising proportion of aggregate activity (with shares of 45.2 per cent and 57.5 per cent respectively in 1991–95), the equivalent share in Germany is approximately 17 per cent at the end of the period considered. In France the share of foreign-owned firms on the overall total is 25.6 per cent for the 27 years.

Considerable sub-national differences exist across the four EU economies. In the case of the UK, France and Italy, a very strong concentration of the overall technological activity carried out by MNEs (both indigenous and foreign-owned) in the period 1969–95 is found in just a few regions. Whilst both foreign-owned and indigenous firms concentrate their research in the same region in the UK (the South East, with 47 per cent of the total), in France (Île de France, with 58.3 per cent) and in Italy (Lombardia, with 52.8 per cent), the same does not hold for Germany, where Nordrhein Westfalen hosts the highest share of indigenous activity (29 per cent), but only represents the second most popular location for foreign-owned research after Baden Wuerttemberg. Turning to the technological profiles across the sub-national economies, a very clear dichotomy between *higher order* and *intermediate* regional research locations emerges from the regression analysis. In the South East region of the UK, Lombardia in Italy, Bassin Parisien in France and all six German regions, statistical support was found for the thesis that these regions attract foreign-owned firms not because of their existing indigenous technological specialization. Foreign firms are attracted there because of the wider technological competencies and infrastructural supports available (i.e. banking, finance, insurance and business services, degree of openness to foreign-owned investors, business climate, corporate and enterprise culture, and so on). The technological activity of foreign-owned and indigenous firms in these regions is typically broad ranging in nature and extends across a spectrum of, often general purpose, technologies. These regions have been labeled as *higher order* locations in the geographical hierarchy, both within the national boundaries and even in the wider European context. In the case of Île de France, which emerged as the main technological core of France both for foreign and indigenous research, the aggregate patenting activity of foreign firms located in the region is dependent upon the technological specialization of indigenous firms. This result is puzzling only at a first glance, because the regional comparative advantages of Île de France lie, in the main, right in leading these general purpose technologies: thus, this is consistent with the region being included in the higher order group of regional cores. The results for the other Italian, British and French regions suggest the presence of *intermediate* research locations. Technological specialization profiles of foreign-owned firms were found to be closely correlated with those of the indigenous technological expertise, also over time, so that knowledge spillovers are likely to be intra-sectoral in nature. This is consistent with the hypothesis that *intermediate* regions attract the innovative activities of foreign-owned MNCs because of a very particular set of sectorally-specific expertise on offer in that region. By basing research facilities in such locations, foreign-owned MNCs may be able to upgrade their own technological capabilities in particular technological fields which may be sub-sets of their own major areas of technological interest.

Source: Adapted from Cantwell and Iammarino 2002.

geographical hierarchy, and upon its gradual change over time. Indeed, as we will see in later chapters, these global-local interactions are likely to further exacerbate the differences between localities. In the 'first order' regional locations, MNEs' strategies for innovation aim at exploring local knowledge and expertise, which will then be integrated within the firm to widen the technological competence at the corporate level through the intra-firm network. When MNE research has a pronounced exploratory nature, it is likely to be attracted by 'first order' regions, treating them as a source of general expertise and skills (Cantwell and Janne 1999). In such regions, technological spillovers operate through exchanges across multiple knowledge sources in and around core technological systems, creating linkages between actors in quite separate alternative fields of specialization (Cantwell and Iammarino 1998, 2003).

A region at the top of the hierarchy is more likely to attract a broad range of foreign innovative activities, as MNEs will generally try to extend their established lines of specialization through intra-firm networks (Cantwell and Iammarino 1998, 2002, 2003; Cantwell et al. 2001; Cantwell and Noonan, 2001). However, these interactions are rather complex. For example, as already noted, it has been shown that in the industries in which the host location is technologically strong, the vibrant presence of indigenous companies tends to deter foreign-owned firms of the same industry from conducting substantial local levels of technological activity in the region's primary technology (Cantwell and Kosmopoulou 2002). At the same time, the strongest firms of other industries or technologies which are not direct competitors of the local leaders may be attracted to that location specifically because of those technologies which represent possibilities for diversification, while a comparative local strength in an industry encourages outward investment in foreign-located technological activities. It also discourages the inward investment of foreign-owned companies in the same industry. The outcome of this is that foreign penetration in clusters of domestic technological excellence tends to be low in most countries (Cantwell and Kosmopoulou 2002). On the other hand, intermediate locations characterized by a narrower scope of technological advantages are seen by MNEs as sources of specific capabilities in some particular fields. These regions may therefore be negatively affected by the innovative strategy of foreign affiliates, which in these cases follow a logic of exploiting indigenous expertise, with the possible aim of out-competing local rivals. In other words, as the position of a region in the hierarchy falls, so the technological specialization of foreign-owned firms in that region becomes more closely related to the equivalent pattern of specialization of the local indigenous firms (Cantwell and Iammarino 2003).

Indeed, at an aggregate level, most such MNE networks for innovation and technology have run between the US, Western Europe and East Asia (particularly Japan, Korea, Singapore and Taiwan), suggesting that the globalization of innovation has, until recently, rather to be seen as a process of 'triadization' (Archibugi and Iammarino 2002; OECD 2008). However, in the last two decades, more sophisticated innovative efforts have gradually become geographically spread across national boundaries too. What has emerged is that inward foreign investment by MNEs has not preceded the occurrence of the earlier stages of innovation in economic catching up, but rather that new locations that had already built up sufficient absorptive capacity in the form of basic capabilities have benefited from the virtuous circle spurred by MNEs operations (Athreye and Cantwell 2007; Iammarino et al. 2008; Mudambi 2008; Fifarek and Veloso 2010). An empirical illustration of cumulative causation and interdependence between indigenous capabilities and MNE regional embeddedness in the context of the emerging economy is reported in Box 4.4.

Furthermore, as mentioned in Chapter 2, vertical disintegration and international outsourcing (offshoring) have recently emerged as further modes of control and coordination of MNE activities, giving rise to what has been labelled the 'concentrated dispersion' of geographical production and knowledge networks of MNEs (Ernst 1997, 1998; Ernst et al. 2001; Ernst and Kim 2002). Global production networks (GPNs) capture the spread of the value chain across firm boundaries and geographical borders (e.g., Dicken 1994, 2003a, b, 2004, 2005, 2007; Ernst and Kim 2002; Henderson et al. 2002; Dicken and Henderson 2003; Coe et al. 2004; Hobday et al. 2005; Hess and Yeung 2006). The GPN approach combines together the insights of various similar perspectives, in particular those of global commodity chains (GCCs) and global value chains (GVCs) (see, among others Gereffi and Korzeniewicz 1994; Gereffi and Kaplinsky 2001; Gereffi 2005; Gereffi et al. 2005). Although there is a substantial similarity among the different concepts (GPNs, GVCs, GCCs), there are also some crucial differences. In the words of Coe et al. (2008, p. 272): 'First, GCCs/GVCs are essentially linear structures, whereas GPNs strive to go beyond such linearity to incorporate all kinds of network configuration. Second, GCCs/GVCs focus narrowly on the governance of inter-firm transactions while GPNs attempt to encompass all relevant sets of actors and relationships'.[18]

GPNs integrate the dispersed supply and customer bases of large MNEs (so-called *network flagships*), that is their subsidiaries, affiliates and joint ventures, suppliers and subcontractors, distribution channels and value-added resellers, as well as their R&D collaborations and different kinds of cooperative agreements. MNEs break down the value chain into a variety of discrete functions, operations and transactions, and locate them where

BOX 4.4 TECHNOLOGICAL CAPABILITIES AND GLOBAL-LOCAL INTERACTIONS: A CASE STUDY ON THE ELECTRONICS INDUSTRY IN TWO MEXICAN REGIONS

In FDI-led high-technology manufacturing industries in less advanced countries, such as the electronics industry in Mexico, MNEs are critical actors in the creation of technological capabilities. Their interactions with, and indirect impact on, the other components of the regional system are crucial. The micro-level analytical survey among manufacturing firms in the electronics industry was conducted by means of a detailed questionnaire administered to a representative sample of locally-owned firms and foreign subsidiaries (overall 80 firms). Different indicators were used to construct a firm-level technological capability index which provided a composite measure of technological capabilities to assess inter-firm differences. The regions considered for the purpose of this study are both characterized by a strong presence of the electronics industry and FDI inflows: Jalisco, in particular the metropolitan area of Guadalajara was chosen in central Mexico; the state of Baja California was selected among the border states. To assess capabilities at the meso level, face-to-face semi-structured interviews with key personnel from 30 innovation-oriented regional organizations were also carried out.

Among the most important factors associated with technological capabilities at the firm level we find size, human capital, and the use of external sources of technology and exports. The analysis also shows that the most important mechanism of technology transfer for developing capabilities in locally-owned firms is the knowledge acquired by local personnel when they worked for foreign subsidiaries. At the meso level, four endogenous (indigenous human capital, local innovation-oriented organizations, the intensity of interactions among the regional system components, and public sector engagement) and three exogenous factors (origin of capital, an industrial sub-sector in electronics and the historical path) are identified. The electronics industry in Jalisco has stronger capabilities than the same industry in Baja California. Firm-level capabilities, in foreign subsidiaries and locally-owned companies, are higher in Jalisco:

advanced product-centred and process capabilities are more common and links with external sources of technology are stronger and more frequent. Similarly, attributes of innovation-oriented organizations (i.e. regional capabilities) are also much sounder in Jalisco, including human resource formation, basic and applied research, government initiatives and private organizations' support.

The two studied regions offer a clear illustration of the different trajectories of development and types of global-local interactions in terms of indigenous capability building. While foreign MNEs active in Baja California seem to act mostly as *enclaves*, in Jalisco foreign subsidiaries have active relationships with other firms and local organizations, and have entered a virtuous circle of rising technological capabilities. Spurred by strategic innovation and technology policies designed at the local level, the process of cumulative causation involving MNE subsidiaries as a major actor has stimulated the upgrading of technological capabilities and of the regional knowledge base as a whole. In turn, higher local capabilities have attracted more technologically advanced foreign activities.

A few critical implications can be drawn from the case study. First, in line with the theoretical literature, FDI should be seen as a complement to, and not a substitute for, local efforts to develop technological capabilities. The attraction of FDI in high-technology industries is not a sufficient condition to develop advanced regional technological capabilities. On the one hand, the presence of MNE subsidiaries does not guarantee technology transfer, and on the other the absorption and assimilation of imported technology is not a passive activity; rather it demands large, continuous and concerted efforts from all components of the regional system. Second, this illustrates that, for developing and emerging regions with FDI-led industries, more than investing on one or another critical asset, it is rather the combination and integration of all of them that is key to achieving stronger regional technological capabilities. Finally, the importance of idiosyncratic social structures and historical paths cannot be overlooked when considering learning processes and knowledge generation and accumulation mechanisms across space. Indeed, Jalisco and Baja California show very different historical paths. The first region includes old colonial cities dating from the sixteenth century and has a long tradition as a state with abundant industrial activity. Conversely, Baja California's main cities are relatively new; migrants from central Mexico represent a high proportion of its population, and its industrial development has been strongly linked to FDI attraction schemes launched in the mid-1960s.

Source: Adapted from Iammarino et al. 2008

they can be carried out most effectively, improving the firm's access to resources and capabilities, and facilitating the penetration of new markets. The main purpose is to gain access to localized capabilities that are complementary to the firm's own ones, at the same time broadening its own capacity of knowledge transfer to individual nodes of the GPN. Such knowledge linkages open up new development and upgrading opportunities for the various regions and the firms involved. Indeed, it has been argued that GPNs in particular industries – such as in electronics – are actually shifting to global innovation networks (GINs), with the integration of functions such as engineering, product development, design and research within inter-firm networks which are located for the most part in emerging locations and in regions within new-comer economies (Ernst 2010). As we will see in Chapters 7 and 8, while there is clear evidence that US, European and Japanese MNEs still exert a tight control on this changing geography of innovation, mainly through their scientific and intellectual property dominance, the offshoring via GINs is generating a handful of new innovative hubs in Asia (Ernst 2010, p. 5).

The precise form and effects that these knowledge and production networks take and provoke depend upon many factors. These factors include the knowledge base of the MNE and the local economy, the industry structure and sectoral composition of the local economy, the nature of local agglomeration forces, the type of transactions and relationships between global and local firms, the intra-region absorptive capacity, and the extra-region linkages and degree of openness (Bell and Albu 1999; Giuliani 2005). All these aspects will be explored in Chapter 5.

4.7 CONCLUSIONS

In this chapter, the sources of innovation at the micro-level of the firm were examined from the perspective of traditional economic theory, the Penrosian and the Resource-Based View (RBV) of the firm, and the economics of technological change and evolutionary views. We next considered the issues of technological competence, capabilities and learning, and these were discussed also taking into account both the nature and the dynamics of technological progress.

The chapter then examined the meso-level of the analysis of the sources of innovation, focussing initially on sectoral patterns of innovation, as well as the issues raised by firm heterogeneity, industrial dynamics and diversity, and technological regimes. These are in fact the foundations for understanding the nature of systems of innovation, defined with respect to their geographical, sectoral and technological dimensions.

These two levels of analysis have helped us to provide the reader with the basic building blocks for understanding the phenomenon of the globalization of innovation and for interpreting the evolution of MNEs production and innovation networks. Indeed, the geography of knowledge sources is crucial for explaining the locational and organizational behaviour of what are nowadays the most common types of MNEs as categorized in the Behrman-Dunning classification of Chapter 2. These most common types of MNEs are those which are seeking specific technical and non-technical capabilities, expertise or skills (i.e. *specific-* and *strategic-asset seeking* MNEs), and/or seeking optimal forms of integration and restructuring of both production and innovation operations (i.e. *scale and scope economies seeking* MNEs). Locational advantages change over time, and these changes relate to technology and MNE organization and behaviour. They result in, and are also the result of, a mix of sectoral and systemic features which bring together the technological, organizational and relational capabilities for interactive learning. As we will see in the next chapter, an understanding of these features at a regional level can greatly help explain the dynamics both of firms – and particularly MNEs – and also the spatial-industrial configurations of regions that we observe.

NOTES

1. See the classification in Section 2.2 in Chapter 2.
2. The contractual paradigm that developed from the seminal work of Coase and Williamson pushed the argument even further forward, defining the firm as a 'nexus of contracts'. Jensen and Meckling (1976) argue that 'a firm is the nexus of a set of contracting relationships among individuals' and that 'it makes little or no sense to try to distinguish those [contractual relationships] which are 'inside' the firm from those . . . that are outside of it' (Jensen and Meckling 1976, p.215).
3. This is also clearly acknowledged by Rugman (2000), who instead highlights the different normative premises of modern resource-based views of strategic management and Penrose's work.

4. Among the few contributions reunifying Penrose's view with the MNE theory and highlighting the importance of external sources of knowledge, see Pitelis (2002a, 2007a).
5. Historical contingency refers to the actual existence of selection mechanisms in socio-economic processes: change is neither solely based on structural elements subject to general rules, nor purely driven by random effects. At each point in time in a system's evolution, a number of paths are theoretically possible, but only a few will be chosen by the actors because each path must conform to the logic of socio-economic dynamics (Schwerin and Werker 2003).
6. Teece et al. (1997) specify that *dynamic* refers to the capability to renew competencies to accomplish the requirements of the markets. It is about innovative responses when time to market is critical, technological change is rapid, and there is uncertainty with respect to the pattern of future competition. *Capabilities* emphasize the role of strategic management in adapting, integrating and reconfiguring internal and external skills, specific assets and functional competencies to match the requirements of a changing economic environment.
7. See Section 4.6 below for some case studies on capabilities of MNEs and regions in different economic development contexts.
8. Incidentally, the confirmation of the Gibrat's law, which is valid if we remove the effects of excess mortality among small firms and the excess variability of growth of the smaller firms that survive, suggests that there are some problems for capabilities-based theories, because the independence of growth and size implies that capabilities are not so persistent after all. Turbulence as an emergent property dominates (Dosi et al. 1997). However, whilst more empirical work is needed on capability persistence in MNEs vis-à-vis other firms, the concept of interactive dynamic capabilities can actually be useful in explaining invariance in the larger firms.
9. The emergence of new technological paradigms (Schumpeter 1939) refers to the rise of major related and pervasive radical innovations that drive the dynamics of the economy as a whole in any historical time (Freeman et al. 1982).
10. Pavitt himself predicted the shrinking of the 'supplier dominated' category on the basis of a tendency for such firms to adopt 'scale intensive' or 'information intensive' strategies (Pavitt et al. 1989).
11. See, among others, Rosenberg (1976), Freeman et al. (1982), and Freeman (1986) on the demand-pull and supply-push debate.
12. The concept of 'development blocks' refers to synergistic clusters of firms and technologies within an industry (or a group of industries) giving rise to new business opportunities. At the core of a development block there is a core technology, i.e. technological paradigm (see Dahmén 1988; Dosi 1988).
13. Other approaches have addressed localized structural and institutional factors shaping the innovation capacity of specific geographical contexts. Among the many sub-national typologies considered by the literature we find, for instance, milieux innovators (Aydalot 1986), new industrial districts (Becattini 1987), technological districts (Markusen 1985, 1996; Storper 1992), and learning regions (Asheim 1995; Florida 1995; Morgan 1997).
14. Embeddedness refers to a set of characteristics appropriate for systemic innovation and reflecting the extent to which a social community operates in terms of shared norms of cooperation, trustful interaction and untraded interdependencies (Cooke 2001; Fritsch 2001; see also Dosi 1988).
15. Exceptions in looking at regional and local systems of innovation from a historical perspective are, among others, Rantisi (2002), Schwerin (2004), Boschma and Wenting (2007).
16. Physical technologies entail specific routines to deal with physical artefacts, and specific knowledge that can be more easily understood, identified and analysed. On the other hand, social technologies entail specific routines rather difficult to understand, identify and analyse, although specific knowledge is still required (i.e. sociology, social

behaviour, and so on). The co-evolution of both has been addressed as the engine of economic growth (see Nelson and Nelson 2002; Nelson 2003).

17. The issue of diversity will be extended in Chapter 5, where the argument of related variety is treated in relation to evolution of regional and local systems.
18. We prefer here to refer to GPN, as the approach that has shown evident linkages with both the innovation systems literature and their sub-national (regional) dimension (see also Coe et al. 2008).

PART II

Multinationals and the changing economic geography of globalization

5. Multinationals, variety of geographies and evolution

5.1 INTRODUCTION

Orthodox economic approaches largely consider both the multinational firm and the region or local system as *black boxes* whose characteristics, behaviour and innovative features are determined by exogenous factors. Conversely, endogenous perspectives, such as those in the management literature, tend to explain structure and growth mechanisms primarily as the result of forces which are purely internal to the firm. Hence, neither of these approaches is per se appropriate for investigating the links between the external and internal sources of innovation, and the issue 'global versus local' in the creation of new knowledge. Rather, the structure, features, behaviour and evolution of both the MNE and geographical space need to be considered within the context of their increasing interdependencies and in the light of their interactions between a variety of different knowledge transfer mechanisms. In the words of Dicken, 'global' and 'local' are not fixed scales; rather, they represent the extreme points of a dialectical continuum of complex mutual interactions' (Dicken 1994: 103).

This chapter is devoted to integrating the micro-level of the firm – focusing in particular on MNEs' technological capabilities and strategies – and the meso-level of the regional system or industrial cluster, in order to explore the new combinations of internal and external sources of innovation in the modern age of globalization. By definition, a cluster is neither a homogenous entity, nor a static and secluded one. Hence, it will become apparent that even firms and clusters that are labelled on the basis of industry-specific characteristics may have very different technological bases, use different sources of innovation, and display very different evolutionary patterns. This degree of complexity is all the more evident when considering multi-product and multi-technology firms, such as is the case with most MNEs, as these are likely to follow a variety of technological trajectories in multiple locations.

Building on the micro foundations of firm location and innovative behaviour – as discussed primarily in Chapters 3 and 4 respectively – and integrating them with the meso perspectives of regional and local industrial systems as discussed in Chapter 4, we will discuss the dynamics of different spatial agglomerations by taking into account the ways in which firms, and in particular MNEs, may interact with the local industrial environment, as defined in technological and, ultimately, social terms. On the one hand our final aim here is to expand a schematic transactions costs-based taxonomy of spatial clusters so as to include some of the basic properties of innovation and learning processes. This will allow us to depict possible models of spatial evolution. On the other hand, the organizational and spatial structure of MNEs also changes over time. As we will see, the interaction between MNEs and industrial locations, and the degree of openness and attractiveness of regions and clusters, is essential for understanding their dynamics and the directions of global competition.

5.2 AGGLOMERATION ECONOMIES IN THEORY

In Chapter 3 we used the Weber, Hotelling and Salop models to examine the interactions between firm location behaviour, factor and output prices, product variety, transactions costs, and firm strategy. One of the major issues which began to emerge from these discussions is that many aspects of firm location behaviour have a dynamic nature. In the Weber model these evolutionary outcomes are a result of new supply or market opportunities in response to factor price changes; in the Hotelling model they are a strategic response so as to prevent a competitor firm from dominating the market from an advantageous location; and in the Salop model they are a result of the effects of product variety which mitigate potential price wars and also promote spillover effects.

This type of dynamic behaviour can lead to firms clustering together in space. However, there is immediately a potential problem associated with industrial clustering, and this is the effect on land prices. If firms tend to cluster in a particular location there is automatically a pressure on local land inputs due to the competition for land, the result of which is that local land prices rise. In order to maintain a sufficient labour supply at that location, labour prices will also have to rise accordingly, otherwise local workers will tend to move away. As such, it immediately becomes clear that industrial clustering is therefore associated with rising local land and labour prices. From our knowledge of the Weber model, these congestion effects mean that the location now becomes less profitable, holding everything else constant. As such, the original rationale for the geographical clustering of firms disappears. That is, of course, unless there are some countervailing local forces or mechanisms which somehow more than compensate for these adverse congestion effects. In the next paragraphs we deal with the mechanisms by which such compensation effects may operate, and these are known as agglomeration economies.

5.2.1 Marshall's Agglomeration Economies

The current thinking regarding the importance and advantages of industrial clustering centres on the concept of agglomeration economies. These agglomeration economies are the location-specific economies of scale which accrue to a geographically clustered group of firms and people, but which do not accrue to, or more specifically do not spill-over to, other firms or people in other locations. In modern terminology, such agglomeration effects are positive externalities. The local boundedness of these externalities implies that agglomeration effects are not evenly distributed across space, but rather they tend to only operate in specific regions. Moreover, from the above argument, such agglomeration effects will only be evident if they more than compensate for the congestion costs associated with industrial clustering.

The nature and advantages of agglomeration economies for localities was first discussed in detail in the 1851 British Parliamentary Inquiry into the expansion of the railway system into Manchester, which considered the types of mechanisms operating with 'manufacturing' districts. Some forty years later, Alfred Marshall (1890) then picked up these specific themes in his discussion of 'industrial districts'. In his analysis he provided three explanations for the existence of economies of scale in situations of local industrial clustering.

Marshall's first observation concerns the existence of what today we would refer to as *knowledge spillovers*. Localized knowledge spillovers have mostly a tacit and informal nature and refer to the advantage that social actors accrue in accessing and using knowledge that freely spills over, both intentionally and unintentionally, from other co-located actors. Marshall's observation was that knowledge spillovers tend to operate specifically at the level of a local area which, following the 1851 inquiry, may in some cases be a neighbourhood, a city, or even a group of cities in a region.

Marshall's second explanation for local external economies arises due to the presence of *specialist local inputs*. These inputs may be non-traded inputs such as infrastructure, while others may be traded locally, such as specialist local services or intermediates. The key issues here are the economies of scale afforded locally by the provision of such inputs or services specifically because of the size of the local market. The provision of many non-traded or traded inputs require huge capital investments to be undertaken, and a large local market allows these costs to be defrayed across a high number of local consumers, thereby reducing the risk associated with these investments faced by the suppliers. At the same time, the local consumer firms benefit from the fact that the existence in situ of such specialist services, intermediate inputs and infrastructure, reduces the costs faced by all the local firms.

Marshall's third observation relates to the fact that industrial clustering facilitates the rise of *specialized local pools of skilled labour*. A region which is not only industrially specialized but which also has an established tradition in a particular sector, over time will develop a large local labour force with the technical and organizational skills appropriate for that industry. In this case the major advantage of geographical proximity is therefore that it allows for an efficient local job search and hiring-matching process within the labour market, as there will be both many firms and many workers with appropriate skills. The search costs associated with employment are therefore systematically reduced in a clustered location. At the same time, in a locality which exhibits a high degree of industrial clustering in a high variety of industries, such a diverse local industrial structure allows both firms and workers to adjust effectively to shocks associated with changes in economic conditions, as long as the shocks are not correlated across the local industries (Mills 1972a, b). This third observation of Marshall therefore also suggests that industrial clustering allows both firms and workers to reduce the risk associated with investment. As such, this effect can be understood as a portfolio effect and the evidence for this can be easily found in the analysis of the risk premia associated with different types of the real estate markets (Mills and Hamilton 1994).

The Marshallian mechanisms which underlie agglomeration processes are neatly described by Duranton and Puga (2004) as reflecting processes of learning, sharing and matching. Taken together the combined local existence of these processes implies that the rates of return on capital will tend to be maximized in areas characterized by industrial clustering. This is because clustering itself provides a mechanism for circumventing many market failure characteristics which are endemic in real-world markets.

5.2.2 Localization and Urbanization, Peripherality and Competition

Other aspects of these clustering processes have been discussed by a range of commentators. Three different perspectives which are important for our analysis, namely those of Ohlin-Hoover, Vernon and Porter, will be shortly summarized here below.

The first additional perspective to agglomeration and clustering which we need to consider follows the work of Ohlin (1933) and Hoover (1937, 1948). The Ohlin-Hoover schema ascribes agglomeration economies on a firm or sector basis into three categories, namely *internal returns to scale*, *localization economies*, and *urbanization economies*. Internal returns to scale refer to the fact that some firms or organizations grow and become large due to economies of scale in their own production processes, and these firms often are located at one principal location. As such, a single geographical place is the location for a large quantity of both workers and capital investment. Obvious industry examples here are aircraft manufacturing, automobiles, electronics and even major software houses. Strictly speaking, in Marshallian terms these are not agglomeration effects, as the scale effects are internalized within the boundaries of a firm, rather than being location-specific externalities which accrue to a co-located group of business companies. We will return to this point in subsequent chapters, as this is very important for our understanding of the behaviour and organization of multiplant and multinational firms, many of which are large.

The internal returns to scale exhibited by large firms point to the fact that scale and co-location are often interrelated phenomena. Indeed, the presence of large firms also often generates other externalities via several channels and linkages – such as for example purchasing, sub-contracting, outsourcing, joint ventures, and so on – on other firms in the same sector in the same locality. These are referred to as localization economies; that is, the agglomeration effects which accrue to a particular sector in a particular location and the fact that they operate between firms means that they can be genuinely regarded as external effects. Moreover, in many cases, such localization economies also work even without the presence of a major dominant large firm, and this often occurs in clusters of small firms in the same region and in the same industry.

Finally, the third form of agglomeration economies in the Ohlin-Hoover schema are urbanization economies. These are the agglomeration effects which accrue to a diverse group of industries all of which are co-located. This is basically – although not exclusively – a city phenomenon, and exists because in an urban environment people from a diverse range of firms and organizations interact, thereby facilitating the processes of learning, sharing and matching.

Another important approach to industrial clustering phenomena relates to the work of Vernon (1960) and his emphasis on the product cycle. The issues raised by Vernon's product cycle model as applied to multinational business and location have already been discussed in Chapter 2. The insights on agglomeration economies were originally derived from Vernon's work with Hoover (Hoover and Vernon 1959) on the economies of cities and regions, and in particular on the economy of the New York metropolitan area. Their original work on the spatial structure of the New York city economy observed that in the central core areas of the city were located firms and activities related to the production and delivery of products which were substantially new, thereby in the early stages of their product life cycles. Over time, as the products became more mature and standardized, their production was increasingly moved out of the core city area to more decentralized locations around New York, while the core locations focussed on the provision of other new products at the early stages in their product life cycles.

Combining the work of Hoover and Vernon (1959) with the work of Vernon (1966) these observations were further developed by Markusen (1985), who argued that within corporate multiplant and MNE structures there tends to be a regular centre-periphery type of spatial structure which is built around the product cycle logic. As mentioned in Chapter 2, such a spatial structure tends to have the primary headquarter and research and development functions all located in the core areas, which are dominated by the types of knowledge spillovers described by Marshall, and tends to focus on the production of goods in the early stages of their product life cycles. Meanwhile the production of the more mature or standardized goods which are at later stages of the product life cycles will tend to take place in lower cost and more geographically peripheral locations. Within the organizational structure of the multiplant and multinational firm, there is a continual process of movement, whereby over time the firm moves the production of each good from the core to the periphery of the corporate geographical structure, according to the stage of its product life cycle. As such, the types of international differences in multinational production implied by the Vernon (1996) analysis are also largely replicated at a sub-national interregional level within multiplant and multinational corporate organizational structures.

The other relevant approach to industrial co-location phenomena follows the work of Michael Porter (1990). Porter's approach focuses on the arrangements of firms, organizations and institutions which are clustered in a region, and the particular combinations of organizations and mechanisms by which competitive processes between firms promote the overall competitiveness of the group. The emphasis of the Porter approach is on the role played by innovation, whereby firms continuously strive to improve their innovative and competitive performance. The link here between innovation and geographical proximity – which was examined in Chapter 4, is that being clustered in the same location generates demonstration effects, whereby firms can clearly see if their competitor firms are gaining a competitive advantage. This mutual transparency afforded by proximity drives the competition between the individual firms within the clustered group, promoting the competitiveness of the whole cluster. The spillover effects which Porter emphasizes are therefore what are referred to as demonstration or observational effects. The most important interactions among clustered firms are based on rivalry, and it is rivalry that generates differentiated products, expands domestic demand and makes it more sophisticated.

The agglomeration arguments of Marshall, allied with the additional insights of Ohlin, Hoover, Vernon and Porter, provide us with a rich variety of reasons why industrial clustering is advantageous in many situations. By these various externalities, spillovers and demonstration and interaction effects, the co-location or clustering of firms often overcomes the negative outcome of rising local land and labour prices, and more than compensates for any adverse congestion effects.

5.2.3 Agglomeration Economies, Related Variety and Regional Growth

Much of the literature on agglomeration economies has dealt with the question of whether a specific regional industrial structure enhances knowledge diffusion, innovation and local economic growth. The core question is whether firms learn more from local firms in the same industry – that is, cluster or regional specialization – or from local firms in other industries – that is, cluster or regional diversity. In other words, are the most innovative and fast-growing regions and clusters sectorally specialized or diversified? A conclusive answer to such a question has not (de Vor et al. 2008), and indeed *cannot*, be found. In fact, an evolutionary interpretation of these issues indicates that there is no unequivocal causal relationship between spatial industrial structure and economic growth, and that the links between diversity and innovation at the spatial level are far more complex than what is assumed by a mechanistic 'structure-conduct-performance' type of view.

Contrary to the Marshallian tradition seen above, according to which agglomeration externalities stem mainly from specialization, and thereby accrue to spatially proximate firms that operate in the same industry, Jane Jacobs (1961, 1969) puts the emphasis on the role of the urbanization economies in the Ohlin-Hoover schema, and therefore the role of sectoral diversity in promoting local economic growth. She argues that urban diversity in terms of productions, facilities, skills, tastes, needs and cultures, generates the cross-fertilization of ideas, and that ideas developed in one sector can be fruitfully applied to many other economic activities co-located in the same urban area. Whilst inter-firm knowledge spillovers are essentially intra-industry in the Marshallian tradition, they are inter-industry in Jacobs's view, according to which it is the exchange of complementary knowledge and experience across different sectors that yields greater economic returns (Iammarino 2011).

What Jacobs describes in her books are essentially large metropolitan centres, but her argument on diversity spurring knowledge transfer and diffusion, and therefore economic growth, has been applied to geographical scales other than the urban area.[1] There are obviously differences between urbanization and diversification economies, as the former often imply the latter, but the reverse does not generally hold. For example, both London and the region of Baden Wurttemberg are considered to be highly industrially diversified, but the forces of spatial agglomeration and the mechanisms for spillovers follow different channels. Moreover, the issue is further complicated by the fact that while there is evidence to suggest that the link between innovation and cities can be strong, particularly with respect to certain sectors (e.g., Acs 2002; Duranton and Puga 2005), there is also much support to the fact that cities do not always seem to be innovation cores (Shearmur 2011), nor does innovation necessarily concentrate on cities (e.g., Simmie and Sennet 1999).

Some empirical studies have advocated dynamic agglomeration externalities across industries as a primary source for regional economic growth (e.g., Glaeser et al. 1992; Feldman and Audretsch 1999; Paci and Usai 2009). Some others have found dynamic agglomeration externalities to be particularly strong within the same sector, supporting the view of localization economies and intra-industry spillovers as a major engine for localized growth (e.g., Henderson et al. 1995; Forni and Paba 2001). Other scholars have obtained mixed results for different sectors within the same region, with sectoral diversity having positive (e.g., Combes 2000, for services) as well as negative (e.g., Combes 2000, and Blien and Suedekum 2005, for manufacturing) effects on growth, also in dependence on the industry life cycle (e.g., Audretsch and Feldman 1996). Recent contributions have shown that the results on the supposedly causal connection structure-performance at the regional level are actually highly dependent on methodological choices, varying sensitively to the adoption of different geographical units of analysis, estimation methods, variable specification, sectoral aggregation, etc. (e.g., Coombs 2000; Van Oort 2004; Mameli 2007; Mameli et al. 2008).

Yet, what emerges in the actual world is the coexistence of different regional specialization patterns, any of which may lead to success or failure in the competitive struggle. To understand the links between diversity and growth it is therefore necessary to go more in to depth on the meaning of 'regional diversity'. As recognized by Jacobs herself, knowledge spills over effectively only when complementarities exist among industries in terms of shared competencies and capabilities. Such complementarities are captured by the notion of *related variety* (Boschma 2004, 2005; Frenken et al. 2007; Boschma et al. 2009; Boschma and Iammarino 2009). To understand diversity both intra-region and interregion (or cluster), it is essential to distinguish between different forms of regional variety, because they involve different economic effects. Knowledge will only spill over from one sector to another when they are complementary in terms of competencies and capabilities.[2] Hence, related variety is needed in order to enable effective connections and innovation governance. This is because some degree of cognitive proximity is required in order to ensure that effective communication (of both information and knowledge) and interactive learning takes place, avoiding knowledge and information mis-alignment and/or cognitive lock-in (Swann 1997; Nooteboom 2000). Thus, it is neither regional diversity (which may involve too wide a cognitive distance) nor regional specialization (which may result in a too narrow cognitive base) per se that stimulate innovation and growth. Rather, local specialization in related variety is more likely to induce effective interactive learning and to facilitate the collective coordination of innovation processes (Boschma 2005). As such, the concept of related variety goes beyond the traditional dichotomy of localization versus diversification externalities. It is not diversity per se, but diversity 'in what' that matters in accruing benefits to the region (Boschma and Iammarino 2009).

Furthermore, the agglomeration literature generally does not account for inter-sectoral linkages across regions, overlooking the fact that new variety may be brought into the region or cluster through the establishment of extra-local linkages. The economic geography literature is paying increasing attention to the importance of extra-regional linkages as a way of preventing regional lock-in (e.g., Bathelt et al. 2004). Some recent contributions have shown that what seems to matter is to have relationships that bring new knowledge into the region through a wide range of links (Boschma and Wenting 2007; Boschma et al. 2009; Boschma and Iammarino 2009; Neffke et al. 2009). This is highly relevant for our purposes here, because as we have seen in Chapter 4, MNEs are one major channel of new knowledge and capabilities across space. In the long run, related variety is also a key input for the evolution of the industrial structure of the region or cluster: new industries grow out of technologically related industries, in which new firms recombine and exploit the knowledge and skills taken from local related industries.

5.3 MNEs AND SPATIAL CLUSTERING

As is seen in Chapter 3, the type of leader-follower behaviour suggested by Alchian's framework finds ample empirical support in the common observation of the clustering of small and medium sized enterprises

(SMEs) around large ones. Interestingly, although the behavioural critique has been so popular in economic geography approaches to location behaviour,[3] it actually appears to be relatively less problematic for the types of multiplant MNEs than for almost any other type of firm. The reason is that MNEs by definition tend to have not only the best information resources, but also the best information-gathering resources of all types of firms. As such, multinational firms ought to be able to engage in location strategies which most closely mirror optimal location behaviour. Moreover, as well as relocation, locational adjustment is also a potential spatial factor adjustment strategy for MNEs. Multinational corporations reallocate resources within existing multiplant spatial structure specifically in order to reduce relocation costs. In contrast, such a strategy is rarely open to SMEs. As such, MNEs are relatively less likely to exhibit the Alchian follower behaviour than other firms, except where they follow other multinational firms for the first time into new foreign markets. However, once an MNE has become established in a new market, then its information gathering facilities mean that its location behaviour will follow the profit maximizing principles as far as possible. Although the 'conflicting goals' argument seen above applies largely and particularly to the case of MNEs, it is still legitimate in a variety of contexts to employ profit-maximizing models of location behaviour, such as those framed in the Weber, Hotelling and Salop traditions, in order to consider aspects of the spatial behaviour of large multinational corporations.

The additional aspect of Alchian-type follower behaviour, as far as it relates to MNEs, implies that other smaller firms will tend to cluster around large multiplant and/or multinational firms. If agglomeration effects operate on the upstream supply chain elements of the MNE firm's industry, then such clustering will itself maximize the chances of multinational firms benefiting from such externalities. For these reasons, within the business school and management literatures, where industrial clustering is observed, a key aspect of a location's attractiveness for a firm is perceived to be its potential for enhancing the competitive advantage (Porter 1985) of the firm (Porter 1990, 1998a, b) from such externalities and spillovers. Moreover, the implication of the line of thinking based on the agglomeration arguments outlined earlier, is that clusters, once formed, have a strong element of irreversibility, and firms therefore have much to gain from locating in such agglomerations.

As is remarked in Chapter 2, Porter's thinking has recently pervaded all areas of the international business literature, because it appears to provide a way in which the 'L' component of the OLI paradigm can be discussed in more geographical detail. Rather than simply defining it in terms of a country, L can be now identified in terms of particular clustered locations at the sub-national level. However, the particular way in which this Porter (1990) argument has generally been interpreted in the international business literature is both positive and normative (McCann and Mudambi 2005). The positive conclusion is that MNEs have a great deal to gain from locating in clusters, and on such a basis, the additional normative conclusion is that MNEs should generally locate facilities where other similar establishments are concentrated. For example, following a combination of the Marshall, Vernon, Porter and Alchian's arguments, 'knowledge-intensive' MNE operations should be located in 'knowledge-intensive' regions characterized by other similar knowledge-intensive activities and establishments. In contrast, rather more routine and conventional factor-intensive activities should simply locate in lower wage areas along with similar types of activities.

There is evidence to suggest that knowledge spillovers do operate in certain locations. For example, it is well known that R&D-intensive industries tend to be highly spatially concentrated (e.g., Castells and Hall 1994; Saxenian 1994; Audretsch and Feldman 1996; Almeida and Kogut 1997), and that this spatial concentration has tended to persist even in the face of rising local labour, land and other local input costs. Recent empirical studies have compared 'first-nature' (e.g., market size, comparative advantage, and trade costs) and 'second nature' (various kinds of externalities, e.g., knowledge spillovers, capital-market externalities, vertical production linkages) factors driving the geographical agglomeration of MNEs (Alfaro and Chen 2009; see also Mariotti et al. 2010). Positive externalities not only play a significant role as determinants of MNE location, but relative to domestic plants, knowledge spillovers and other externalities seem to exert a much stronger impact on MNEs in terms of the traditional Marshallian centripetal forces (Alfaro and Chen 2009).

However, the involvement of MNEs in clusters is not ubiquitous. There is also evidence suggesting that the potential advantages for MNEs of being located in clusters are very sensitive to the nature of the industry structure in which the firm operates (Cantwell and Kosmopoulou 2002). These rather ambivalent findings reflect an analytical weakness of many of the clustering-agglomeration arguments as applied to MNEs. In particular, much of the recent international business literature seems to suggest that MNE location decisions are quite straightforward, in that MNEs should simply locate in clusters in order to benefit from localized spillovers. However, these decisions are anything but simple, as they also depend on organization and knowledge internalization issues. Moreover, a careful assessment of the empirical evidence on clustering advantages is required, particularly when applied to MNEs, because observations of clustered MNEs are not necessarily indirect evidence for local spillovers.

There are two reasons for this. Firstly, even without any agglomeration or clustering effect, a random distribution of activities across space (Ellison and Glaeser 1997) means that some activities will appear to be clustered while other activities will appear to be dispersed. Observations of spatial industrial concentration are not necessarily evidence of clusters (McCann 2007). Secondly, in the clustering-agglomeration type of literature the definitions of what is 'local' and what is 'spillover' are often very vague, in that the critical geographical dimension over which any such knowledge spillover-based competitive advantage is assumed to operate is never specified. Indeed, it has been argued that the characteristic of tacitness commonly associated with knowledge, together with the free, unintentional and disembodied nature of *pure* knowledge externalities, have been often misinterpreted, giving rise to a mystification of the concept of spillover which is applied indiscriminately to indicate both deliberate and unintended exchanges, flows and dissemination channels, regardless of the actual transmission mechanisms (Breschi and Lissoni 2001a, b, 2003, 2004; Breschi et al. 2005a, b). In the case of MNEs this is particularly problematic, because there is much empirical evidence to suggest that the information and knowledge transfer mechanisms of MNEs extend well beyond the dimensions of the individual industrial cluster or metropolitan areas (Audretsch and Feldman 1996; Suarez-Villa and Walrod 1997), and may also extend well beyond a regional or even national level (Arita and McCann 2000; Cantwell and Iammarino 2000, 2002). Therefore, how important it is for a multinational firm to be located either within or immediately adjacent to a particular industrial cluster or metropolitan region is often very unclear.

In order to discuss these issues in a coherent manner, it is necessary to reverse the logic regarding the potential advantages of clusters and to consider another question: when should an MNE *not* locate in a particular industrial cluster or region? As we have seen in Chapter 3, the central rationale for the firm – and in particular the MNE – is as a means of internalizing information and knowledge transactions costs within the individual firm. At the same time, as we have discussed earlier in this chapter, one of the key rationales for industrial clustering is to internalize knowledge and information transactions costs within the group of clustered firms, rather than within an individual firm. As such, there may be cases where the rationale and objectives of the MNE are in conflict with that of the cluster. Therefore, from the perspective of an MNE, the location decision must depend on the balance between the opportunity costs and benefits of locating in a cluster and the opportunity costs and benefits of not locating there, allowing for the organizational features of the firm. By adopting a transactions-costs approach to understanding the types of inter-firm relations which exist within a specific cluster or region, we will see shortly that while there may be conditions under which it could be advantageous for an MNE to locate facilities within a cluster, there are also many situations in which this will not be so.

5.4 ANALYTICAL TYPOLOGIES OF SPATIAL CONFIGURATIONS: IMPLICATIONS FOR MNEs AND SPILLOVERS

5.4.1 The Transaction-Costs Spatial Typologies

If we adopt a transactions-costs perspective we can define three distinct types of industrial clusters, according to the nature of firms located there, and the nature of their relations and transactions within the cluster ore region (Gordon and McCann 2000; McCann et al. 2002; McCann and Sheppard 2003; Simmie and Sennet 1999). These three distinct types of industrial clusters are the *pure agglomeration*, the *industrial complex*, and the *social network*. It has to be highlighted that the difficult analytical problems relating to the diverse features of spatial agglomerations are compounded by severe problems of identification and definition.[4] Here we provide a classification of spatial types which is independent of either the sector or the location, but instead is based on the microeconomic behaviour and objectives of the co-located actors, and on the transactions evident in the cluster or region. The key feature which distinguishes each of these different ideal types of spatial industrial configurations is the nature of the relations between the firms located there. The characteristics of each of the cluster types are listed in Table 5.1, and as we see, the three ideal types of clusters are all quite different.

In the model of pure agglomeration, inter-firm relations are inherently transient. Firms are essentially monopolistically atomistic, in the sense of having almost no market power, and they will continuously change their relations with other firms and customers in response to market arbitrage opportunities, thereby leading to intense local competition. As such, there is no loyalty between firms, nor are any particular relations long-term. The external benefits of clustering accrue to all local firms simply by reason of their local presence. The cost of membership of this cluster is simply the local real estate market rent. There are no free riders, access to the cluster is open, and consequently it is the growth in the local real estate rents which is the indicator of the cluster's performance. This idealized type is best represented by the Marshall (1890, 1920) model of agglomeration, and is the notion of clustering underlying models of new economic

Table 5.1 Industrial clusters: a transactions costs perspective

Characteristics	Pure agglomeration	Industrial complex	Social network
firm size	atomistic	some firms are large	variable
characteristics of relations	non-identifiable fragmented unstable	identifiable stable trading	trust loyalty joint lobbying joint ventures non-opportunistic
membership	open	closed	partially open
access to cluster	rental payments location necessary	internal investment location necessary	history experience location necessary but not sufficient
space outcomes	rent appreciation	no effect on rents	partial rental capitalization
notion of space	urban	local but not urban	local but not urban
example of cluster	competitive urban economy	steel or chemicals production complex	new industrial areas
analytical approaches	models of pure agglomeration	location-production theory input-output analysis	social network theory (Granovetter)

geography (Krugman 1991a, b; Fujita et al. 1999). The notion of space in these models is essentially urban space in that this type of clustering only exists within individual cities or tightly spatially grouped concentrations of firms in the same sector.

The industrial complex is characterized primarily by long-term stable and predictable relations between the firms in the cluster. This type of cluster is most commonly observed in industries such as steel and chemicals, and is the type of spatial cluster typically discussed by classical (Weber 1909) and neoclassical (Moses 1958) location-production models, representing a fusion of locational analysis with input-output analysis (Isard and Kuenne 1953). Component firms within the spatial grouping each undertake significant long term investments, particularly in terms of physical capital and local real estate, in order to become part of the grouping. Access to the group is therefore severely restricted both by high entry and exit costs, and the rationale for spatial clustering in these types of industries is that proximity is required primarily in order to minimize inter-firm transport transactions costs. Rental appreciation is not a feature of the cluster, because the land which has already been purchased by the firms is not for sale. The notion of space in the industrial complex is local and regional, but not necessarily urban, in that these types of complexes can exist either within or outside of an individual city. This complex model is actually the single explicitly spatial element in the transactions costs approach of Williamson (1979), where the focus is on the types of flow-process scale economies which firms can realize by being part of vertically-integrated production complexes.

The third type of spatial industrial cluster is the social network model. This is associated primarily with the work of Granovetter (1973), and is a response to the hierarchy model of Williamson (1975). The social network model argues that mutual trust relations between key decision making agents in different organizations may be at least as important as decision-making hierarchies within individual organizations. These trust relations will be manifested by a variety of features, such as joint lobbying, joint ventures, informal alliances, and reciprocal arrangements regarding trading relationships. However, the key feature of such trust relations is an absence of opportunism, in that individual firms will not fear reprisals after any reorganization of inter-firm relations. Inter-firm cooperative relations may therefore differ significantly from the organizational boundaries associated with individual firms, and these relations may be continually reconstituted. All of these behavioural features rely on a common culture of mutual trust, the development of which depends largely on a shared history and experience of the decision-making agents. This social network model is essentially aspatial, but from the point of view of geography, it can be argued that spatial proximity will tend to foster such trust relations, thereby leading to a local business environment of confidence, risk-taking and cooperation. Spatial proximity is necessary but not sufficient to acquire access to the network. As such, membership of the network is only partially open, in that local rental payments will not guarantee access, although they will improve the chances of access. The geographical manifestation of the social network is the so-called 'new industrial areas' model (Scott 1988), which has been used to describe the characteristics and performance of areas such as Silicon Valley and the Emilia-Romagna region of Italy (Piore and Sabel 1984; Scott 1988; Storper 1997; Castells and Hall 1995). In this model space is once again local or regional, but not necessarily urban.

In reality, all spatial configurations will contain characteristics of one or more of these ideal types, although one type will tend to be dominant in each cluster or region. Yet, as we see there are some elements of each of these particular cluster frameworks which are mutually exclusive of the other spatial typologies. Therefore, in order to understand the advantages to the firm of being located in any particular cluster, it is first necessary to determine which of the ideal types of industrial cluster, described in Table 5.1, most accurately reflects the overall characteristics and behaviour of the firms in the cluster.

5.4.2 MNEs, Clustering and Spillovers

For our purposes, what is important is to understand how the above transactions-costs descriptions of spatial types can inform our discussion of the attractiveness of cluster for MNEs and multiplant domestic enterprises (MPDEs). As we have already seen, the central rationale for the MNE is as a means of internalizing knowledge and information transactions costs within the individual firm, whereas the rationale for industrial clustering is to internalize knowledge transactions costs within the group of clustered firms. Therefore, from the transactions-costs discussion here it becomes clear that to understand the relationships between MNEs and industrial clusters it is necessary to consider how the organizational characteristics and objectives of the MPDE and the MNE relate to the knowledge spillover characteristics and inter-firm behaviour of the other clustered firms. In order to do this we must reconsider the firm's perceptions of the benefits of knowledge spillovers. In particular, we must distinguish between spillovers which result in knowledge inflows from those which result in knowledge outflows, and also we must distinguish between unintentional and intentional knowledge flows.

It is safe to assume that all firms regard knowledge inflows positively, irrespective of whether they are intentional or unintentional. However, a firm's perceptions of the benefits of knowledge outflows will depend on the structure of the industry in which the firm competes and also on the technology embodied in the industry. This is because unintentional knowledge outflows have both a positive and negative effect on the individual firm (McCann and Mudambi 2004, 2005). On the one hand, the private effect of an unintentional knowledge outflow on the owner-originator firm is a leakage of its valuable intellectual capital, which would be viewed negatively by the firm (Grindley and Teece 1997). Meanwhile, on the other hand, the potential positive effect of an unintentional knowledge outflow is the public good aspect of knowledge (d'Aspremont et al. 1998). Such outflows can contribute to a virtuous cycle by strengthening the knowledge base of the region and making it a more attractive location for other knowledge-bearing firms. This, in turn, should generate larger future knowledge inflows to all the firms in the group. As such, the individual firm's view of the net benefits of these unintentional knowledge outflows will therefore depend on its assessment of the relative costs and benefits of these two opposing effects. These costs and benefits will depend on both the industry structure and also the technology and knowledge characteristics of the industry. For the moment we will just comment on the issue of industry structure, and in the next section we will deal with the effects of technology and knowledge characteristics.

In the case of a competitive market structure characterized by a large number of firms, each with a relatively small market share and profits, such firms probably have little to lose from unintentional knowledge outflows and more to gain from inflows stemming from a strong clustered location. As such, in the case of a broadly competitive market structure the public good aspect of knowledge would appear to dominate, with the local knowledge outflows being viewed as generally positive both for the firms themselves and for the local region (e.g., Jaffe et al. 1993; Saxenian 1994). On the other hand, an oligopolistic industrial structure is characterized by a few large firms, each with a large market share and considerable strategic interdependence. Often, oligopolistic firms realize that unintentional knowledge outflows to industry rivals can be extremely costly in terms of lost competitive advantage, because the private good aspect of knowledge is one of their dominant considerations. In these situations where any unintentional knowledge outflows from a firm are more valuable to a firm's competitors than any potential knowledge outflows from these competitors to the firm, the overall effect of such outflows is perceived to be negative. From the perspective of firm location strategies, the clustering of oligopolistic firms would appear to jeopardize their proprietary knowledge assets by exposing themselves to the possibility of unintentional outward knowledge spillovers. If this is so, this will lead such firms to decide not to locate in clusters, unless they can find a way of avoiding such knowledge leakages. We can therefore use this argument concerning the avoidance of unintentional knowledge outflows to reconsider the attractiveness of a cluster for MPDEs or MNEs, given that most of them are oligopolistic firms.

In terms of our spatial typologies above, the possibility of unintentional knowledge outflows is associated most obviously with the model of pure agglomeration. Knowledge can be shared between two parties, but if there is little or no inter-firm loyalty within the system, this knowledge will also be passed on to third parties who are beyond the control of the originator of the information. As such, a pure agglomeration-type of location will create concerns for an oligopolistic MNE establishment. Similarly, in the case of a social network type of environment, where the non-opportunistic relations between the firms are built upon longstanding mutual trust and shared experience, an immigrant oligopoly MNE firm will benefit little from such features, as these trust systems are generally based on networks of small firms aiming to collaborate together.

On the basis of these arguments, if we apply Akerlof's (1970) market-for-lemons model, many industrial clusters and regions which include large oligopolistic competitors will generally be characterized by adverse selection and should either fail to form, or become concentrations of mediocrity (McCann and Mudambi 2004, 2005; Mariotti et al. 2010). This will be particularly so in the case of clusters which are characterized by the types of inter-firm relations typical of pure agglomeration or the social network. The reason for this is that the knowledge and information internalization logic in favour of the MNE is largely inconsistent with either the externality argument in favour of the pure agglomeration or the interpersonal and inter-organizational relations argument of the social network. Similarly, the clear organizational boundaries of the large firms typical of an oligopoly are inconsistent with the organizational forms assumed by either pure agglomeration or social networks.

These arguments therefore provide a powerful counter-argument to the simple clustering logic and appear to explain the empirical observation that many of the largest firms – either MPDEs or MNEs – do *not* co-locate their knowledge creation activities with those of their competitive rivals (Simmie 1998; Cantwell and Santangelo 1999; Shaver and Flyer 2000; Alcácer 2006; Verbeke et al. 2009). Moreover, in situations where they do so, the organizational aspects of the firms are designed specifically to avoid the sharing of knowledge (McCann 1993; Arita and McCann 2002).

The one major clustered form of industrial organization which is almost entirely consistent with the strategies and objectives of oligopolistic MNEs is that of the industrial complex. In some situations, inward investing MNE firms will find it optimal to locate facilities close to other similar firms, specifically in order to carry out particular types of long-term inter-firm transactions. In these cases, the intentional sharing of knowledge and information between the firms is a mutually planned process with knowledge inflows and outflows being carefully managed within a system of bilateral monopoly frameworks. This type of clustering is commonly observed in industries such as chemicals, pharmaceutics, petroleum, automobile manufacturing, as well as many sectors of the electronics industry (McCann 1997). Yet, the inter-firm relations embedded within this type of system are entirely different to the types of relations assumed to operate in the clusters models based on simple descriptions of knowledge spillovers. As such, observations of MNE clustering of a type consistent with an industrial complex model may often be misinterpreted as a cluster based on an agglomeration-spillovers model. Some apparently more sophisticated work (Devereux et al. 2000; Barrell and Pain 1999) has fallen into this trap, by simply assuming that groupings of FDI investments by MNEs within an individual country must be clear evidence of agglomeration economies, irrespective of the geographical location and spatial scale of either the country or its internal regional and urban system. It appears that many of the mistakes made by the original international business literature are being repeated when it comes to the modern empirical analysis of the location behaviour of MNEs.

There is another aspect in which the industrial complex model is also particularly important when considering the MNE location behaviour. In a context where knowledge transactions can be largely defined (and also protected) via legal contracts and where both the production function and also the spatial transactions costs relationships of the firm are largely predictable and stable over the medium term, then the Weber model and the logistics-costs model (McCann 1993, 1997) become particularly relevant. The reason is that in order to move away to another location more distant from the markets and suppliers in search of lower input factor prices, the production function relationships and the spatial transactions costs incurred must be predictable and stable over the medium to long term. Otherwise the opportunity costs of moving away to more distance locations become too high, because the risks involved become so great. This is actually the converse of a point made by Scott (1988) who argued that where firms operate in an highly unstable trading environment without many formal legal and organizational safeguards, and here the extreme case are firms in the informal sector, then they are forced to cluster together. As such, the more predictable, stable, formal and protected are both the production function and transactions costs characteristics of the firm, the more readily will a firm relocate in order to take advantage of spatial costs variations.

It is not possible to have a multiplant spatial structure in which the relationships between the firm's individual establishments are of a tacit or informal nature. Moreover, this argument is applicable to both MPDEs and MNEs. However, multinational firms are the most powerful example of firms which formalize and protect their knowledge and transactional features precisely in order to foster their geographic mobility across countries. As we will see in this book, the relationships between the individual MNE establishment and its immediate locality may have elements of any of these three types of spatial-industrial arrangements described above, or indeed any of the four types described below. Yet, while the industrial complex is a very common organizational form at the establishment level, the increasing autonomy and capabilities of MNE affiliates and subsidiaries suggests that both of the other two forms are also becoming more important in terms of the relationships between a subsidiary establishment and its local environment. However, *within* an individual MNE corporate structure the relationships between the dispersed MNE establishments *must always* be arranged in a tightly organized manner, as represented by the industrial complex. The tightly organized MPDE and MNE structure permits strong control over the flows of knowledge, finance, goods and people, between the individual MNE establishments, and the maintaining of the control over these relationships is the job of the headquarters. Most crucially for our purposes, all of the international business evidence suggests that it is the very tightness of the headquarter control over the O and I dimensions which itself allows the multinational firm to geographically disperse its establishments across countries. As well as having clustered establishments, the organizational and transactional forms represented by the industrial complex model therefore also provides firms with a high ability to disperse or relocate subsidiary and affiliate establishments to areas which are distant, dispersed or not locally clustered. O and I are strongly related to L via formal channels and via a formal logic which can be readily analysed using models such as the Weber (1909) or the logistics-costs models, both of which now become powerful analytical instruments for thinking about the location and relocation behaviour of MNEs.

5.5 KNOWLEDGE, TECHNOLOGY AND THE VARIETY OF SPATIAL CONFIGURATIONS

5.5.1 Variety and Dynamics of Industries and Spatial Systems

As seen in Chapter 4, the literature on industrial dynamics and technological regimes has produced valuable stylized facts on both firm and industry heterogeneity and characteristics and their impact on growth paths. Yet, this literature has, until recently, largely neglected the spatial dimension of innovation and technological processes.[5] On the other hand, the literature on innovation systems, with its strong emphasis on the links between organizations and institutions, has discussed some of the main determinants of system variety, including various levels of geography. However, that literature has largely overlooked the evolution of spatial systems themselves over time in response to innovation processes.

Following the evolutionary tradition, which aims to explain economic growth in terms of technological and institutional change along with firms' strategies and industry dynamics (Freeman 1982a; Perez 1983; von Tunzelmann 1995; Verspagen, 2000), some key concepts can be adapted from this literature in order to build an analytical framework for interpreting the evolution of industrial clusters. Note that, within this approach, a conceptual distinction has to be made here between growth and the evolution of a cluster or region. 'Growth' is defined in terms of a quantitative and qualitative increase of the cluster/region's existing resources and structures, including its mechanisms of specialization and diversification. In contrast, 'evolution' implies structural and institutional changes that take place within the cluster or region, leading eventually to the partial or complete transformation of the existing structures, spatial settings and economic and innovative advantages of the local system, resulting in either regional regeneration or decline. Although the two phenomena are clearly interrelated, our focus here is more on the evolution of clusters as a result of globalization forces and technological change embodied in MNEs.

Few observations are worth mentioning in this respect. On the one hand, despite the high microeconomic turbulence of industries, the overall degree of spatial clustering seems to remain fairly stable over time, although a bigger variation is observed in the location of such agglomerations (Dumais et al. 2002). On the other hand, as shown by Bottazzi et al. (2005), intersectoral differences in agglomeration patterns can be at least in part accounted for by differences in the regimes of technological learning across industries. All this suggests that in terms of technology there are two overlapping levels of dynamic features that link industrial and spatial structures. Firstly, there are cross-sectoral location-specific advantages that underlie the intra- and inter-industry heterogeneity of spatial patterns. Second, there are the sector-specific knowledge conditions that help explain regularities in industrial patterns across different regional locations. One of the greatest challenges in analysing innovation processes and technological change is therefore dealing with this issue of the interdependences between sectoral diversity and spatial diversity (Frenken and Boschma 2007). In response to these challenges, here below we further expand and develop the typologies of industrial clustering initially outlined in Table 5.1 specifically so as to link the major technological trajectories and innovation patterns followed by firms and industries to the major forms of spatial industrial clustering evident.

In line with what has been argued so far, industrial cluster and regional dynamics can be seen as a function of three main factors:

- firm-level dynamic capabilities, which refer to the intra-organizational technological and organizational capabilities of both local firms and MNEs;
- technological regimes, which refer to industry dynamics and the prevailing knowledge conditions in the specific industry structure of the cluster/region;

- systemic integration, which refers to the evolving functions in the cluster or region as a whole, driven by inter-organizational capabilities, networks, institutional learning, and the overall system's degree of openness.

The dynamics of the process of the accumulation of capabilities at both firm and system level are critical determinants of the vertical structure of firms and industries. Moreover, they also have a major influence on spatial structures. In history-friendly simulation models (Malerba et al. 2006), the degree of vertical integration or disintegration of firms in a specific industry depends on the actual distribution of capabilities across firms in that industry and in the upstream and downstream industries. In addition, the degree of heterogeneity and the distribution of capabilities are also affected by market structures and selection. As such, the presence of MNEs with wide-ranging capabilities which are superior to those available on the local market will spur processes of either specialization or vertical integration in backward- and forward-linked industries located within the region, with major effects on the spatial patterns of industries. In turn, the level of capability accumulation on the part of the cluster's firms is influenced by the degree of vertical integration, because specialized firms accumulate knowledge and capabilities differently from vertically integrated firms (Malerba et al. 2006).

The variety of firm-level capabilities, technological regimes and systemic integration account for the observed variety of spatial configurations and the different sources of innovation in each particular cluster/region. This is the case, even allowing for the fact that the innovation processes will change over time, following changes in each of the three factors. However, whilst capabilities at the firm and system levels can be considered to be largely endogenous to the cluster or region, the features of the technological regime are largely exogenous to the cluster or region. This fact is what justifies our choice to introduce them as the critical dimension for expanding the transactions costs cluster taxonomy presented above.

As broad examples of such industry-geography interdependencies, traditional sectoral systems with highly scattered (cross-country) spatial concentrations will often be connected to technological regimes with low degrees of opportunities. Meanwhile, capital goods and machinery systems with strong sub-national agglomerations and localized knowledge bases will usually be allied with technological regimes offering medium to high opportunities. At the same time, advanced sectoral systems such as microelectronics, biotechnology and software, exhibit technological regimes characterized by new and fast-growing opportunities. These sectoral regimes will usually be associated with high spatial concentrations of activity, both within and across national boundaries, which are also characterized by a high degree of integration between the local and global knowledge bases (Breschi and Malerba 1997a, b). In such fast-changing fields, innovation processes are critically informed by new developments occurring outside of any individual firm. Therefore, technological progress benefits greatly from the active participation in the particular technological community or context where new developments take place (Frost 2001), even if these are geographically outside of the region. In turn, new technological opportunities will affect the overall rate of technological growth, and global innovation flows will privilege those locations offering the best and fastest-growing breaks. Technological regimes characterized by new and fast-growing opportunities therefore are those which are also most expected to generate the greatest turbulence and the largest changes in the locational or sectoral hierarchies.

5.5.2 The Limits of the Transactions Costs Spatial Classification

One of the advantages of the transactions costs classification discussed above is that it provides an organizing framework which is capable of dealing with the diversity of spatial concentrations that we observe. At the same time, a weakness of this framework is that, in the transactions costs perspective itself, hierarchies, and particularly but not exclusively firm structures, are reduced to being a consequence of the types of transactions costs. As such, the transactions costs approach underlying the classification of cluster types in Table 5.1 is essentially static in nature, and exhibits a very narrow definition of knowledge and technology. There are three aspects to these limitations.

First, while the classification scheme presented above is essentially static, as we have discussed, we know from all of our previous discussions that the relationship between firm location and technology is mainly dynamic (Bottazzi et al. 2008). This is true both in terms of industrial demographic features such as firm entry, exit, growth, and relocation, and also in terms of cluster life cycle features such as cluster birth, growth, decline, openness, and attractiveness. The transactions costs classification scheme does, however, offer possibilities for the inclusion of technology and innovation components, such as learning, capability accumulation, and knowledge creation, as additional dynamic explanations for such diversity.

Second, the transactions costs arguments applied to the perceived impacts of knowledge inflows and outflows by all firms including MNEs, depend to a large extent upon a narrow definition of knowledge. In the definition of knowledge which is employed, knowledge is regarded as a purely public good, and is therefore assumed to be non-excludable, non-depletable, and free to be used without limits by anyone, at any time, and anywhere across geographical boundaries. In such an approach, it is clear that knowledge and information are treated as being largely synonymous. However, once tacit knowledge is taken into consideration, it becomes clear that technology as a whole cannot be easily traded or exchanged, and only the potentially public knowledge component is, at least partially, liable to be accounted for in terms of transactions costs analysis.

When narrowing the notion of technology to something akin to information, and concentrating primarily on the organization of the exchange of such information, there is a tendency to overemphasize the appropriability issue (Winter 1987, 1993). In contrast, the returns on innovation to a firm may well be mainly returns to its creation of capabilities, a process supported by, but not simply reducible to the generation of new potentially public knowledge. In addition, knowledge can at the same time be both 'sticky' within the organization or firm boundaries, while also being 'leaky' or mobile, generating outflows in the environment immediately external to the firm (von Hippel 1994; Wernerfelt 1984). As such, ideas, inventions and practices which are difficult to move within organizations under some circumstances under other circumstances may prove to be quite capable of moving outside of them (Winter 1987; Steinmueller 2001; Seely Brown and Duguid 2001), an observation which thereby questions the centrality of the appropriability issue.

The third limitation of the transactions costs view is related to the sources of innovation in the industry, or to the technological regime prevailing in the cluster. The transactions costs view alone is unable to account for how the innovation and knowledge accumulation processes themselves contribute to the emergence and evolution of clusters and regions. It is therefore necessary to map the knowledge and technology characteristics discussed in Chapter 4 onto the simple transactions costs framework presented. The reason for this is primarily that the multinational firm is the focus of this book and, as we have seen, in recent decades changes in the nature of knowledge and capabilities have been responsible for transforming MNEs simply from technology transferors to technology creators. As such, changes in O, L and I advantages will all determine how the logic of an MNE itself, as well as the type of spatial agglomeration in which it locates, will evolve over time.

5.5.3 An Extended Technological Classification of Spatial Types

In order to discuss the evolutionary dynamics of the variety of spatial cluster types we observe, the transactions costs-based classification scheme reported in Table 5.1 is now extended here in Table 5.2, in a manner in which the main underlying knowledge conditions of the cluster are now

Table 5.2 Technological classification of spatial types

Characteristics	Pure agglomeration	Industrial complex	Social network	
			Competence-based SN	Trust-led SN
Nature of technical knowledge	codified, explicit and mobile, transmitted by way of information	specific, systemic, routinized, R&D-intensive; based on non-transferable experience	tacit, generic, non-systemic, sticky and leaky; transmitted within cognitive networks	codified, sticky, mature; transmitted within geographically localized networks
Technological trajectory	oriented to processes, problem-solving, organizational innovation	oriented to complex products, cost-cutting	oriented to radically new products	oriented to processes, customer-driven
Dynamics	stochastic	strategic	disruptive, uncertain	incremental
Sources of innovation	external to the firm, multisectoral	internal to the firm and the sector, institutionalized R&D in incumbents, relevance of basic science	external to the firm and the sector, relevance of applied science and cooperative R&D	external to the firm, internal to the sector, relevance of machinery acquisition
Appropriability of innovation returns (and market structure)	low (perfect or monopolistic competition); based on standards, copyright, design	high (oligopolistic competition); based on patents, secrecy	low (low entry barriers); based on patents, copyright	low (collaborative monopolistic competition); based on trademarks
Technological opportunities	variable	medium/low	very high, uncertain	low
Degree of cumula-tiveness	low	high	low	high
Knowledge-base	broad, diversified	narrow, targeted, specialized	broad, research-based	narrow, specialized along the filière

Table 5.2 (continued)

Characteristics	Pure agglomeration	Industrial complex	Social network	
			Competence-based SN	Trust-led SN
Modes of governance	market	hierarchies	relational and cognitive networks	social and historical networks
Examples of industrial specialization	finance, banking, insurance, business services, retailing	steel, chemicals, automotive, pharmaceuticals, machine tools, medical instruments, ICT hardware	high-tech, general purpose technologies	customized traditional goods, textiles, footwear, furniture, tourism
Example of cluster	Global cities; 'Silicon Valley' (California)	'Silicon Glen' (Scottish Electronics Industry)	'Silicon Fen' (Cambridge UK)	Italian industrial districts in *made in Italy* industries
Pavitt classification	Information Intensive, Supplier Dominated Firms	Production Intensive Firms (Scale Intensive and Specialized Suppliers)	Science-Based Firms	Supplier Dominated Firms

made explicit. This extended knowledge-based taxonomy is developed by linking the transactions-costs spatial classification scheme to the well-known Pavitt (1984) taxonomy of the sources of innovation, and then by modifying it to take account of the nuances of the knowledge and innovation features of different cluster types. This allows us to take account of the ways in which firms, and in particular MNEs, may interact with the local industrial and technological environment, and the multiple linkages which exist between knowledge conditions and regional economic structure and growth. This will also help us to understand how spillovers occur and change over time. Again, as with the transactions costs classification scheme already described, the knowledge framework we now develop here assumes that in each locality, one typology is likely to be dominant, but that does not necessarily imply that it is at all meant to be exclusive. As before, this allows for geographical configurations which will exhibit features of more than one category at the same time.

In the knowledge and innovation taxonomy described by Table 5.2, we see that in the model of pure agglomeration the bulk of knowledge is explicit and codified, is available to any local actor and organization, and is generated outside firms' boundaries, being largely created in other private and public organizations. Variety and promiscuity are distinctive features of cities, in the sense of there being an absence of long term loyalty between agents. The combination of different streams of knowledge therefore occurs across a broad range of sectors (Jacobs-type externalities), and individual and organizational innovation linkages or relations are unpredictable due to the low degree of knowledge cumulativeness. However, even though in many cases the critical distance over which urban agglomeration externalities operate may be that of the broad city-metropolitan area (Gordon and McCann 2005), as is assumed by many theoretical models of agglomeration, there is also much evidence to suggest that for many firm-types and industries the critical distances over which agglomeration externalities operate may be very much larger, and as large as that of the city-region (Suarez-Villa and Walrod 1997; Simmie 1998; Arita and McCann 2000; Caniëls 2000; Cantwell and Iammarino 2003). These considerations come from observations of regional innovation systems, the location and performance of MNE R&D facilities, the behaviour of local and regional labour markets, and the structure of transportation networks, particularly air-transport systems.

The combination of diverse kinds of knowledge into an interdependent economic and technological base crucially needs a plurality of knowledge sources and networking among those sources. These features of economic systems, and particularly their opportunities for communication, play a major role in determining the conditions of the production of new technology (Patrucco 2001; Antonelli 2000, 2008). In this respect, urbanized and metropolitan regions have been demonstrated to provide highly positive institutional environments for technological progress, due to the variety and availability of complementary economic activities, business services, science and technology, institutions, and communication and network infrastructure mechanisms. As we will see later in Chapter 7, these considerations largely account for the extreme spatial concentration of MNE headquarters in the major world cities, where the range, depth and variety of the local urban transactions and networks are strongly information- and power-intensive, and institutional connectivity is dense enough to ensure that the MNE's corporate global control functions operate properly (Friedmann 1986; Sassen 2001).

In knowledge terms the industrial complex is quite different to the agglomeration model. The industrial complex model is instead associated primarily with highly cumulative learning from sources inside the industry and the firm, such as in-house R&D, and on the basis of knowledge that is specific to industrial applications. In analogy with the Schumpeter Mark II type of regime, such cases generally exhibit low entry rates (high barriers) and high industry concentration. Large incumbents, often MNEs, account for most of the sector's innovative activity, and these firms can profit from their innovations in part because they have the potential to exclude rival firms from using the new product and processes they have generated. In these situations, knowledge based on specialized non-transferable experience is the major input in local innovative activity, and the incumbent firms tend to have a major innovative advantage over new firms because innovation is largely routinized and processed within the existing hierarchical bureaucracies. As such, leading firms, particularly large MNEs, play a crucial role, and power asymmetry is central to the value chain and the governance of innovation in the cluster or region (Cooke 2001).

In knowledge terms the social network model is the third industry-space typology. However, once the different types of spatial configurations are distinguished in terms of technological regimes, structures and governances, as well as their transactional relations, it becomes clear that the social network model ceases to be a relatively homogenous and consistent single analytical category. Thus, the single transactions costs-based social network typology previously discussed in Table 5.1 can now be split into two subcategories, namely the 'competence-based social network' category and the 'trust-led social network' category. Such a distinction is based on the dominant features of the technological regime which are likely to prevail in each of the two social network types. In the case of the competence-based social network the technological regime is entrepreneurial, science-based and technologically widening. In the case of the trust-based social network the technological regime is routinized, technologically cumulative and deepening. Here, we therefore move well beyond the innovation adaptation of Table 5.1 described Iammarino and McCann (2006) which treated the social network model as a single category, whereas all of our arguments here point to two very distinct categories of social networks.

In the competence-based social network model high technological opportunities come primarily from sources outside of the firm and the industry sector, such as university academic research. In this kind of technological environment the type of knowledge tends to be both generic and non-systemic, with high rates of market entry and exit, a strong degree of volatility of market shares, and low levels of market concentration. In such an environment, the tacit and sticky nature of knowledge requires geographical proximity, particularly in the early stages of the industry or cluster life-cycle. On the other hand, the openness of the innovation system, the relative 'leaky' character of new knowledge and consequently the high potential for spillovers, and the related emergence of new rules, standards, blueprints and verification procedures, all point to the importance of external sources of technical knowledge. Moreover, the sources of technical knowledge may or not necessarily be localized. As such, innovation is frequently associated with a high level of uncertainty regarding both the nature and levels of technology and demand, and a high degree of market turbulence. As a consequence, a lower survival rate for new firms is likely to be associated with higher levels of innovativeness and growth among the surviving firms (Alchian 1950). Innovations also therefore mainly come from knowledge that does not have a routinized nature, with new firm start-ups playing an important role, and SMEs accounting for a large share of innovative activity. There is normally either no role or at best only a limited presence of large MNEs *in situ* in the first stages of the cluster life cycle. In later stages of development, however, where both industry and spatial patterns have reached a certain stability, such clusters may become attractive to MNEs, but in the early stages of the competence based cluster, MNEs are rarely part of the story.

In the trust-led social network type, there is not necessarily any clear hierarchical structure, and the overall coordination of the innovation system is left to a mix of cooperation and competition. Knowledge is largely codified and mature and develops along stable trajectories which are mainly oriented towards process innovation. Knowledge is transmitted essentially by way of informal personal contacts which are strongly embedded in a common cultural matrix, comprised of elements of social and political lobbying, and various backward and forward linkages. These backward and forward linkages sometimes are intensified by the presence of MNEs. If this is the case, then the cluster evolution will also in part depend on the nature of the transactions established by the MNEs with both small and large local firms. As such, the structure of vertical and horizontal linkages between MNEs and local firms may be crucial in determining the nature and extent of the local spillovers. Even in the same industry, clusters may be characterized either by highly vertically integrated firms, or by stage production with significant subcontracting linkages. The trust relationships and the role of reputation will differ considerably in such different spatial and industrial settings (Guy 2009).

Reinterpreting the simple spatial types in terms of technology, knowledge and innovation characteristics, allows us to uncover more subtle and detailed aspects of firm location behaviour. In particular, in the case of MNEs, any simple assumed correlations between local knowledge spillovers, the spatial concentration of activity, and the advantages of a particular location for firm investment, do not necessarily hold. For example, in the case of pure agglomeration, while the overall links between innovation and cities are very strong (Acs 2002), the arguments relating MNEs to knowledge spillovers and innovative cities are rather more complex. This is because the knowledge objectives and the innovation performance of multinational firms are not necessarily centred on cities (Simmie and Sennet 1999). As a consequence, the model of pure agglomeration presents the biggest difficulties in being traced back to any of the archetypical kinds of technological regimes which are identified in the literature. The reason is that the heterogeneity of spatial patterns which can be included under this heading, and ranging from the city to the regional innovation system, depends critically on the idiosyncratic local capabilities and the degree of system integration. Agglomeration therefore allows for the coexistence of different technological regimes characterizing the same geographical unit of analysis, but the sense in which it is driven by 'pure' externalities can be unclear.

In contrast, the industrial complex will often be characterized by regions with a highly specialized industrial structure. For modes of innovation which are primarily based on major R&D and capital investment and long development lead times, in many cases the industrial complex may provide the most suitable environment for technology creation and experimentation. Where evident, the composite and advanced local knowledge-production basis is also likely to display a strong concentration at the spatial level. The industrial complex structure is also often a suitable structure for the development of multilateral networks of dissimilar but complementary relations between MNEs and local actors (Patrucco 2001, 2003). Where MNEs invest in a region, the system will often exhibit the characteristics of the complex. On the other hand, as seen above, in the industries in which the host region or cluster is technologically strong, the major presence of indigenous firms tend to deter MNEs in the same industry from carrying out considerable innovation activities in the primary technological fields related to the local industry strengths. As such, when MNEs choose to invest in these types of regions it generally implies that the focus of the local relations is mostly related to supply chain issues, rather than technology or knowledge sourcing. However, MNEs operating in other industries will still be there attracted for innovation and knowledge purposes in those technologies whose lines of development represent diversification from the primary technologies of their own industry (Cantwell and Kosmopoulou 2002).

Finally, from the perspective of geography, there is also a fundamental difference in the particular modes of innovation governance between the two social network models (Simmie et al. 2004; von Tunzelmann 2003, 2009a, b). While trust-led social networks are generally geographically embedded, and also rooted in historical experience based on strong social proximity, competence-based social networks may rely on various kinds of communities of practises that do not necessarily require a spatial dimension, and where instead a strong role is played by relational and cognitive proximity (Boschma 2005).

Some empirical analyses have shown that inter-industry knowledge spillovers are likely to become more intense in centres of technological excellence where spillovers seem to operate mainly through exchanges in and around core technological systems (see also Crozet et al. 2004). This is the case where spillovers are primarily rooted in 'general purpose technologies' such as, for instance, background engineering, mechanical methods, new materials, electronics and ICTs, and the additional spillovers which these create between actors in quite separate alternative fields of specialization. These types of centres of excellence – which correspond to the 'first order' regions discussed in Chapter 4 – exhibit inter-industry technology spillovers, and are most likely to be classified either as pure agglomerations or competence-based social networks. However, as we have seen, these two different types of clusters tend to offer different possibilities with regard to the local role of MNEs, and the contribution that they make to local spillovers. Yet, it is these two types of clusters which tend to experience a faster process of convergence between old and new technologies, and potentially therefore also a greater degree of competitiveness.

In the light of the arguments presented so far, it becomes clear that all industrial clusters can be characterized both in terms of the transactions costs and relations characteristics described in Table 5.1, and also in terms of technological regimes and knowledge characteristics summarized in Table 5.2. As in all taxonometric attempts to classify units of analysis which are based on a stylized reduction of the complexity of the whole population, differences between the categories are necessarily maximized. However, as Pavitt himself said about his own taxonomy, the main weakness of our attempt 'is the high degree of variance still found in each category' (Pavitt 2000, p. xi). This is also true here in our case, although there is a subtle difference between the Pavitt schema and ours. Pavitt's approach was inductive and based on detailed empirical observations of individual units of analysis such as firms (Archibugi 2001), and even though his classification scheme was the most parsimonious yet developed, he still observed enormous variation even within his individual taxonometric categories. In contrast, our taxonomy is deductive, and based on the different theoretical streams of literature which exist regarding the economics and geography of innovation, and the various attempts within this literature to classify composite units of analysis such as clusters and regional systems. However, fundamentally these various theoretical approaches are all aiming to provide the most parsimonious and empirically and theoretically defensible classification system for explaining complex sets of observations. As such, it is to be expected that in our theoretically constructed taxonomy our individual categories should also exhibit significant internal variation.

Yet, what is important in terms of economic geography is to identify the dominant form of spatial-industrial structure which a locality exhibits. Once again, many observed cases will exhibit aspects of more than one, and sometimes even all, of the individual categories, depending also on the spatial scale which is the specific unit of analysis. However, in order to make sense of the overall workings of the local region and also the possibilities and rationale for MNE investment in that region, it is necessary to identify the dominant transactions, relations, knowledge and innovation characteristics of the region. It is an awareness of these knowledge and innovation features that clearly demonstrates that many regions will never be appropriate locations for particular types of cluster policies aimed at promoting either Silicon Valley-style agglomerations or new economy 'creative' clusters, even though there have been multiple attempts at branding places as such.[6] In terms of technology, knowledge and innovation, the developmental features of 'mature' regions are very different indeed to those of newly emerging regions (Farshchi et al. 2009). Moreover, shifts in technology can lead to processes which drive both the rise and decline of different industrial clusters and regions (Cantwell and Iammarino 2003; Boschma and Lambooy 1999).

However, the existence of spatially localized interdependencies and relations does not necessarily imply that innovation depends principally on them. Knowledge circulation within the region is complementary to that which occurs across regions, and to the linkages which operate between local and non-local actors (Bathelt et al. 2004; Wolfe and Gertler 2004). Being able to build new competencies and capabilities involves the ability to establish links at all levels, from the 'global' to the 'local'. Such linkages can be established through many different channels: trade exchanges; inward and outward foreign direct investment; involvement in global production networks; technological alliances; integration in knowledge networks that, beyond firms, may encompass a variety of regional organizations. As we have seen, the extent to which a region is both able to attract innovative resources from elsewhere and also able to transmit new ideas and innovation outwards – so spurring its external integration, depends first and foremost upon its extant absorptive capacity and knowledge base (e.g., Simmie 2003; Morgan 2004).

The notion of regional absorptive capacity, in conjunction with that of openness and attractiveness, has been effectively applied to explain diversity in the context of industrial clustering (e.g., Bell and Albu 1999; Arita and McCann 2000, 2002; Garofoli 2003; Owen-Smith and Powell

2004; Giuliani et al. 2005; Iammarino and McCann 2006; Morrison 2008; Belussi and Sammarra 2010) and regional innovation systems (e.g., Cooke et al. 1998; Cooke 2001; Asheim and Isaksen 2002; Cantwell and Iammarino 2003; Cantwell and Piscitello 2005; Doloreux and Parto 2005; Iammarino 2005; von Tunzelmann 2009a, b).

In the cluster literature global firms can assume a critical role in bridging capabilities at both the micro and the spatial system level. As we have seen in Chapter 4, MNEs can act as gatekeepers that import knowledge that may (or may not) diffuse to other co-located firms, depending on the regional absorptive capacity and on the complementarity between the knowledge inflows and the local knowledge base. As discussed in Section 5.2 above, the related variety brought into the region through extra-regional channels may enhance diversification into new applications and new sectors. At the same time, this increased related variety may allow for more extending of the local knowledge base, thus helping to produce the type of diversity which is conducive to growth and eventually to the evolution of new spatial structures. In addition, as already emphasized in Chapter 2, there is now evidence of increasing spatial agglomeration at the functional level, and this is particularly visible in service industries which are far more affected by functional rather than by sectoral features (Defever 2006). This may provide an additional angle to look in dynamic terms at the taxonomical exercise provided above. For example, although it has so far only been examined at the national level in the context of the enlarged EU, the co-location of the different stages of the value chain of MNE affiliates shows that, rather unexpectedly, MNE headquarters do not display any pull effect on the location of any other MNE function (Defever 2006). This observation casts further doubt on the identification of knowledge spillovers on the basis of general location choices of MNEs.

5.6 TOWARDS THE EVOLUTION OF MNE AND SPATIAL TYPES

Clusters and regional systems are not necessarily static in that they may evolve over time. This evolution often involves the blending of the various features of the different knowledge-technology-cluster typologies into new and different combinations, and possibly also involves shifting from one dominant type to another, according to the relative stage in their life cycle. As we have already argued, in order to relate the issue of the local economic geography to the location behaviour of the MNE, what is important is to identify firstly which of these idealized spatial types best approximates the dominant characteristics of the locality being considered, and secondly also the technological, knowledge and organizational characteristics of the MNE.

5.6.1 Evolution of Spatial Structures

The arguments developed in the previous sections allow us to ask questions about how observed industrial clusters themselves may evolve over time and how such features may relate to the types of multinational firms that locate there. The reason is that, following our original Alchian (1950) arguments in Chapter 3, because of global scale, global network effects, and knowledge and technological advantages, the location of multinational investment into a region can stimulate processes of change within that region. In this respect, the interdependence of industrial and spatial dynamics in the current era of multinationalism and globalization depends on the balance between the internal and external sources of innovation or, in other words by the degree of *closure* versus *openness* of the innovation models at both the firm and spatial system level (Belussi et al. 2008). For example, on the one hand, in the case of the industrial complex and the two types of social networks, the extent to which firms within the cluster or region rely on internal (intra-firm) or external (extra-firm) innovation sources is largely determined by the technological regime conditions within the cluster, whereas in the case of pure agglomeration it is essentially space-driven. On the other hand, the extent to which the cluster or the region as a whole is inward or outward looking is largely influenced by the system's capacity for learning at the boundaries, and also by the system's potential for developing dynamic interactive capabilities and spreading its knowledge base outside its boundaries. The way in which the firms belonging to a particular cluster modulate their strategy, combining local and global sources for innovation and building intra-regional, interregional and international knowledge networks, is an indicator of the openness of the spatial system as a whole. An MNE's local presence can substantially alter such innovation and production models at the spatial level (Belussi and Sammarra 2010). This is because even in a largely closed innovation model in a particular location, multinational intra-firm knowledge sourcing will be external to the cluster, thereby influencing its evolutionary paths.

All of the taxonometric categories presented in Table 5.2 assume that local knowledge spillovers operate in some form, although they may vary significantly in both nature and scope across the cluster typologies. As such, the knowledge-technology-cluster categories presented in Table 5.2 therefore offer a framework for examining the interactions between the global and local location behaviour of MNEs, and how both inward and outward MNE investment into and from a locality itself may foster changes over time in the local economic environment. This is not to say that all industrial clusters and regional systems are directly involved in the creation of new knowledge or that the evolution of clusters is a simple linear and deterministic process, as these issues have to be established by observation and analysis. Indeed, as we have already maintained, innovative firms are not necessarily clustered together in space, and neither are clusters and regions necessarily innovative systems. Moreover, in reality it is likely that various mixed, diversified, and idiosyncratic patterns of cluster evolution will be observed, even within the same region (see also Menghinello et al. 2010). However, the development over time of industrial clusters and regional systems is crucially related to issues of knowledge, technology and innovation. The various possible evolutionary trajectories of these spatial transitions are depicted in Table 5.3.

A few brief well-known examples can be used here to illustrate the possible directions of spatial evolution (Box 5.1).

Once we examine the knowledge, technology and transactions-costs characteristics of each of the spatial typologies of our taxonomy, four broad insights emerge from the observation of these various selected examples:

- First, some particular types of evolutionary or transition patterns are much more common than others, some of which are very rare indeed. It seems thus possible to argue that the probability of some evolutionary patterns is higher relative to others, though this may vary over historical periods.
- Second, what also becomes apparent is that such evolutionary patterns are not specific to, or characteristic of, particular industrial sectors. Observing Tables 5.1, 5.2 and 5.3 together demonstrates that industries such as the semiconductor and electronics industry exhibit multiple possible cluster types, and multiple possible evolutionary trajectories. At the same time, depending on their industrial structure, clusters may well be engaged in a prevalent but not exclusive trajectory at any given point of time.
- Third, some simple stylized and popular analytical classifications have little or no explanatory power in terms of the evolution of cluster geographies or characteristics. For example, the traditional technology content dichotomy often employed in the literature which distinguishes between 'high-technology' sectors and 'low-technology' sectors turns out to have almost no explanatory power whatsoever. It is clear from Table 5.3 that the observed evolutionary features of many well known high-skills and high-technology sectors

Table 5.3 Evolution trajectories of spatial configurations: some examples

Origin	Destination			
	Pure agglomeration	Industrial complex	Competence-based SN	Trust-led SN
Pure agglomeration	Global cities	US and UK automobile industry	– Broadway Theatre – West End Theatre – Montreal video-game industry – Bangalore software industry	UNCOMMON Eighteenth century Scottish banking
Industrial complex	– Boston Route 128 electronics industry – Seattle aircraft industry – Los Angeles defence industry	– Global semiconductor (wafer and assembly) industry – Scottish electronics industry	UNCOMMON Nokia's networks in Finland	UNCOMMON West London media industry
Competence-based SN	Silicon Valley semiconductor industry (stage II)	– Toulouse Airbus industry – Sophia-Antipolis life-science industry	– Formula 1 London Hinterland – Emilia Romagna robotics and mechatronics industry	– Italian and Spanish ceramics industry – Sophia-Antipolis IT industry
Trust-led SN	Birmingham and Connecticut gun-making industry	Some *made in Italy* industrial districts in leather industries (Veneto and Tuscany)	– Silicon Valley semiconductor industry (stage I) – Cambridge (UK) electronics industry – Emilia Romagna life-science industry	Some small *made in Italy* industrial districts in clothing industries (Marche and Abruzzo)

BOX 5.1 EXAMPLES OF EVOLUTION OF SPATIAL TYPES

(1) As we will see in Chapter 7, the rise of the so-called 'global cities' – a topic that, as discussed in Chapter 2, was first touched on by Hymer – is associated primarily to widespread agglomeration effects. These cities have longstanding agglomeration advantages, which are actually reinforced under the current era of globalization. As such, over time the evolutionary transition process is one of a shift from pure agglomeration reaping the home-market effects within the national domestic economies, to pure agglomeration reaping home-market effects which operate across a multi-country or global scale. As we have seen above, the largest MNEs' headquarters are strongly and increasingly concentrated in the major world cities, where the local urban transactions and networks are strongly information- and power-intensive, and where the degree of connectivity is maximized in order to guarantee an efficient functioning of corporate global control functions.

(2) The early development of the global automobile industry on both sides of the Atlantic approximated to being something akin to pure agglomeration (Hall 1998; Boschma and Wenting 2007), but over time this system evolved to represent the classic industrial complex model we now see (Best 1990), which is dominated by large oligopolistic producers, clustered in particular localities, with complex and highly organized input-output supply chain systems (Markusen 1996). The evolutionary process of this location-specific sector has therefore been primarily from pure agglomeration to industrial complex.

(3) The transformation from a pure agglomeration to a competence-based network system as an evolutionary transition may probably be best represented by knowledge-intensive service industries within 'creative cities'. Some of the numerous examples here are the theatre industries associated with Broadway, New York, and the West End of London, and the video-game industry in Montreal (Grandadam et al. 2012). The initial advantages provided by the large urban agglomeration of Bangalore in attracting both MNEs and indigenous firms in the software industry, and the subsequent upgrading of both local and global competence, may be at least partially seen in this specific evolutionary trajectory (Athreye 2005; Chaminade and Vang 2008).

(4) Conversely, the transition from a pure agglomeration to a trust-based network is a rather uncommon evolutionary pathway. A possible example of this is the development of banking systems within Scotland, which emerged from heterogeneous small-scale financial competition into a locally-institutionalized system with many local and culturally specific characteristics.

(5) The shift from an industrial complex to something more akin to a pure agglomeration may be characteristic of the transition of industries previously dominated by an oligopolistic protective contract-based organizational system to a more openly competitive environment. Some observers contend that global competition and technological change will drive these types of transitions in the Boston Route 128 electronics industry (Saxenian 1994), and the same may also be taking place in the Los Angeles defence contracting and electronics industry (Scott 1991).

(6) The evolution from an industrial complex to even larger scale industrial complexes is obviously characteristic of industries such as petrochemicals and pharmaceuticals. However, it is also characteristic of many sectors of the electronics and semiconductor industry such as those based primarily in Japan (Arita and McCann 2002; McCann et al. 2002; McCann and Arita 2006) and 'Silicon Glen' in Scotland (McCann 1997).

(7) The move from an industrial complex to a competence-based network is a very rare evolutionary trajectory. One example is perhaps represented by the case of Nokia's domestic networks in the south of Finland (Ali-Yrkkö 2001).

(8) Similarly, the transition from an industrial complex to a trust-based network is also an uncommon type of spatial/industrial dynamics. However, a possible illustration of this is the development of a highly localized TV, film and media industry in West London, the logic of which is driven primarily by competition for contracts with the BBC. Many of the executives of the companies emerged from the BBC itself after the UK TV industry was deregulated during the 1980s onwards.

(9) The most recent stages of development in the Silicon Valley semiconductor industry – with particular reference to some subcomponents of semiconductor and electronics – would appear to be characteristic of the shift from primarily a competence-based

network to something largely characteristic of the pure agglomeration model, with highly competitive and flexible local market arrangements, which have emerged in quite different ways. For example, the Silicon Valley elements of this industry which have tended to focus on semiconductor design activities. Although the early post-war features of such activities in Silicon Valley were mainly typical of the trust-led social network model (Saxenian 1994; Hall, 1998), this industry developed during the 1970s along the lines of a competence-based social network, and has now emerged into something which is akin to a pure agglomeration (Arita and McCann, 2000, 2004; Saxenian 1994), exhibiting the supplier dominated characteristics of Pavitt's classification. The majority of the design innovations developed in Silicon Valley are made possible essentially because of the miniaturization innovations generated in the wafer processing and wafer assembly parts of the industry which are primarily located elsewhere, as components of global value chains with large MNEs flagships. As such, the evolutionary transition of the Silicon Valley cluster has been from trust-led to competence-based social network, and more recently to pure agglomeration.

(10) In terms of the evolution from a competence-based network to an industrial complex, one possible example of this transition is that of the Airbus industry based around Toulouse in France (Longhi 2005). This might also represent the evolutionary path of the life-science industry in the Sophia-Antipolis industrial cluster (ter Wal 2010). Sophia-Antipolis is one of the archetypes of successful European high-tech clusters and, for our purposes here, offers a very interesting case of how different evolutionary trajectories can coexist even within the same spatial structure. As shown by ter Wal (2010), over time the cluster has progressively specialized in information technology (IT) and, to a much lesser extent, in life-science. The development of the IT industry – after the crisis of electronics in the early 1990s – has been characterized by a shift from foreign MNE-led growth to mainly locally-based growth driven by technological-intensive spin-off and start-up SMEs established by the highly qualified labour left in the region by the relocating MNEs (Quéré 2007, cited in ter Wal 2010). Conversely, the life-science industry has remained through time dependent of foreign MNE subsidiaries, scarcely embedded into the region and in its dense localized knowledge

and learning networks (ter Wal 2010): the industry has not shown high dynamism since the mid-1990s, going through a shrinking process at the advantage of the IT sector, and displaying increasingly the features of an industrial complex model.

(11) Probably one of the most illustrative examples of a local competence-based network which has emerged into a global competence-based network is that of the Formula One racing car R&D cluster in the hinterland regions of London. Further evidence about this 'steady transition' can be found in some successful Italian industrial district cases, such as that of the Montebelluna sport industry (Belussi 2010), or the Reggio Emilia mechanical district (Ciapetti 2011). In this latter case, the district's transformation has recently occurred as a result of the rising specialization of local firms in the mechatronic industry – that is, the integration of mechanical, electronic and information-based components. Here the process of technological upgrading and integration in MNE networks has been described as being competence-enhancing and based on incremental and architectural innovation (Ciapetti 2010).

(12) The evolution from a competence-based cluster to a trust-based typology may be illustrated by some of the features of the post-War development of the ceramics industry in Sassuolo (Italy) and in Castellón (Spain). These two cases have shown evidence of inter-district trust-based relationships and mutual interdependence of social networks (Molina-Morales and Martínez-Fernández 2008). Also, as seen above, in the IT industry located in the Sophia-Antipolis cluster in the south of France, the emergence over time of collective learning mechanisms and the significant increase of local-local knowledge interactions (ter Wal 2010) may give rise to a shift towards a trust-led social network kind of spatial/industrial configuration.

(13) The transition from a trust-based system to something approaching a pure agglomeration model, although not really common, is characteristic of the changes experienced in the nineteenth century by the gun-making and small-arms industries in Birmingham in the UK and in Connecticut in the USA (Best 1990).

(14) The evolutionary path from a trust-based system to an industrial complex is characteristic of the changes exhibited – as a

consequence of globalization processes – by some traditional Italian industrial districts operating in *Made in Italy* industries, such as the leather *filières* located in and around the cities of Verona and Florence (Belussi 2010; Bacci et al. 2010).

(15) The transformation from a trust-based to a competence-based social network is typical of certain sub-sectors within the global semiconductor and electronics industry such as – as seen above – the early post-war development of Silicon Valley, the development of Silicon Fen around Cambridge UK, and of some electronics and life-science industries in Emilia-Romagna. Indeed, one particular aspect that emerges from the copious empirical evidence on the Emilia-Romagna case – one of the most celebrated examples of regional innovation system's upgrading and opening (e.g., Castells and Hall 1994; Braczyk et al. 1998; Hall 1998; Cooke et al. 1997; Belussi et al. 2008, 2010) – is the presence of multiples evolutionary trajectory that have blended together the two typologies of social networks identified by our taxonomy.

(16) Example of clusters which have maintained their trust-based system over time include again some of the small Italian industrial districts, such as the clothing and shoes sectors in and around the cities of Ancona and Teramo (Sammarra 2010).

Source: Adapted from Iammarino and McCann 2010.

do not point to any particular pattern of cluster characteristics or evolutionary trajectories. Indeed, the technological, knowledge and transactional characteristics of the clusters are what determines cluster and regional types, rather than technology content dichotomies, and it is the changes in these technological, knowledge and transactional features which determine spatial and industrial evolutions.

Similarly, the well-known cases of the Italian 'industrial districts' are far from being straightforward or homogeneous. For example, if we consider the transformation of the Italian industrial districts in the light of the process of internationalization in which outsourcing and offshoring becomes increasingly widespread, we may indeed argue that 'the industrial district has often proved to be rather a "stage" in one of the possible different paths of industrialization' (Becattini

1987, p. 32). As such, although the industrial district model is argued to come from the transformation of an original competence-based social network spatial typology (Paniccia 1998; Dei Ottati 2003), such a model now provides examples of evolutionary transitions from a trust-led social network to something which exhibits many of the relational characteristics of the industrial complex model. The only major difference here is the greatly reduced level of geographic localization of many of the local input-output linkages, which are increasingly transferred into global production networks (Garofoli 2003; Belussi and Sammarra 2010). On the other hand, there is some recent empirical evidence which seems to indicate that there are possibilities for mature industrial regions characterized by spatial agglomerations of the trust-led social network type to undergo a process of revitalization of technological capabilities towards new and research-intensive productions (Heidenreich 2005; Belussi et al. 2008, Farshchi et al. 2009). Where this transformation takes place it represents a shift away from a trust-led social network type of model to something which is closer to a competence-based social network.

- Fourth, there are some industrial cluster cases which are as yet rather difficult to classify. These are the cases of the newly emerging and rapidly changing industrial sectors such as biotechnology and multimedia (e.g., Swann and Prevezer 1996; Baptista and Swann 1998; Brouwer and den Hertog 2000; Coenen et al. 2004; Fuchs and Koch 2005). Many of the innovations within these sectors actually take place in large MNEs whose locational criteria primarily reflect those of the industrial complex. Yet, in situations where activities in these sectors are geographically concentrated amongst small and medium firms, they appear to correspond most closely to the competence-based social network type of system. The inter-industry spillovers that emerge from the local integration of a variety of knowledge sources, both internal (intra-firm and intra-cluster) and external (to firms and cluster), and of input-output networks, may be in these cases highly spurred by the presence of foreign firms. As such, the local competence-creating processes depend on the embeddedness of the MNE's local subsidiaries (Cantwell and Piscitello 2005; Tavares and Teixeira, 2006; Rabbiosi 2008; Kramer et al. 2011).

5.6.2 MNEs, Related Variety and Regional Hierarchies of Technological Excellence

Another illustrative example of co-evolution of technological profiles of firms and space can be found in global-local interactions for the creation of new knowledge. As is argued in Chapter 4, MNE networks for innovation seem to conform to a hierarchy of regional locations, both *within* and *across* national boundaries; a hierarchy which is subject to slow change over time. In such a context, as we have seen, if the imperative for the MNE to create global R&D networks has grown more pressing over recent decades, for industrial clusters and regions to reap the best technological opportunities it is necessary to be able to attract external knowledge flows related to the regional knowledge base (see also Menghinello et al. 2010), as well as to produce outward knowledge flows. The complexity of MNE activities is therefore associated with the variety of industrial clustering possibilities, and MNE firms' dynamics are increasingly intertwined with the technological and organizational characteristics of local contexts.

In line with the evolutionary economic geography framework, the sectoral patterns of regional technological activities gradually change as new industries develop and new technological linkages are forged between sectors. Specialization and diversification tend to move along a cumulative path in which the creation of new technological competencies depends on the pattern of advantages previously established. In other words, along with the related variety argument rehearsed above, diversification is one of the possible forms of incremental change in the composition of the regional innovation base, whilst in other cases regional profiles may be reinforced and concentrated in their established areas of technological expertise. Such forces will be a major cause underlying the evolutionary shifts from one main spatial typology to another, as depicted in our dynamic framework above.

Depending on the strategy followed by the MNE its interaction with the local environmental conditions for innovation may give rise to different paths of evolution of regional specialisation patterns. If the strategy of the large firm is to tap into localized specific capabilities in order to upgrade its own technological competence, this might add substantially to the strength of the sector in question in the host region and the technological specialization of the location is likely to be reinforced (higher concentration). If the MNE has developed a strategy for technological diversification through tapping into specific competencies in a range of different 'first order' locations (Cantwell 1994),[7] some of the latter may turn out to be comparatively advantaged by general external economies, therefore experiencing both an impulse to centripetal forces (geographical agglomeration) and a decrease in the degree of sectoral concentration (higher diversification) (Cantwell and Iammarino 2001).

As such, many MNEs may fit into regional profiles of specialization, thus supporting the process of local technological concentration. In contrast, depending upon the initial regional technological profile, MNEs may spur the diversification of the regional knowledge base. Yet, in spite of a high degree of technological diversity, some 'first order' regional innovation systems, and especially those characterized by the presence of more mature clusters of activity, which have become more narrowly specialized in their technological activities, might experience a slower process of convergence between old and new technologies because of a lock-in trend due to both the initial technological profile and the institutional environment. Other 'first order' regions which have broadened their profiles by moving into complementary trajectories and achieving institutional integration, might experience a faster process of convergence between old and new technologies, thereby reaching a greater potential for 'diversity for growth' (see Box 5.2 for an empirical example).

To sum up, local technological profiles in the principal regional systems of innovation are distinctive, and the fact that only some regions are able to adjust to the highest technological opportunities confirms that the location-specific nature of technological change implies heterogeneity and diversity, and eventually the rise and the decline of technological poles within a country or an area. As a consequence, a 'competitive bidding' between regional systems of innovation is likely to occur (see also Basile et al. 2008; 2009). If this is indeed the case, the advantages and disadvantages of competition and gains in competence may be more or less balanced out, depending on the capabilities, structures, institutional alignment and integration of the regional innovation system. In other words, it depends on the region's overall capacity and speed to react to major technological changes.

5.6.3 Evolution of MNE Organization

At this point we should consider why it is that different MNEs adopt different organizational structures and in which cases would particular organization structures be more appropriate. From the spatial typology arguments above it is clear that if an MNE wishes to internalize almost all of its own knowledge and to control the generation and exploitation of that knowledge very tightly, it is appropriate for the MNE to adopt a very hierarchical organizational structure such as is depicted in Chapter 3 in Figures 3.1 and 3.23. Where subsidiaries engage with other local customers or suppliers, these engagements will be framed within a system of primarily long-term legal contracts. In other situations, it may be the case that the parent firm wishes to maintain tight control over certain international operations relating to particular subsidiaries, while also allowing for more local autonomy and discretion in other subsidiaries which undertake specific local knowledge-based activities. A resolution to these

BOX 5.2 EU REGIONS AND MNEs: CHANGE, STABILITY AND STRENGTHENING OF TECHNOLOGICAL COMPARATIVE ADVANTAGES

The empirical analysis of the evolution over time of both technological capacity and sectoral specialization of different EU regional centres provides support for the hypothesis that a geographical hierarchy of regional locations can be established also *across* national boundaries in the European Union area. We test statistically the extent of stability in the sectoral profiles of technological specialization of the major regional centres in Europe, and examine what changes have occurred in these profiles between the 1970s and the 1990s.

Bearing in mind that the regions examined here are 'first order' locations – therefore showing a broad spread of innovation across a wide range of fields of activity – a regional trajectory may simultaneously follow different directions as between sectors (consolidating established strength in certain fields, while seeing it eroded in others), according to the initial geographical and sectoral conditions. For comparative purposes, the underlying sectoral composition of regional specialization has been considered in relation to shifts in the cross-sectoral distribution of comparative advantages as a whole. We have thus devised a taxonomy based upon two dimensions of regional technological evolution: stability and change, concentration and diversification.

The eight European core regions chosen for the analysis, with their respective regional shares in their own country of large MNEs USPTO patents due to local research facilities, are as follows: the South East in the UK (47 per cent), Lombardia in Italy (52 per cent), Baden Wurttemberg in Germany (19 per cent), Île de France in France (58 per cent), Flanders-Brussels in Belgium (79 per cent), South Netherlands in the Netherlands (63 per cent), the Stockholm-Ôstra Mellansverige region in Sweden (50 per cent), and the Basel region in Switzerland (58 per cent) (of which only the latter is not an EU region). Thus, the basic criterion for the selection of the 'core' technological region for each country was that it account for close to or greater than 50 per cent of the total US patents granted from 1969–95 to large firms for research in the country in question. This criterion was applied in every case except Germany, in which the cross-regional dispersion of patented research is much greater, such that no single region attains 30 per cent of patenting from national sources (Cantwell and Noonan 2001).

Overall, the results of the statistical and regression analysis support our hypotheses, indicating that trends towards a reinforcing or a broadening of specialization may simultaneously occur in the same large area (Europe) depending upon the regional location, its structure of comparative advantages and the underlying agglomeration forces. The regional cases show that three locations have registered broader patterns of technological specialization over the 27 years here considered (Baden Wurttemberg, Flanders-Brussels and South Netherlands), while for the other European regions the sectoral composition of their technological profile has followed a strong cumulative path (South East, Lombardia, the Basel region, Île de France and Stockholm-Ôstra Mellansverige).

In conclusion, the empirical findings show that in the European area, which has been influenced by a process of strong economic restructuring as a consequence of integration processes, two different paths may be observed over the last decades:

- 'First order' regions, especially those characterized by more mature clusters of activity, which have become more narrowly specialized in their technological activities, might experience a slower process of convergence between old and new technologies because of a lock-in trend due to both the initial pattern of specialization and the institutional environment;
- 'First order' regions, especially those in which geographical agglomeration of general economic activity has been strongly affected by EU integration, being also closer to the EU institutional core, which have broadened their specialization, might experience a faster process of convergence between old and new technologies, reaching a potentially greater competitiveness. Such a positive evolutionary path has involved catching up in some fast-growing and 'general purpose' technological fields, which are likely to lie at the heart of spillovers between indigenous and foreign-owned firms (e.g., ICT sectors – such as telecommunications, image and sound equipment, semiconductors and office equipment – and General mechanics).

Interestingly, in the regions which have experienced a reinforcement of their technological profiles over time, a greater contribution to this pattern of strengthening is due to non-EU MNEs, which are more prone than their EU-owned counterparts to tap into local fields of innovative expertise, hence being relatively more oriented towards asset-seeking, local competence-acquiring strategies. MNEs from other EU partner countries seem instead to have a relatively higher propensity to expand their own lines of technological specialization, assuming a more sectorally-oriented perspective in the process of rationalization of their innovative operations within the integrated area.

Source: Adapted from Cantwell and Iammarino 2001, 2003.

complex needs may imply a decision-making structure of the type represented by Figure 3.24 in Chapter 3, which allows for the possibility of a high degree of local autonomy on the part of the subsidiary located at L_{C2}, in order to take advantage of the knowledge sourced by participating in the regional social networks. At the same time, in the modern global economic environment knowledge is developing rapidly, and many aspects of knowledge are localized in particular competitive regions. Therefore, in order to rapidly respond to and best learn from local knowledge capabilities, firms may employ heterarchical organizational structures, as depicted in Figure 3.25 in Chapter 3. Such structures, however, are not necessarily the ultimate response to the current global economic environment, as they also imply very flat autonomous decision-making systems, which can themselves lead to tensions in parent-subsidiary relations (Asakawa 2001a, b). These tensions arise because of the conflicts associated with the fact that the parent firm and headquarters operations will often wish to retain the scale advantages of a hierarchical organization, while at the same time also reaping the benefits from the local knowledge gained via the relatively more autonomous subsidiaries. Following the arguments reported in Chapter 3,[8] under such conditions of conflicting goals and organizational stresses (Simon 1952a, 1959) MNEs may adopt satisficing (Cyert and March 1963) strategies, even though these may be sub-optimal from the point of view of the MNE as a whole. In other words, while on the one hand subsidiaries embedded in leading technological centres of competence (Cantwell and Janne 1999) may be sources of potential competitive advantage for the MNE, many of these advantages may actually remain unrealized due to the internal political structure of the MNE. As such, even if in purely geographical terms the spatial structure of the MNE is optimal, the overall ownership-location-internalization (OLI) structure of the MNE may not be optimal, unless the MNE's organizational and decision-making structure allows the firm to best take advantage of its geography.

The particular organizational structure exhibited by an MNE is also in part dependent on the time period in question. The arguments so far suggest that the organizational and spatial structure of an MNE often emerges over time, as the firm locates establishments in new places thereby expanding its own degree of multinationality. As we have seen, the reason why firms locate establishments in particular places is due to a complex mix of strategic issues relating to prices, outputs, risk and market competition, as well as due to the issues relating to knowledge generation, monitoring and diversification. This particular mix of issues implies that the spatial structure of an individual MNE often emerges in an evolutionary manner. As well as this, however, MNEs in general have also tended to exhibit different organizational structures in different time periods, depending on the technological and institutional context. During the emergence of modern multinational enterprises in the later decades of the nineteenth century and the early part of the twentieth century (Jones 1996; Chandler and Mazlish 2005), firms tended to exhibit fairly simple hierarchical organizational structures, as represented by Figures 3.1 and 3.23 in Chapter 3. For instance, the international emergence of US multinational manufacturing subsidiaries in Europe, and particularly within the UK, at the beginning of the twentieth century,[9] tended to be characterized by tightly-controlled hierarchical decision-making systems. In these organizational systems, the costs involved in coordinating activities over space and across national boundaries meant that each European country hosting such subsidiaries usually exhibited a national headquarters which was responsible for all the other subsidiaries within the same country, and which reported directly to the US parent headquarters. However, as the number of European establishments of US MNEs increased during the years after the Second World War, many of these firms developed more complex organizational systems, as characterized by Figure 3.24 in Chapter 3. These organizational developments were undertaken in order to allow for more autonomy on the part of the local European establishments, so that multinational firms could better take advantage of, and also respond directly to, the local specificities of each national market (Casson and McCann 1999). The high degree of heterogeneity in the economic and institutional conditions across the European countries meant that their specific market characteristics differed significantly, and that the effective management of such heterogeneity required a great deal of local knowledge, expertise and decision-making capability. Moreover, where MNEs attempted to develop a European-wide strategy within this diversified environment, the required complexity of the decision-making capability became magnified to an even greater extent. As a result, MNEs tended to move away from the simple hierarchical decision-making structure of the early decades of the century, towards more complex quasi-hierarchical and, in some cases, heterarchical structures.

From a theoretical point of view there are good reasons for these changes. As a firm's market becomes more specialized and its geography becomes more complex, the overall performance of a firm depends on its ability to appropriately respond to its various local economic environments. The ability of a multinational and multiplant firm to respond optimally to changes in its local economic environment will depend not only on its ability to successfully monitor factor cost and market price changes but also to react to them rapidly. This in turn depends on the level of subsidiary decision-making autonomy. As such, a firm's responsiveness to local environmental changes tends to increase as the decision-making structure becomes flatter and more decentralized. The desire to respond effectively to local environmental variations in part underpinned the general tendency for MNEs in the post-war years to change from simple hierarchies to more complex and in some cases flatter organizational structures.

In addition, the improvements in both transportation and communications technologies during the second half of the twentieth century, which will be discussed in Chapter 7, also facilitated the developments of these less hierarchical organizational systems. Indeed, over the last two decades, there is evidence to suggest that many MNEs now exhibit very flat structures (Birkinshaw 2001), as depicted in Figure 3.25 in Chapter 3. Obviously, certain types of multiplant MNEs are also systematically more disposed to flatter decision-making structures than others. For example, modern multinational firms in service industries such as retail (Kotabe 2001) or international real estate investment (Ball et al. 1998), which primarily undertake local market-seeking activities and rely to a much lesser extent on the inter-country coordination of corporate activities, would often be characterized by very flat corporate decision-making and organizational structures. This would also be true in cases where different establishments within a corporate multinational grouping operate with fundamentally unrelated technologies, such as is often the case within the pharmaceutical, chemicals or electronics industries. In general though, the most common forms of MNE organization structures can be represented by the quasi-hierarchical structures depicted in Figure 3.24 in Chapter 3. Moreover, within this grouping there is almost an infinite number of possible organizational arrangements, particularly where MNEs exhibit hundreds of subsidiaries in dozens of different countries and in scores of different sub-national regional locations.

The recent shift towards new organizational structures in which many subsidiaries increasingly act as critical knowledge-gathering establishments also implies that many of the simple centre-periphery assumptions of the product cycle model may no longer be tenable (Cantwell 1995). However, the fact that more subsidiaries may be gaining a relatively higher level of autonomy does not automatically imply that all MNE establishments will be located in so-called knowledge centres. From the above arguments, the actual economic geography of the MNE will also depend crucially on the organizational structure of the firm and its changes over time (Birkinshaw 1996). For example, a subsidiary may originally acquire a regional mandate where it is responsible for the coordination of activities with regard to a particular class of products, overseeing other subsidiaries in the same region. Eventually, however, it may obtain a global product mandate where its responsibilities become worldwide. Yet, such developments do not necessarily lead to observable location changes. Rather, as we have already seen, it is often the internal logic and organization of the activities within the network of the MNE establishments which is adjusted. Therefore, the geographical logic underpinning the MNE's location choice for any subsequent new investments will also depend on this emerging organizational system.

5.7 CONCLUSIONS

On the basis of the building blocks provided in the previous chapters, here we have proceeded to integrate the different micro- and meso-level perspectives of analysis by focussing on issues related to spatial clustering, technological accumulation, global-local interactions and knowledge networks. The critical role played by research-intensity, international expansion and the globalization of innovation was examined, and the broader relationships between MNE networks for innovation and regional hierarchies were discussed.

In this context we were then able to examine the characteristics of knowledge and technology and the dynamic relationships they create between firms and space. This allowed us to firstly build a clusters classification scheme based on the transactions among co-located firms, and then expand it into a broader and more comprehensive taxonomy defined according to the knowledge, innovation and transactions-costs characteristics of the cluster or region. These observed knowledge-innovation-transactions-costs categories of spatial and industrial configurations consequently provide for a more refined understanding of the rationale and situations in which multinational firms may wish to locate in particular locations. The crucial issue here is the extent to which the rationale and characteristics of the MNE 'fit' with those of the local economy, as it is this degree of fit which will determine the extent to which the immigrant MNE will foster locally-embedded innovative linkages, which lead to the generation of further knowledge spillover effects.

Finally, this extended taxonomy also allowed us to move some preliminary steps forward in uncovering the evolutionary trajectories of different cluster types, all of which may exhibit a major presence of multinational investment. As we see, this approach offers the greatest possibilities for assessing the degree of the fit between the MNE and its local environment, and evaluating the likelihood of the development of local innovative networks and the direction of the co-evolution process of both firm and spatial structures. More generally, the issue of how technology, knowledge, transactions and organizational relations together impact on the degree of fit between multinational investment and the local economy points to the need for, and importance of, a deep understanding of the historical changes underlying such coevolution processes. These historical evolutions are the issue to which we now turn in the next chapter.

NOTES

1. However, Jacobs herself, in recognizing the economic potential of regional diversity for growth, acknowledges that diversity per se does not necessarily evolve in 'diversity for growth' (Jacobs 1961, 1984).
2. Related variety is not defined in terms of input-output linkages among sectors. It is relevant to draw a distinction between the economic dimension and the cognitive one because business networks do not necessarily coincide with knowledge networks (see, for example, Giuliani 2007).
3. See Section 3.6 in Chapter 3 for the behavioural theories of firm location.
4. It is beyond our scope here to engage with the ongoing debate on the relevant spatial scale for economic analysis and, as a consequence, with the definition of 'cluster'. See, for example, Maskell (2001), Martin and Sunley (2003) and Moulaert and Sekia (2003).
5. Some empirical studies have applied industrial dynamics to regional economics, identifying determinants and effects across space of firms' entry (see, for example, Fritsch 1997; Dumais et al. 2002; Acs and Armington 2004a, b) and post-entry firms' growth (e.g., Swann 1993, 1998; Swann and Prevezer 1996; Baptista and Swann 1998; Audretsch and Dohse 2007; Boschma and Wenting 2007; Capasso et al. 2011; Neffke et al. 2011).
6. www.netlingo.com/word/siliconia.php.
7. See Box 4.3 in Chapter 4.
8. See Section 3.6 in Chapter 3.
9. See Chapter 6 in this book for a historical account of globalization and MNE expansion.

6. Globalization and multinationals in a historical process

6.1 INTRODUCTION

Any discussion dealing with issues relating to globalization faces the problem of what exactly is meant by the term. 'Globalization' is used on a daily basis in the popular press and all forms of cultural dialogue, and with regard to almost every social and economic issue we currently face. The term can be interpreted in a variety of different ways, carrying either or both positive and normative connotations, depending in large part on our perceptions and the vantage point from which we are considering these issues. It is not a single phenomenon or process, but a catch-all concept to describe a wide range of forces at work (Steger 2003; MacGillivray 2006). It has been defined very differently according to the social science within which it is applied. Already, almost two decades ago, Paul Streeten (1996) had, half in jest, collected the various definitions in the literature. One largely applied is that provided by Giddens (1990, p. 64): 'the intensification of world-wide social relations which link distant localities in such a way that local happenings are shaped by events occurring any miles away and vice versa'. Thus, one may think of globalization as a high (and increasing) degree of interdependency and interrelatedness among different and geographically dispersed actors and places. In principle, therefore, there may be a higher level of globalization even keeping constant the level of international transactions and operations. In other words, globalization and internationalization are not perfectly synonymous (Archibugi and Iammarino 1998, 2002; Cantwell and Iammarino 1998).

A first point of dispute has been to establish the overall importance of global forces in social life in the sense of the question – does true globalization exist or not? This in principle would require the identification of the different types of globalization and an estimate of their weight according to geographical location, industrial sectors, social groupings, and so on. In a nutshell, since the late 1980s/early 1990s there have been those maintaining that globalization has come effectively to dominate the greater part of economic life (Ohmae 1990; Chesnais 1994; Barnet and Cavanagh 1994; Brecher and Costello 1995; Perraton et al. 1997) and those who have been rather sceptical about its importance in quantitative terms (Ruigrok and van Tulder 1995; Michie and Grieve Smith 1995; Hirst and Thompson 1996). Another major controversy has opened up on the value-judgements attributable to globalization, such as whether a 'global' society is a good or a bad thing. Contrary to what has happened in popular debates whereby the nature of the issues at stake is left largely undetermined and vague – answers to this type of question could (and should) only be provided by clarifying the actors of reference. For example, for many people globalization immediately conjures up images of a form of international capitalist exploitation (Klein 2000, 2002), mediated by rapacious multinational companies greedily acquiring the resource base of poor communities. According to this largely negative view, globalization is therefore seen primarily in terms of the extent to which such communities are increasingly disadvantaged by the globalizing nature of capitalism. For other people, globalization immediately points to issues of global branding (Levitt 1983; Douglas and Wind 1987; Day and Reibstein 1997), whereby dominant corporations increasingly embed themselves in the cultural life of heterogeneous communities and identities, therefore encouraging a form of global cultural homogeneity dominated mainly by Western values. Once again, this view of globalization tends to be largely negative. For others, globalization conjures up images of the 'rise of Asia', and a gradual shifting of political and economic hegemony eastwards (Ferguson 2008). In terms of value-judgements, this view is based on an understanding of historical and political processes and is not bound to any particular normative assessment of those processes and their outcomes. Meanwhile, for those who enjoy or earn a living from travelling and tourism, globalization means cheap airfares and easy access to places that would have been unimaginable three decades earlier. This view tends to see globalization as being positive. Finally, for others, globalization immediately implies the communication possibilities afforded by information technologies such as e-mail, phone-text, Google, YouTube and Facebook, in which access to people, information and news sources across all locations around the globe can be instantly facilitated by electronic media. This view of the world converging to something akin to a 'global village' tends also to see globalization in largely positive terms, merging together the power of new technologies and the enlargement of markets and the intensification of all types of exchanges. The pace of globalization and that of technological change have indeed been strictly interrelated and, from a long-term perspective, it appears less important to establish which one should be considered responsible for triggering the other rather than to establish that they mutually enforced each other. For many decades there has been a circular process in which new technologies act as a 'lubricant' for economic and social globalization. In turn globalization, while facilitating the circulation of people, goods, capital and above all, ideas and knowledge, allows for the sustenance of a historically unprecedented rate of technological change (Archibugi and Iammarino 2002).

As we have discussed above and in the previous chapters, MNEs are in many respects the most important actors in globalization processes. Yet, the relationship between multinational firms and globalization is a difficult one to analyse. This is because the intra-firm connections and inter-firm relationships which constitute the essence of multinationals represent simultaneously both complementary and interdependent ways of allocating and organizing resources and assets across national borders, thereby contributing at the same time to the growth and change of both the firm and the external world. The problem here is thus to disentangle what should be considered exogenous from what is indeed endogenous to the firm. The traditional perspective on the 'home' and the 'host' countries of the multinational corporation implies a 'sectioning' of the firm, which divides its multifaceted nature and impact and separates it from the overall context of its activity as a multinational firm (Wilkins 1994, 2001). In what follows we will present a picture of the historical trends in globalization as a series of inter-reliant economic and technological *activities* carried out across borders, keeping in mind that the main agents performing such activities are multinational *firms*, which are the focus in this book. Such a historical overview is helpful in order to understand how firms, activities and their geographical locations *co-evolve* over time.

6.2 THE 'FLAT WORLD' THESIS

The view of globalization which focuses primarily on technological changes such as the Internet, mobile phones, and thereby also sees the world as rapidly becoming a global village, has been the focus of increasing popular discussion in recent years. This view of globalization was given an enormous boost with the publication in 2005 of Thomas Friedman's influential bestselling book *The World is Flat*, and subsequently updated editions (Box 6.1).

Even though it is never explicitly stated in his book, the motivation for Friedman's argument is actually primarily an economic geography argument. The idea that technology is making the world flat derives from the notion that geographical space and location are becoming increasingly irrelevant as arbiters of economic and social activity. The implicit belief underlying the general argument that the world is becoming flatter is essentially that of a dramatic fall of the transactions costs associated with

BOX 6.1 GLOBALIZATION ACCORDING TO FRIEDMAN

For Friedman globalization is the world becoming 'flat' as a natural and direct outcome of the technological changes associated with global information and communications technologies (ICTs). These technologies have not only changed the ways in which people are able to communicate with each other, but also have increased the number and the global distribution of people with whom we are easily able to communicate. This enormous increase not only in the level of mass-communication possibilities across the globe, but also in the speed of such communications implies that we increasingly observe a greater degree of similarity and an increasing degree of cultural homogeneity between people in different parts of the world. The reason is that increasing numbers of people are able to engage in dialogue with one-another. Friedman bases his argument on ten epoch-making technological and political phenomena which he interprets as having fundamentally and irreversibly changed global society. These ten exceptional phenomena are: the fall of the Berlin Wall; the advent of the world wide web; the development of work-flow software; the advent of uploading and file sharing; the advent of outsourcing; the rise of offshoring; the development of supply-chaining; the rise of third-party dedicated internal logistics operations; the rise of information availability via search engines; and the development of wireless technology. Friedman argues that, since the Millennium, the combined impacts of these ten phenomena have rapidly been re-shaping the world in a way which is both qualitatively and quantitatively different from previous eras of globalization.

In Friedman's hypothesis, the three eras of globalization are characterized and distinguished primarily by the behavioural emphasis on the country, the firm and the single individual respectively. Through the lens of these apparently-changing behavioural perspectives, which Friedman sees as being made possible by technological and institutional changes, he then argues that we have already embarked on a new era of globalization by which the whole world is increasingly becoming flatter. As such, we would move towards a world with rapidly diminishing differences between people in different locations, the outcomes of which would be greater similarity, greater homogeneity, more equality, and more individual-based democracy.

Source: Friedman 2005, 2006, 2007.

overcoming space, as a result of technological and institutional epoch-making phenomena.

Yet, Friedman's argument is not entirely new. The idea that technological improvements mean that geography and distance have become of little or no importance goes back to the early 1980s (Warf 1995; Gaspar and Glaeser 1998), where commentators assumed that the overriding likely impacts of the emergent information technologies would be to promote dispersal away from areas of population concentration. During the 1980s these arguments were increasingly bolstered by widespread contemporaneous observations of a process of urban-rural shift, whereby increasing numbers of people and firms appeared to be moving away from major cities and into smaller cities (Fothergill et al. 1985). In 1992 O'Brien (1992) announced the 'end of geography' and by 1997 Frances Cairncross (1997) had also proclaimed the 'death of distance' (Cairncross 1997); they were followed, among others, by the 'weightless world' (Coyle 1997) and the 'vanishing of distance' (Reich 2001). As we have already noted, although to most readers of Friedman's book these assumptions and arguments immediately strike a chord and therefore seems completely reasonable, from the perspective of economic geography many of Friedman's assumptions and arguments about globalization are entirely wrong. In particular he makes one critical error in his thesis. For Friedman it is unambiguous and obvious that technology has reduced the costs of engaging in activities across space, thereby flattening the world. For urban economists and economic geographers, however, things are far from being so straightforward. Spatial transactions costs have not fallen over recent years, but instead they have changed, and in many cases they have actually increased. These changes are transforming the relationships between firms and geography. In particular, the relationships between multinational firms, cities, regions and countries are changing both dramatically and rapidly, and from the perspective of economics these transformations are far more profound than any changes brought by YouTube or Facebook (Christopherson et al. 2008; Stiglitz 2002; Storper and Venables 2004, Steingart 2008). Moreover, as we will see over the following sections, the available empirical evidence suggests that many of these changes are actually diametrically opposed to the 'death of distance' conclusions. In particular, rather than becoming flatter, in many aspects globalization is nowadays making the world more uneven, more curved (McCann 2008), or even more spiky (Florida 2005a, b; Rodríguez-Pose and Crescenzi 2008).

In terms of economic geography and multinationals, why is it that the 'flat world' thesis could be so wrong, given that so many people accept it and are influenced by this line of reasoning? As we will see, the answer is that while technological changes have indeed affected the behaviour of individuals and corporations, the changes actually occurring are in many ways very different to those posited by the supporter of the flatness thesis, and there are two main reasons why this is so. Firstly, one basic problem with the flatness argument is that the analytical perspective and the objective of the analysis are often somewhat conflated in this logic, leading to many problems of both analysis and interpretation. For example, as we will see shortly, while institutional and technological changes are indeed causes of evolution in the nature of globalization, phenomena like outsourcing and offshoring are actually responses to such technological and institutional transformations, and not in themselves causes of changes in globalization. Secondly, many of the types of behavioural responses considered in these arguments are simply the responses of individual people, and in particular their leisure responses to the possibilities afforded by technology. As we will also see shortly, however, the employment responses associated with these technological changes are fundamentally different to these posited leisure-related observations.

In order to understand these changes, and to see how and why they differ so profoundly from the flatness characterizations, it is first necessary to understand the main features of economic and technological globalization as being the result of a long-standing historical process, the characteristics of which have evolved over different periods. The reason for this is that in order to get a sense of where we are today, and the outcomes of this process in terms of the current logic of our worldwide exchange architecture, our international economic order, and the global geography of production and consumption, it is necessary to understand these historical processes and to consider the major features of contemporary globalization from a historical perspective. Our analysis of globalization will therefore focus on the interrelationships between technological, institutional and organizational changes, as they impact on the roles played in globalization processes by countries, regions, cities and multinational firms. One of the aspects which will become apparent is that there was indeed a change in these relationships during the inter-war years of the twentieth century and, in that sense, Friedman's historical characterization of the eras of globalization has some credence. However, we will also see that, in terms of economic geography, the processes of fundamental change in the economic relationships associated with the most recent era of globalization are leading to almost completely opposite outcomes of the flat world thesis.

6.3 GLOBALIZATION AS A HISTORICAL PROCESS: INDUSTRIAL REVOLUTIONS AND TECHNOLOGICAL LEADERSHIP

Globalization is not a new phenomenon, in that the processes of investment internationalization and economic interdependence have been taking place over several centuries (Bordo et al. 2003; Steger 2003). However, the date at which globalization began is debateable, and there are various possible departure points which we may point to as being critical junctures in this long-standing process (McCann 2008). Early developments along the road to globalization include: the invention of double-entry book-keeping and the banking systems which emerged in the fifteenth century Italian city-states of Florence, Venice and Siena; the Dutch capitalist dynamic (Roubini and Mihm 2010) following the invention of the joint-stock company in the Netherlands in the early seventeenth century (Ferguson 2008); the growth of underwriting and insurance markets in the seventeenth century financial markets of London; the advent of industrialization in eighteenth century Britain; and the subsequent industrialization of other parts of the world (Ferguson 2008).

Variations in the dates by which we might define stages of globalization are not of themselves a problem. Different historical characterizations are essentially a result of different theoretical perspectives. Moreover, differences are likely to be driven not only by the variety of analytical methodologies but also by that of the objectives of analysis. Different commentators have defined the process of globalization using different time markers (O'Rourke and Williamson 1999; Crafts and Venables 2003; Steger 2003; Crafts 2004; Baldwin 2006; MacGillivray 2006; Maddison 2006), depending on the particular aspects of globalization they were concerned with. For example, scholars focussing on global migration behaviour (Hatton and Williamson 2005) would perceive the 1850s and 1890s as critical break-points in globalization eras, driven fundamentally by individual concerns rather than any questions related to country or company. Banking historians would argue that the validity of the network between individuals was paramount right up until very recent times, with issues of country or company being of little relevance. Similarly, monetary historians would stress the ways in which changes in the nature of credit have led to distinct eras in the globalization processes. As such, the Darien adventure (Schama 2006), the South Seas Bubble, the abandonment of the Gold Standard, the 1930s liquidity crisis, the 1960s rise of the Eurodollar markets, and the 1984 Big Bang in London (Casson and McCann 1999), would all point to pivotal changes in the eras of globalization, none of which are fundamentally about either countries or companies. Other vantage points from which to view the history of globalization, such as the history of the oil, transport and energy industry, the history of technology, the history of the nation-state, the history of imperialism, slavery, human rights, gender, culture, politics, not to mention Marxist, structuralist or military histories, would also suggest quite different critical break-points which define stages of globalization.

From our perspective of multinationals and economic geography there are three major aspects to the historical processes of globalization that we need to consider. These are the relationship between countries and globalization, between multinational firms and globalization, and between regions, cities and globalization. In a sense, each of these issues relates to the question of scale, and globalization is fundamentally a question of scale. Yet, what exactly we mean by scale is itself not so obvious when we consider the relationship between MNEs and globalization, and the reason is that the existence of multinationals implies that factors are able to move across national borders. The issue arises when we consider the recent citation for Paul Krugman's 2008 Nobel Prize in economics. Krugman's research focussed on two aspects of economies of scale. The first aspect is the role played by economies of scale in determining the nature and pattern of international trade in situations where production factors are immobile between countries; the second aspect is the role played by agglomeration economies in shaping the economic geography of production in a situation where production factors are mobile between locations. From the perspective of MNEs and globalization, these two different aspects of economies of scale lead to another fundamental question: in terms of economic performance, is the scale of a country more important than the scale of a city (region) or is the scale of a city (region) more important than the scale of a country? In order to deal with these questions we need to understand the processes which have driven and facilitated globalization historically. In the following sections we will first consider the technological and institutional changes which have driven globalization from a historical perspective from the early seventeenth century up until the late 1980s; then we will move on to discuss the main transformations that have characterized the most recent phase of globalization.

6.3.1 Nations and Empires

In many ways, from our perspective of MNEs and economic geography, the most appropriate date to consider as the start of the modern processes of globalization – in the sense of the rise of modern global capitalism – was the beginning of the seventeenth century with the formation of the first multinational corporations of the Dutch, British and French East and West India companies (Carlos and Nicholas 1988). Many scholars of the early globalization eras would perceive the trading company as being of paramount importance in understanding globalization, with the country playing instead a very limited role. In the case of the formation of the first joint stock company – the Dutch East India Company (VOC) in 1602 – it can be argued that the Dutch state played a major role (Ferguson 2008), whereas this was to a much lesser extent so in the case of the founding of the British East India Company in 1603. In general, however, once the monopoly trading powers were granted to these trading companies, their actual conduct during the seventeenth and eighteenth centuries was almost entirely outside of the remit and control of the state. These arguments would suggest that the role of the financial power of the very first genuinely multinational joint-stock corporations, in terms of their ability to marshal people, resources, and weapons, was far more important in driving globalization during the first two centuries of modern globalization than the role of the nation-state. As such, MNEs have been at the very heart of the globalization process right from the beginning of the modern era. The critical role of these companies in spearheading economic globalization continued apace through the eighteenth and early nineteenth centuries as international hegemony was slowly transferred from the Netherlands and Spain to France and Great Britain. During this period the expansion of the areas under their control by the empires of Great Britain, France, Russia and Austro-Hungary, as well as the US (Ferguson 2004), provided increasing resources and inputs into the national economies. The period spanning from the sixteenth to the nineteenth century represents the era in which almost all parts of the world came to be dominated to some degree by the large European powers of France, Netherlands, Spain, Austria-Hungary, Britain and Russia (Findlay and O'Rourke 2007; Maddison 2006, 2007a). In the case of Great Britain, by the end of the nineteenth century and the beginning of the twentieth century, eventually one quarter of the globe was contained in its Empire, in which the supply of raw materials was provided for by its colonies in return for manufactured goods produced primarily by Britain itself (Ferguson 2003).

During this period the modern notion of sovereignty which governed the international relations between these powers first arose, the establishment of which is usually attributed to the Treaty of Westphalia in 1648. The principles established by this treaty were that the state was viewed as exercising comprehensive (over all issues), supreme (no superior authority), unqualified (by any other state), and exclusive (no joint sovereignty) control (Easton 2007). Importantly, these principles were established at a time when – in comparison to today – the prevailing environment was one in which economic activity was primarily contained within an individual nation's own colonial systems (Findlay and O'Rourke 2007). This nation-empire structure is in marked contrast to the modern notion of the nation state, which is a relatively recent concept, and primarily a result of the nineteenth and twentieth century breaking up of the very empires which had driven the globalization processes for some three hundred years (Ferguson 2006). That this is so can be seen from the fact that only about ten of today's 190 or so nation states existed largely in the form they do now at the beginning of the nineteenth century, and only twenty or so existed largely in their current form even in the mid-nineteenth century (Easton 2007). Yet, this is not to imply that states have necessarily become smaller over time because of globalization processes. For example, the micro-state model of the German confederation of 35 monarchies and four free cities in the period 1815–1866 became obsolete by the mid nineteenth century, primarily due to the falling costs of distance and increasing regional trade (Easton 2007). Similarly, countries such as Italy, Australia, Russia, Canada and the United States all underwent processes of enlargement as access to the raw material inputs required by industrial systems became increasingly urgent. Institutional changes giving rise to the newly-unified and integrated countries such as Germany, Italy, Australia, Canada and the United States, can in many ways be a seen as a rational reaction to nineteenth century forces of globalization, which in some contexts called for larger states, and in others for smaller states.

Between the middle of the eighteenth century and the end of the nineteenth century it was the empire expansion of Great Britain which more than any other nation played the dominant role in spearheading globalization. The British economy above all was transformed by technological change, and only at the dawn of the twentieth century did the US economy overtake that of the UK in scale, as the westward expansion of the US provided for the resource and material inputs required for the American East Coast and Mid-West manufacturing industries. Yet, as the two dominant economic superpowers, it was the technological, financial and institutional interrelations between Great Britain and the USA which uniquely moulded the emerging forms and patterns of MNE business and globalization during the late nineteenth and early twentieth centuries.

6.3.2 The Emergence of Multinational Business

Unquestionably, the British economy was the first modern industrial system. The British industrial revolution of the mid eighteenth century was led mainly by mechanized textile production and the associated development of machinery and chemicals industries. By the first decades of the nineteenth century it was based primarily on the power of steam engines and the mechanization opportunities that steam offered which radically transformed the British economy. By the mid nineteenth century, Britain's economy was the workshop of the world with British output across all sectors accounting for approximately one third of global output (Cameron 1993). The raw materials required for this productive effort originated primarily in the colonial parts of the British Empire such as the West Indies, North America, and many parts of Africa. This resulted in a two-way flow of trade between Britain and its colonies, with Britain importing raw materials and exporting finished and semi-finished goods. The initial development of this global trading system exhibited three key interrelated but distinct features, without which the growth of this system would have been impossible: rapid technological change, the development of sophisticated forms of financial instruments, and the expansion of market areas.

The industrial revolution of the eighteenth century heralded an era of significant technological progress in which production innovations arose from both the employment of new forms of machinery and the increased specialization of labour activities. Modern commentators frequently discuss these features of economic growth from the rather narrow technical perspective of Adam Smith's observations of labour organization within the firm. However, these developments were taking place within a much larger context, which included the growth of input and output market areas (North 1981). This growth was necessary to provide both the input raw material required for the production process and an enlarged market for British industry's increased output. However, the expansion of these market areas itself also encouraged the development of more sophisticated financial instruments, which provided the debt and insurance capital required to facilitate colonial expansion by both individuals and trading companies. Early in the internationalization of the British economy, the London merchant banking sector had therefore played a key role in the development of global markets.

Historically, as mentioned above, the UK economy performed also a very particular role as the major link between the US economy and those in Europe. Against a background of common linguistic, cultural, and historical ties, the establishment of strong and stable economic ties between Great Britain and the United States was brought about both by increased mutual trading linkages and also by major flows of foreign direct investment between the two countries. The close economic integration engendered by these two-way flows of FDI meant that each country traditionally acted as an incubator for the industrial innovations developed in the other (Casson and McCann 1999). Yet, although the linkages between Britain and the US are critical for understanding modern globalization, during the nineteenth century Great Britain was the world's largest exporter of FDI capital in general (Jones 1996), and for most of the century the pattern of trade between the United States and Britain largely reflected the global trading patterns which were internal to the British colonial system. Raw material imports to the United Kingdom – such as cotton – were exchanged for industrial exports to the United States. As the US economy grew, however, its trading relationships with Britain began to change, and by the mid nineteenth century many of the technical and financial innovations of the British industrial revolution had been imported into the United States. These imports took place both by the straightforward imitation of British innovations in areas such as accounting practices and the formation of stock exchanges (Kaplan and Johnson 1987), and by the direct implementation of British techniques and industrial practise via British foreign direct investment in the United States (Casson and McCann 1999).

The American adoption and adaptation of many of these British industrial innovations was initially concentrated primarily in the north-eastern states, and the resulting productive capacity in no small measure contributed to the eventual outcome of the US Civil War. Yet American industry was also beginning to exhibit its own distinct character, based on a series of domestic industrial innovations. As such, an important marker in the emergence of modern multinational business is 1851, the year of the Crystal Palace exhibition in London. Following the Crystal Palace exhibition, the system of manufacturing arms on the basis of standardized interchangeable parts was the first American production innovation with a major influence in Britain (Best 1990, 1998). The subsequent American civil war economy, which required large-scale munitions production, not only encouraged the widespread adoption of these principles but also led to many further technical innovations within American industry. The two-way flow of ideas and innovations between British and American industry increased during the middle years of the nineteenth century. However, the British industry, which was based primarily on craft-based principles of production in which workers were paid according to skills rather than to the nature of the activity, would progressively become the imitator of American innovations during the second half of the nineteenth and early twentieth century (Hounshell 1984). As a result, the flow of ideas and industrial innovations became progressively more unilateral than bilateral, with Britain adopting and adapting American innovations within its own industrial context (Casson and McCann 1999).

Many of the ideas originated in the US refining, distilling, and tobacco industries, but their worldwide impact on production organization arose primarily due to their application in metal working industries, which dominated goods production in the early twentieth century (Guy 2009). The automotive manufacturer Ford first developed the automated production line at the turn of the twentieth century based on techniques already adopted in US slaughter houses, in which the continuous flow of materials was ensured by a conveyor belt. Soon other US industries adapted these ideas to their production, and the assembly line became the standard production model for many industries. Central to this system was the need for managerial coordination to ensure the smooth running of internal operations. This is because the whole system was vulnerable to individual breaks in the chain of activities. The requisite coordination of these large-scale activities therefore contributed to the second major industrial innovation with US origins, namely the development of a large managerial hierarchy. This came both as a response to the technical economies of scale afforded by increasing firm size in sectors such as railroads (Chandler 1977) and automobiles, and also as a means of establishing market control over production factors and inputs, particularly in sectors such as oil and steel. The principle of managerial control extended well beyond merely coordinating activities within the plant to include the control of the flow of inputs into the plant and, wherever possible, to integrate production, inventory, and marketing operations. The internal integration of all these activities was intended to ensure a stable market environment for a firm's inputs and outputs and to exclude any other external influences (Casson and McCann 1999).

The American principles of standardized production based on interchangeable parts required a change in labour practices, from the employment of traditional craftsmen to ones based on specialization, precision working, and measurement. During the last decades of the nineteenth century and the beginning of the twentieth century, however, the changes in production techniques and labour practices went much farther. The third major industrial innovation was the development of flow-line principles of production, in which groups of machines were laid out according to the order of the machining activity required, instead of according to function. This process innovation, which reduced internal inventory handling and transport costs, required inventory flows to be balanced at all points in the production process, affording major economies of time by increasing the speed of production throughput. However, genuine mass production only came about with the advent of the continuous flow principle, in which no two products were ever at the same stage of production on a single production line (Best 1990, 1998).

The development of the mass production system and the managerial hierarchy required to facilitate this model of production organization in many ways marks the second major industrial revolution of the modern era. Yet, these technological and organizational developments not only transformed production, but also transformed global organizational and investment possibilities. During the first era of globalization from the early seventeenth century through to the early twentieth century, large multiplant corporations operated almost entirely within the confines of their own national and colonial systems. Many British, French and Dutch commercial enterprises had multiple facilities and operations scattered across their respective empires as well as across their home countries. Similarly, by the second half of the nineteenth century many US firms were also rapidly expanding facilities and operations across the North American continent as the US empire spread west (Ferguson 2004). Yet, up until this period the economic engagement between different countries and empires was overwhelmingly dominated by exporting and importing operations, rather than by actual direct investment in each other's economies, which was still very limited by today's standards (Findlay and O'Rourke 2007). In the latter years of the nineteenth century and the eve of the twentieth century the growing income per capita of the dominant powers, which supported a steady expansion of product variety and substantial technological advances in terms of new products, processes, and improvements in transportations and communication systems, spurred both innovation activities and MNE expansions enormously. The extent of international operations went increasingly beyond simply commercial exchanges and capital flows embedded in foreign direct investment. The *degree* of multinationality started to extend both in terms of geography – that is, overcoming the boundaries of colonial empires and involving large corporations headquartered in a wider number of countries – and also in terms of functions – that is, involving more complex forms of control and business association with foreign operations, such as, for example, the diversification of innovation and technological activities in multiple locations outside the company's headquarters. The base for firms' long term profits began to shift away from market growth, trade growth, and capital accumulation, and became increasingly grounded on technological accumulation and organizational innovations, which gradually transformed production methods and systems. As highlighted by Schumpeter, such transformations relied upon the creation of locally specific technological competencies, embodied in the collective corporate capabilities of the large multinational companies (Cantwell and Fai 1999). Indeed, while during the nineteenth century the early profile of technological accumulation and corporate leadership in the most advanced economies was still, at least partially, mirroring national resource endowments (for example, US large firms showed advantages in woodworking and oil-related technology; UK firms in metal working and coal-related technology; German firms in artificial dyestuffs), at the turn of the twentieth century technological accumulation in large business companies started to follow a less predictable path, showing ever-weaker links with the natural resource assets of the country of origin (Rosenberg 1976; Cantwell 1989; Cantwell and Fai 1999).

Yet, in this changed economic scenario, it is largely incorrect to assume the (still nowadays) common view that the internationalization of production activities from firms originating in countries other than the UK and the US was a post-Second World War event (Jones and Schroeter 1993; Wilkins 2001). The evolution of the *modern* multinational corporation dates back to the last decades of the nineteenth century, although multinational firms from earlier times showed in some cases very similar characteristics to the modern MNE. Many continental European companies – a significant share of them representing still today the European version of the twenty-first century multinational corporation – had been active in international production well before the outbreak of the First World War (Franko 1974). These companies, operating mainly in chemical, pharmaceutical, electrical equipment and food processing industries, were primarily German (e.g., Siemens, AEG, BASF, Hoechst, Bayer) and Swiss (e.g., Ciba, Geigy, Brown-Boveri, Nestlé), locating their foreign operations principally, but not exclusively, in other European countries. In the period between 1870 and 1914 the ranking of the countries headquartering larger numbers of companies conducting business abroad saw the UK in the first position, followed by Germany and the US. Immediately after these three leaders there were several European economies competing for the next position, including France, Sweden, Belgium, Switzerland, and the Netherlands, while further down in the rankings were Canada, Austria-Hungary, Italy and Russia (Wilkins 1998). In the inter-war period MNEs with headquarters in European countries further strengthened their presence on the global business scene (Franko 1974).

One of the major impacts of the scale economies generated by mass production was that the role which large companies, and in particular large US companies, played in globalization processes, started to change. In the early decades of the twentieth century multiplant investment in manufacturing evolved from being confined within a broadly colonial type of transnational system, to one with a more genuinely international nature (Jones 1996; Chandler and Mazlish 2005). At the same time, the growing success of the American firms led to much imitation on the part of Britain, as well as the other rapidly industrializing economies in Europe and elsewhere (e.g., Japan). The three American firm innovations of standardized production, managerial hierarchy and flow-line principles, which eventually culminated in mass production techniques, were widely copied. These technological and organizational developments were also reflected in terms of the global rise of British, US and other large European manufacturing firms, most of which for the first time began expanding outside of their own national and colonial systems during the first three decades of the twentieth century (Dunning 1988c; Jones 2005), thereby emerging as genuinely global corporations and becoming truly multinational enterprises.

As we have seen above, British imitation of American industrial innovations had started from a small scale in the mid-nineteenth century, and increased during the following decades, resulting in a bilateral process of technology transfer. By the inter-war years of the twentieth century, however, the emergence of the United States as the world's technological leader implied that the flow of technology transfer had reversed and had become largely unidirectional. British industry's imitation of US innovations also took place via two distinct means. The first was the copying by British firms wherever possible of the characteristic features of the American model of industrial organization, as had been the case for small arms manufacturing in the previous century. The second and more direct method of technology transfer was the introduction of American industrial characteristics into Britain by the immigrant US firms, which adopted a variety of FDI strategies, from mergers and acquisitions to direct greenfield investments. The relative importance of greenfield FDI as the primary mode of US–UK technology transfer increased enormously with the huge expansion of US-owned subsidiaries into the United Kingdom during the inter-war years. This expansion was initially spearheaded by the US automobile firms, but was also closely followed by US consumer goods producers in many manufacturing, food processing, and pharmaceuticals sectors (Dunning 1988c). As well as linguistic, cultural and institutional ties, the laissez-faire attitude of the UK government towards investment and labour relations was a key attraction of Britain as a base for US investment in Europe, as many American firms sought to establish facilities in Britain as a first step in becoming genuinely multinational in nature (Dunning 1988c). The scale and technological superiority of US inward manufacturing investment in the UK, which slowly extended to other countries of Western Europe, meant that the developments in international business and multinational operations up until the Second World War were largely associated with US manufacturing industries.

As well as American production systems, other aspects of the US economy including features of the American regulatory institutions, such as the antitrust agencies, were also imitated by the emerging industrial economies of Europe and Japan. Yet, although these economies came to resemble the US model to a certain extent, they were far from replicating the American system. Part of the reason is that, unlike the US, many of these other countries relied fundamentally on world markets for inputs. At the same time, in these economies the state was explicitly involved in protecting certain domestic industrial interests and the sponsoring of preferred industrial alliances. To some extent, these types of interventions were a response to the perceived need for national self-sufficiency, particularly within the military-industrial sphere. Contrary to the US experience, however, they also reflected different views on the role of the state in industry, in that in countries such as the UK, France, Japan, Italy and Germany the state played a key role in shaping corporate relations and behaviour (Casson and McCann 1999). Through the internal integration of all activities within their managerial hierarchies, American firms excluded any external influences that might cause instability in their productive environments, such as from financial markets or government. This meant that in the mass production sectors, from the turn of the century onwards, US corporate expansion into international markets generally took place via internal funding arrangements and without any government assistance. Apart from corporate culture, however, this was also because national governments were simply too small relative to the major US corporations to exert any meaningful influence over them. Although the mass production economies that later developed in other countries were ostensibly imitations of the US economy, their organization and functioning were often quite different, in that both banks and, in some cases, governments played key roles in the development of inter-firm relations (Piore and Sabel 1984).

Indeed, in line with the theoretical views of Ownership (O) advantages discussed in Chapter 2, home country systems and characteristics are critical for explaining the important differences in the evolution of modern MNEs in the pre-World Wars period. For example, German companies – that by 1914 ranked second after the US in world manufacturing (with British firms occupying the third place) – relied mostly on high-skilled and research-intensive productions drawing on a vast supply of highly qualified personnel. This was reflected in the German leadership in high-technology industries such as chemicals and electrical, allied with its relative disadvantage in mass production techniques, which instead required unskilled labour and were far more advantageous to the American widespread substitution of capital for labour. The extent of business opportunities in the domestic economy relative to those abroad is another major difference. In 1914 the US was by far the largest recipient of world FDI, as its potential for internally-generated profits was the largest in absolute terms. This was also mirrored by the fact that large US companies had always much larger shares of their business operations at home than outside the national boundaries; whereas this was not the case for some of the small (and sometimes even of the larger) European headquartered companies (Wilkins 1998). National and regional (i.e. sub-national: Wilkins 1998, pp. 10–11) education and scientific training systems were also at the core of companies' strengths in expanding abroad, as well as, as highlighted before, the financial and banking system. For example, the strong and overarching presence of German banks in economic activities extended and supported business operations far beyond the domestic sphere, thereby playing a key role in the evolution of German MNEs in the first decades of the twenty century. Likewise, large banks had a crucial influence on the internationalization of both French and Belgian companies (Wilkins 1998).

In marked contrast to the experience of manufacturing industries, early multinational developments within the financial service sectors were in general much smaller and very different in nature to those of the manufacturing sector. From the early part of the eighteenth century onwards, London had been the centre of global finance, serving markets all over the British Empire, as well as markets in both North and South America. The worldwide developments in the manufacturing industries were to some extent mirrored in the financial industries, and obvious examples of these developments were the merchant banks, whose roles in global engagement expanded rapidly from the late nineteenth century onwards. As such, the rise of British and American merchant banks, such as Rothschild, Barings and JP Morgan, started to transform capitalism from what had been primarily a mercantilist and colonial system into a genuinely global market-based system transcending the individual national colonial systems (Ferguson 2008). Yet, on the eve of the Second World War, the UK and other European financial systems were still more or less closed to external influences, with almost no foreign ownership of their domestic financial institutions. The tight regulatory systems for domestic financial services evident in the UK and the US were the norm for all countries prior to the Second World War, and the closed nature of many of these systems was strengthened in the immediate post-war era with the Bretton-Woods architecture. As such, the relatively closed nature of the domestic financial sector meant that, while the London markets were used to raising finance for many global investment opportunities, the effects of the globalization process itself on the behaviour and governance of the British financial services sector was still very limited until well after the Second World

War. In fact, almost all domestic service industries in both the US and European countries, and in particular the financial services industry, were largely unaffected by global competitive developments until the 1960s. In other words, globalization only really began to have impacts on US and European services industries almost half a century after those within the mass production manufacturing sectors.

6.3.3 The Post-Second World War Order

The initial stages of genuinely multinational corporate expansion from the late nineteenth century through the early decades of the twentieth century were altered dramatically by the three catastrophes of the First World War, the Depression and, even more fundamentally, the Second World War and the post-war reconstruction era. These disruptive events had profound effects on the world economic order, as is clearly reflected in the data.

In the period 1870–1914, the levels of world trade and foreign assets to GDP rose almost continually, peaking just on the eve of the First World War. The ratio of trade to GDP at the worldwide level was 10 per cent in 1870, reaching 21 per cent in 1914 (Estevadeordal 2003). Such a trend was recorded in spite of the gradual return to protectionism experienced by the major economies in continental Europe since the end of the 1870s. Germany was the first to change its course by adopting the 'iron and rye' tariff in 1879: raising tariffs and industrial protection followed in France, Italy, Sweden, Russia, Austria-Hungary and Spain.[1] Apart from the United Kingdom, only some small countries, such as the Netherlands, Belgium, Switzerland and Denmark, remained relatively liberal. This may be explained, among other things, by the much lower vulnerability of these countries to the price and rent reductions which globalization had brought about (Daudin et al. 2008). In other words, world trade might have grown more slowly after 1914 than it did before even had the First World War not intervened (Daudin et al. 2008, p. 18). Despite such regional trade trends, the period 1870–1914 was certainly characterized by unprecedented integration in international markets for goods and, even more markedly, capital, and labour. According to Obstfeld and Taylor (2004, p. 55), the ratio of the world stock of foreign direct investment to global GDP was 7 per cent in 1870, but it reached nearly 20 per cent in 1914. The same ratio declined to only 8 per cent in 1930, 5 per cent in 1945, and 6 per cent in 1960. The 1914 figure was not attained again until the late 1970s (Crafts 2004)[2]: the ratio of foreign assets shot up to 25 per cent in 1980, reaching 49 per cent in 1990 and 92 per cent in 2000. As pointed out by Obstfeld and Taylor (2004), capital market integration has traced out a U-shape over the past 150 years, with late nineteenth century integration being followed by inter-war disintegration, and a slow move towards reintegration in the late twentieth century. As such, in contrast to the previous three centuries, much of the twentieth century was actually characterized by long periods of non-globalization, a feature which was really only reversed again rather recently.

During this period, the British Empire and the other European colonial systems were largely dismantled, and US industrial hegemony within the Western capitalist world became unquestioned (Ferguson 2003, 2004). Although there had been previous periods of unemployment, and most notably in the post-Napoleonic era, the 1930s Depression had brought with it the first widespread experience of unemployment, as understood in the modern Keynesian sense. Despite the British origins of Keynesianism (Keegan 1993) during the 1930s, the British government had actually been much less proactive in implementing policies to counter the economics and social effects of the Depression than the New Deal initiatives proposed by their American counterparts (Galbraith 1987; Cameron 1993). US policy-makers had concluded that excessive competition within the financial markets had in part contributed to the crisis. Therefore, they encouraged rigid segmentation within their domestic financial markets, in terms of both market activity and market location, in order to promote security over competition, while underpinning the whole system with a government guaranteed deposit insurance program (Coleman 1996).

At the same time as restructuring its domestic financial markets, the US also lead the way in the post-war restructuring of international transactions. The Depression experience had led Western policymakers to a growing belief that a liberal international financial system – such as was the gold standard of the period 1870–1914, where monetary policy was subordinated to the goals of capital mobility and fixed exchange rates (Taylor 2006) – would undermine the new emerging role of the state in influencing domestic economic conditions. As welfare expenditures grew, governments required domestic households and firms to keep their funds at home in order to provide a strong domestic tax base (Helleiner 1994). Thus, policymaking was focussed on expanding world trade without engendering financial instability: capital mobility was sacrificed to keep currencies on 'adjustable' pegs to the dollar and yet preserve the autonomy of monetary policy (Taylor 2006). The Bretton-Woods negotiations of 1944, and the resulting 1946 Articles of Agreement, provided for the setting up of the International Monetary Fund (IMF) and led to the creation of the General Agreement on Tariffs and Trade (GATT) as a negotiating forum for tariff reductions and trade rules.

The key elements of the Bretton-Woods agreements were fixed exchange rates backed by gold and the free convertibility of currencies. These were designed to avoid unilateral competitive devaluations and the implementation of increasing tariff levels. With its many regulated markets at the micro level, institutional labour negotiation at the national level, and Keynesian fiscal policy to control aggregate demand at the macroeconomic level, the reconstruction era economies were almost all characterized by stable growth, low inflation, and negligible unemployment. During this period, US manufacturing and financial firms were largely independent of foreign markets, and the strong performance of their overseas subsidiaries was largely reliant on the buoyant economies of their trading partners (Piore and Sabel 1984). Similarly, the stable and strong performance of the US economy encouraged high levels of inward FDI into the country from other Western economies. However, for the purposes of domestic economic policy and regulation, within the Bretton-Woods system these capitalist economies were still viewed as largely separate entities, each with its own distinct characteristics. In all European countries the major clearing banks, mutual funds and building societies continued to operate within a domestic system of cartels, and most of the Central Banks acted both in a supervisory role and as representatives of industrial interests in the government arena. These primarily domestic formal and informal modes of governance were typical in all Western economies. For nearly a quarter of a century, therefore, the stability of the Bretton-Woods system appeared to insulate the domestic economies of Western countries from the vagaries of international financial markets, thereby allowing a primarily domestic focus for economic policy-making. At the same time, the stability of the system was also providing a stable environment which was ideal for FDI, and during this period American investment in Europe increased rapidly in manufacturing industries.

Following the mass production developments of the inter-war period, the industry structure of the post-war economy in Europe was principally characterized by large industrial firms and a marked division between the public and the private sectors. In the production sector wages in the large firm, particularly in mass production industries, were set through bargaining between corporate management and labour organizations. In the public sector, wage bargaining tended to track these private sector wages. Yet, there were also significant transatlantic differences in labour relations systems. The role of national unions in wage setting was far more marked in Europe than in the US, where the labour movement had always been more fragmented, and the expansion of unemployment and social welfare provisions in European countries along with widespread nationalization had changed European labour relations. Pay bargaining in a wide range of industrial sectors therefore became institutionalized at the national level, and this ensured that the state in European countries retained a very different role in the economy compared to that in the US.

6.4 THE THIRD INDUSTRIAL REVOLUTION OF THE 1970s–80s

During the late 1960s, the stability of the international trading and financial system provided by Bretton-Woods started for the first time to be undermined. The US dollar came under continuing downward pressure, due to increasing domestic inflation caused by the public expenditures associated with simultaneously funding the Vietnam War and increased welfare state measures. This downward pressure meant that the dollar could not continue to act as the world's reserve currency at existing parities. The result was that the Bretton Woods system collapsed in 1971 when the US dollar was floated. Shortly afterwards, the 1973 oil embargo by the Organization of the Petroleum Exporting Countries (OPEC) and the Soviet wheat deal led to worldwide commodity-price inflation (Piore and Sabel 1984). The combination of rapid inflation and floating exchange rates meant that commodity prices became both uncertain and independent of the performance of domestic markets. During the 1970s the inflationary effects of the oil crisis combined with the increased international competition thereby rendered much of the post-war forms of labour organization obsolete in many sectors. This was because the asymmetric effects of differential price rises between sectors meant that national pay bargaining across industries and regions became impossible for both unions and management. Moreover, nationally organized labour activities exacerbated the confrontations between unions and management. Therefore, the stable environment of industrial relations, corporate governance and domestic market demand which had been characteristic of the Bretton Woods era was ultimately shattered.

These changes in the global economy and the increasing influence of monetarist thinking in macroeconomics brought forth a new emphasis on deregulation and competition in both the real and the financial economy (Keegan 1993). This increased emphasis on deregulation and competition also encouraged firms and organizations, irrespective of their sector or nationality, to search for new management, production, and human resource practices. In manufacturing in particular, in order to adapt to the new competitive environment, firms had to find ways to increase the productivity of their workforce. However, this was not at all a straightforward matter. The reason was that in the previously dominant modes of mass production, corporate profits within the expanding markets depended primarily on the implementation of process innovations without any requirement for labour training beyond the necessary repetitive tasks. As such, labour productivity improvements over the previous seventy years had generally come about primarily via the investments in improved capital equipment, with little pressure on finding ways to increase labour productivity by adjusting work patterns or methods.

From the early 1980s onwards, the structure, organization, and behaviour of Western manufacturing and production industries started to change dramatically. These changes were initially driven by the decisions of many large multinational corporations not only to radically re-think the logic of their production systems, but also to undertake the rapid implementation of these changes both domestically and also in their global affiliates. Moreover, MNEs also required their suppliers to adopt similarly radical changes in production thinking.

The modern inspiration for these new ideas is not at all American or even Anglo-Saxon in origin, but Japanese. Although many of the innovations introduced by the Japanese industry had actually originally developed, at least in part, by American engineers who failed to find an audience in the United States (Horsley and Buckley 1990; Fransman 1990), the development and widespread application of these new technologies and organizational innovations was an entirely Japanese phenomenon (Freeman 1987, 1988; Odagiri and Goto 1993). Human resource management strategies and organizational systems involving labour market principles such as 'labour flexibility', 'quality circles' (QCs), 'continuous improvement', and 'total quality management' (TQM), were all imported into Western industry from Japan, and spearheading the importation and implementation of these ideas was the US automobile industry (Kodama 1995). Taken together, all of these ideas generally came under the overall umbrella concept of 'just-in-time' (JIT) manufacturing. JIT principles focus on the drive for production which is characterized by minimum levels of inventory-holding. These ideas originally derive from the US skyscraper construction industry of the inter-war years, and were refined by the US Marine Corps as it moved across the Pacific in the Second World War. During the early 1950s, however, this new production philosophy was adopted and adapted by Toyota as a way of re-engineering its overall production system so as to allow production capacity increases without the need for increased space requirement or land-holdings (McCann 1998; Fujimoto 1999). By the 1960s, however, the principles and practices common in the Japanese automobile industry had become widespread in almost all of post-war Japanese industry (Schonberger 1996; Nishiguchi 1994), and were largely credited with much of Japan's success at mitigating the effects of the oil-price shocks of the 1970s.

During the 1980s, these JIT and TQM ideas came to dominate the MBA academic literature, and consequently the management practices of those North American manufacturing firms which felt most threatened by the rapidly growing competition from Japan. Most of these US firms were MNEs, and their drive to reconfigure their production systems was also transmitted to their subsidiaries and affiliates both within and outside the United States. However, to avoid the problems of opportunism associated with managing supply chains using these minimum inventory production and distribution techniques, the widespread adoption of JIT principles led to a new emphasis on the fostering of more long-term inter-firm relations and a movement away from short-term market contracting based solely on prices (Oliver and Wilkinson 1989). One of the resulting insights brought about by JIT implementation was the increasing realization within business and management circles that economic geography and location were important competitive issues. JIT raised the profile of location as a competitive factor because the innovations introduced required much more inter-firm transactions between suppliers and customers than the more traditional Western production techniques. As such, they had the potential to significantly increase industrial transport and knowledge transactions costs (McCann 1998). This gave firms an incentive to concentrate their input supplier base as much as possible (McCann and Fingleton 1996) and to restructure their geographical shipment linkages as they rationalized their operations. Moreover, the new emphasis on geography as a competitive factor in the non-financial production sectors was a major psychological change in much of US and European industry. The relative stability and domestic protectionism of the Bretton-Woods era had fostered a myopic isolation in many sectors of the economy of Western countries, and it was not until the 1970s oil crisis that the effects of increasing global competition started to become fully felt. Indeed, it was only during the 1980s that many sectors of Western manufacturing industry first began to seriously consider the impacts of the emerging Asian competitive challenge. Very quickly it became apparent that these new competitive pressures at the global level most heavily affected the large American and European firms whose markets were more geographically diverse, and which were therefore more directly in competition with the emerging Japanese firms. Naturally, US and European firms tended to be MNEs. However, in response to the new competitive environment, as well as implementing new production techniques and managerial systems, many traditional vertically-integrated firms which had prospered during the Bretton-Woods era now for the first time in their history became increasingly the subject of mergers, acquisitions, or radical restructuring, as they sought to develop organizational structures which would facilitate the incoming technological paradigm.

The 1980s witnessed major inflows of Asian FDI, and in particular Japanese FDI, into the US and European economies, as these new immigrant MNEs sought increasing footholds in the world's largest markets. Many of the new production models therefore became increasingly embedded in both US and European manufacturing via the growing number of Japanese subsidiaries locating within the US and Europe during this period.[3] The geographical expansion of Japanese FDI was the outcome of complex interactions between economic, technological and political forces both internal to Japan itself and within the broader external global environment. The organizational structure of Japanese FDI showed substantial differences according to the broad geographical area targeted (Dicken 1988). Of all sectors, the firms which had the biggest transformative impact were the Japanese automobile makers, which quite literally brought the industry into a second era of transformation, following on from the first mass production transformations of the early twentieth century. This second era of transformation was led by radical and architectural technological changes such as the use of shared platforms between different vehicle models, the modularization of components, the development of component systems, all based upon the increasing use of electronics (Dicken 1998, 2007; Delbridge and Lowe 1998). In the automotive industry initially Japanese manufacturing investment abroad was initially heavily concentrated in neighbouring countries in East and South East Asia, where a composite intra-firm division of labour was developed. Since the end of the 1980s, Japanese major car manufacturers – Toyota, Nissan and Honda – were increasingly targeting North America and Europe following market-seeking strategies primarily in response to trading frictions (Dicken 1988). Outside of the US, it was the UK which became the major host location for Japanese manufacturing investment, and consequently the major area of implementation of these new production systems (Trevor and Christie 1988). As in the US, labour relations in the UK manufacturing industry operated under much more flexible labour laws than in other European countries. The new deregulated industrial relations environment allowed for both relatively lower manufacturing wages and a greater ease of hiring and firing. More importantly for the Japanese MNEs, however, was the fact that the flexible laws and the deregulated environment for labour relations facilitated the implementation of JIT production techniques based on firm-level, or rather establishment-level, labour negotiations, as was typical in Japan itself. As well as Japan, this new labour environment also became increasingly attractive for immigrant FDI from manufacturing firms from other emerging Pacific Rim countries, such as Korea and Taiwan. In order to maintain employment levels in the increasingly competitive and uncertain environment, UK and European trade unions therefore began to campaign actively for attracting inward investment, decisions about which were therefore becoming increasingly determined at the local level rather than at the national level.

Although the new production, value-chain, and labour management principles were initially adopted by the automobile producers in the early 1980s, they quickly became disseminated across almost all areas of manufacturing. Moreover, during the latter part of the decade the same principles became increasingly embedded in almost all areas of retail, logistics and transportation sectors. By the end of the decade, the manufacturing, retail, logistics and distribution industries of the Western and Asian advanced economies were already fundamentally different in nature and modes of operation and organization to the equivalent sectors in the 1970s.

In terms of international business, one of the major differences between the economic restructuring taking place during the 1970s and the 1980s and the types of restructuring which had occurred in the previous decades of the twentieth century was that the global competitive changes were now also impacting heavily on the financial services sector. As we have already mentioned, from the inter-war years onwards, financial service industries on both sides of the Atlantic were largely insulated from global competition by domestic regulations, and in the post-war era, by the Bretton-Woods arrangements. However, the highly regulated nature of financial markets started to unravel rapidly from the 1970s onwards. Even before the breakdown of the Bretton-Woods system in 1971 when the US dollar was floated, the first modern effects of globalization on the financial services sector of European economies had started to take place with the rise of the Eurodollar markets. However, such changes did not arise from changes within the domestic financial sectors of European economies, but rather from those in the domestic American financial sector itself. By far the major impact of these transformations was felt in the UK, which acted as conduit via which further structural changes in the global financial services industry were transmitted across Europe.

The City of London was a very attractive location for the offshore activities of American financial service firms because of the uniquely informal and permissive supervisory approach of the Bank of England. The combined effect of the need to avoid US domestic regulatory restrictions plus the informal regulatory environment of the City thereby encouraged the widespread immigration of US financial houses into London. The inflows of US firms also then encouraged other overseas-owned financial houses to locate in London. Apart from the financial services inflows of foreign investment from other European countries, by far the most important inward flows were from Japanese banks.

As well as the ending of US capital controls in 1974, the flood of petrodollars into London following the OPEC crisis provided further impetus towards global financial integration, and away from the protected domestic market characteristics of the City and other European financial centres. By the 1980s these challenges to the traditional organization of the financial markets were also accompanied by new regulatory challenges for the supervisory authorities, as the nature of competition in these globalizing markets was rapidly changing. Moreover, these challenges arose not just from the changes in market structure, but also from changes in the financial products themselves. One of the key aspects in that respect was that the boundary between securities markets and banking was becoming increasingly blurred. As such, policymakers in international financial centres needed to modernize and democratize banking regulation and supervision, so as to desegregate credit and deposit markets, to break up domestic financial monopolies, and to modernize investor protection systems (Coleman 1996). However, the restructuring resulted in even greater unanticipated inflows of short-term foreign capital (Miles 1992), which rendered monetary aggregates very difficult to monitor. Existing approaches to monetary policy became largely impossible to implement, forcing governments to focus primarily on the relatively blunt instrument of interest rate adjustments as the primary means of controlling inflation.

6.5 GLOBAL TRANSFORMATIONS SINCE THE 1990s

During the 1980s, all areas of North American, European and Australasian commerce changed fundamentally from earlier decades in terms of their modes of operation, the approaches to management, the nature of labour relations, the control of production and distributions systems, and the methods and approaches to market regulation. In all Western commercial sectors, the newly-emerging organizational, managerial, regulatory and supervisory challenges had been driven by the rapid rise of overseas competition, in particular coming from the Asian region.

Yet, much of the pressure for these changes had been accelerated and exacerbated by the shift to the new technological paradigm of information and communications technologies (ICTs). As we have already seen in Chapters 4 and 5, this new technological paradigm generated an enormous increase in both the complexity and speed of knowledge and information transfers, and of financial transactions. Both require and allow for new corporate organizational structures and relationships which previously had not been possible (Freeman and Soete 1997; Guy 2007), and which entail fundamental changes in spatial transactions, knowledge flows and firm behaviour. The traditional approaches to industry supervision within North America and European markets were therefore progressively becoming out of step with changes in the competitive environment, and these changes themselves called for appropriate governance and institutional changes at a much larger scale than previously envisaged. However, the sheer magnitude of the technological, institutional and organizational changes which were about to take place were far beyond the imagination of any market commentators in the 1980s, and in many ways, as we will see in the next paragraphs, the modern era of genuine globalization commenced right at the end of this decade.

6.5.1 Institutional Changes and Regionalism

In terms of institutional transformations, the years between 1989 and the mid-1990s saw various fundamental transformations in the institutional structure of the global economy. These changes were the entry of a series of countries into the global labour market for the first time, the growth in the number of bilateral investment treaties, and the creation of areas of economic integration at the macro-region level.

In terms of the expansion of global labour markets, following the political crisis in 1989 the process of deregulation and competition in the Chinese economy was massively accelerated, and the country for the first time in over half a century rapidly began to open up to global investment and market opportunities. The opening up of China brought some 760 million workers into the global labour market for the first time (Venables 2006). In the same year the Berlin Wall collapsed and within twelve months the planned-economy system of the former communist arena had almost entirely disappeared. The fall of the Berlin Wall marked the beginning of the end of the post-Second World War period polarized order, and the countries to the East of the former Iron Curtain re-emerged as 'transition economies', due to their process of radical transformation entailing political, economic, cultural and diplomatic transition (Rodríguez-Pose 2002). Taken together, the emergence of the transition economies brought some 260 million workers into the global labour market for the first time in over half a century (Venables 2006). Meanwhile, the second wave of fundamental economic reforms in India commenced in 1991, facilitating both competition and international investment in the Indian economy: such liberalization process added up another 440 million workers (Venables 2006).

Further liberalization and institutional reforms in Brazil, South Africa and Indonesia bolstered these global shifts. In total, around 1.5 billion new workers effectively entered the global labour market for the first time, implying an increase of the worldwide labour force by more than one third within the three year period 1989–1991.

The second major set of institutional changes which began to take place at this time was the rapid growth of bilateral investment treaties between countries. The number of preferential trade agreements doubled between 2000 and 2006 (UNCTAD). At the same time, there was also a rapid increase in institutional changes which dealt not just with free trade issues, but also with the liberalization of cross-border investment. Globally, the number of international investment agreements (IIAs) between countries – including both double taxation treaties (DDTs) and bilateral investment treaties (BITs) – had reached almost 5500 by 2006 (UNCTAD 2007). This number has increased dramatically from 900 in 1980 and just over 3800 in 1999 (UNCTAD 2000).

The third major institutional change which took place in this period was the establishment and development of areas of economic integration. In terms of global competition and regulatory issues, the development of free trade and common market areas are just as important as the rapid expansion of the global labour market. Indeed, economic integration has moved much faster at the macro-regional than at the global level, and in terms of macro- or super-regional neighbouring countries, the growth of international trade has been higher at the intra-regional than at the inter-regional level since the Second World War (Mansfield and Milner 1999). In this context, the creation of the European Union (EU) internal market in 1992 and the formation of the North American Free-Trade Agreement in 1994 are critical moments.

What was to become the European Union began in 1951 as a six-member organization of modest scope – the European Coal and Steel Community. By 1957 this had given birth to a customs union – the European Economic Community – of the same six states. From 1973 onwards it added new member states, reaching 27 by 2007. From the Single European Act of 1986, it quickly transformed itself into a single market for goods, services, labour, and capital, and took on many of the legal and regulatory functions previously reserved for member states. The initial motives for European integration were geo-political: for all, it was a way to bind the nations of Western Europe together in order to prevent another intra-European war; for some, it was also a way to restore Western European influence in a world then dominated by the US and USSR, the two titans of the Cold War. As the union grew both larger and more closely integrated, however, economic motives became at least as important as geo-political ones, as it became evident at the time of the accomplishment of the internal market in 1992. The main underlying motivations of the EU internal market were in fact the rise of international competition, first from the US and Japan, and then from the Newly Industrializing Countries (NICs) of East Asia and elsewhere, and the consequent fears brought about by a winners/losers outlook; the potential advantages offered by the win/win perspective connected with new technological opportunities; and the belief that market fragmentation at the national level was hampering economies of scale and scope (Rodríguez-Pose 2002).

The international climate at the end of the 1980s and the beginning of the 1990s was indeed very favourable to launching or expanding trade blocs, the result of which was the establishment of NAFTA between the US, Mexico and Canada in 1994, of Mercosur between Brazil, Argentina, Uruguay and Paraguay in 1991, and the very slow but real integration of the economies of the Association of Southeast Asian Nations (ASEAN), which was started in 1967 but only really embarked on a process of engaging economic collaboration in 1991. These institutional developments all responded to a similar logic of promoting trade and investment to rationalize and restructure intra-area markets and resources, spur innovation, and ultimately react to the competitive pressure coming from emerging economies (Hufbauer and Findlay 1996; Hufbauer and Schott 2005; Hufbauer and Brunel 2008).

Several other international organizations and institutional arrangements at the global and continental scales were founded in this period and were also aimed at liberalizing and supervising international trade and investment. Among these, the World Trade Organisation (WTO), which was established in 1995 as a replacement of the General Agreement on Tariffs and Trade (GATT) which had been founded in 1947, is the only really global free trade organization in terms of number of members (153 countries, with some 30 observers potentially future members) and representing more than 97 per cent of total world trade.[4] The Asia-Pacific Economic Cooperation (APEC) was established in 1989 to enhance economic growth and cooperation in the Asia-Pacific region through the reduction of tariffs and other trade barriers and the encouragement of investment across the area. APEC currently has 21 members including the whole of NAFTA as well as Japan and China, and accounts for approximately 40 per cent of the world's population, 54 per cent of world GDP and 44 per cent of world trade.[5] On a much smaller scale there is also the Closer Economic Relations (CER) agreement between Australia and New Zealand which was originally established in 1983 and achieved one of its most significant outcomes in 1990, with the almost total elimination of all tariffs and quantitative barriers to trade between the two countries.

All of these institutional developments promote trade and multinational investment because they allow, at least to some extent, goods, services, capital and labour to move more freely between countries. Today the three largest areas of closely integrated activity – though varying greatly in terms of degree of economic integration achieved – are the three 'super-regions' of the EU, NAFTA and the East Asia region dominated by China, Japan, Korea and Taiwan. The major outcome of such changes has been the increasing importance of these super-regions within the global economic system (Rugman and Verbeke, 2004b, 2007, 2008a, b). As we saw in Chapter 1, by 2005 these three economic super-powers had roughly comparable size economies: the gross income of NAFTA represented 32.6 per cent of global income, the EU 29.4 per cent, and South and East Asia 22.3 per cent; and together, the three regional blocks accounted for 84.3 per cent of global income, and 73 per cent of global trade (World Bank 2007).

Thus, globalization is largely dominated by the advanced countries, and particularly by the three regional blocks, to the point that, as we also noticed in the case of technological and innovative activities in Chapter 4, that many scholars now prefer to use the term 'triadization', instead of globalization, of the world economy. The spatial aspects of these trends have become even more complex and geographical proximity has become increasingly important for trade and FDI (UNCTAD 2000, 2007). Both the trade levels and FDI stocks between the US and Canada, and between the EU nations are much greater than what would be predicted simply on the basis of the scale of these economies (Krugman 2007), as is also true in the case of Japan and the other East Asian economies (UNCTAD 2007). The reason is that international exchanges and investments are highly associated with geographical proximity between adjacent or neighbouring countries. These groups of mutually very open neighbouring countries are here referred to as the broad macro-regions or super-regions, and this tendency for groups of countries located in the same parts of the world to develop stronger interrelations is increasing. The geographical patterns of, principally, double taxation treaties (DTTs) and, secondarily, bilateral investment treaties (BITs), closely resembles the cross-border patterns of FDI (UNCTAD 2003). The most striking case of this is again that of the EU. In terms of bilateral inward investment stocks, in 1995 17 of the top 50 pairs of countries were from Europe, whereas by 2005 this number had increased to 22 (UNCTAD 2007). This reflects the rapid increase in both EU economic integration and its enlargement to Central and Eastern Europe over recent years.

The development of regional institutions of economic regulation, policy coordination and political-economic integration, are also intrinsically linked to the organizational and technological changes that fundamentally affected the technical and spatial properties of production systems (Guy 2010). The degree of interrelatedness among the three types of structural changes is even more evident when the activities of MNEs are taken into account, as we will do in the next sections.

6.5.2 Organizational Changes and New Modes of Production

One of the major features of the emerging twenty-first century international economic order is that of the increasingly important role played by MNEs. The dominant mode of firm organization is steadily changing from domestic firms to international firms to global firms. As we saw in Chapter 1, these recent rapid increases in the weight of MNEs in the global economy reflect long-term trends towards the increasing openness and inter-connectivity of places, pointing to one of the major features of globalization already highlighted: namely that the growing economic interdependence between different locations is being increasingly mediated by multinational enterprises. Indeed, during the 1980s and 1990s, both the gross product of international production and also the gross sales of foreign affiliates increased much faster than either global GDP or exports.

In the short-term, however, multinational growth trends are highly susceptible to the business confidence conditions in the global economy. Foreign direct investment peaked in 2000, while falling sharply in the aftermath of the dot.com bust and 9/11. Global FDI inflows halved (UNCTAD 2003) and FDI flows remained sluggish until 2003–2004, after which their growth once again resumed (UNCTAD 2005). Global FDI grew by 24% in 2004 and by 29% in 2005 (UNCTAD 2006), reaching over $1.306bn by 2006 (UNCTAD 2007). Meanwhile, during 2005–2006, global sales, exports and value-added of the foreign affiliates of MNEs also grew by 18 per cent, 16 per cent and 12 per cent respectively, faster than in previous years (UNCTAD 2007).

Part of the reason why FDI flows are so susceptible to short term business confidence fluctuations is that they are increasingly dominated by service industries and mergers and acquisitions (M&As), reflecting the major changes in the Location (L) advantages pursued nowadays by MNEs. Indeed, empirical studies have shown that access to core markets, clustering, and the internalization of localized innovative capabilities represent the prominent motives underlying M&As in Europe, while institutional factors, such as the EU integration processes or language obstacles, are less significant (Zademach and Rodríguez-Pose 2009). In general, the literature on the spatial effects of M&As have indicated that M&As have not only deeply transformed the organization of business, but have also reshaped the locational patterns of corporate control and management, increasingly concentrated in large metropolitan areas (Taylor 2000; Duranton and Puga 2003; Rodríguez-Pose and Zademach 2006). Although varying significantly across industries, agglomeration economies, metropolitan connectivity, urban concentration, and to a lesser extent also geographical proximity, act all as the driving factors of the link between M&As and industrial and spatial structures (Rodríguez-Pose and Zademach 2006). In the words of Rodríguez-Pose and Zademach (2006, p. 306): 'Corporate takeovers and mergers constitute a key expression of the information-based and globalized economy of the late twentieth and early twenty first centuries'.

Part of the reason for the increasing importance of M&As in FDI flows lies in the relative growth of the service sector. During the last decade of the twentieth century and the first decade of the twenty-first century, the service sector has been the fastest growing component of the global economy (World Bank 2007). M&As are the major mode of international market entry for all forms of service sector FDI. By the end of the 1990s service industries accounted for over 60 per cent of all international mergers and acquisitions, with the EU alone accounting for 60 per cent of all service sector international M&As (UNCTAD 2004). In 1970 services accounted for one quarter of global inward FDI, by 1990 for almost one half (UNCTAD 2004, 2007), whereas by 2005 services accounted for almost two thirds of the total (UNCTAD 2007). At the same time, manufacturing's share of global FDI inflows fell from 41 per cent in 1990 to approximately 30 per cent in 2005 (UNCTAD 2007), whilst the share of primary industries is little more than 6 per cent (UNCTAD 2004). The rapid rise in the importance of service sector FDI has also been accompanied by significant structural changes within the sector. In 1990, 65 per cent of stocks and 59 per cent of flows of global service sector FDI was in trade and finance, whereas by 2002 these shares had fallen to 47 per cent and 35 per cent respectively (UNCTAD 2004), following the rapid growth of foreign investment in infrastructure, power generations and telecommunications industries (UNCTAD 2007).

While, as we have noted, FDI is highly contingent on the levels of international business confidence, most notable here is the fact that the numbers of mergers and acquisitions are very sensitive indeed to global business confidence.[6] Record numbers and values of M&As were achieved in 2000, but fell sharply after the worldwide collapse of business confidence in 2001 (UNCTAD 2003). The value of cross-border M&As dramatically increased in the period 2004–2007, at the end of which they accounted for two-thirds of all FDI inflows (UNCTAD 2007). Over the next few years, it is likely that the growth of FDI will be somewhat reduced from its 2006–2007 levels. Prior to the appearance of the US sub-prime market problems, expectations were positive, with a 54 per cent increase in the value of M&As in the first half of 2007 (UNCTAD 2007). However, global business confidence appears to have fallen markedly within a very short time, as strong uncertainties emerged regarding the levels of exposure of different components of the international financial economy to the US sub-prime mortgage market. In the previous recession following the loss of business confidence in 2001, the decline in FDI was geographically and sectorally uneven. Flows into manufacturing and services declined, whereas those into primary sectors increased (UNCTAD 2003). Similarly, the flows of FDI via M&As fell by more than greenfield FDI, and the equity and loan components of FDI fell by more than reinvested earnings. On the other hand, intra-regional FDI in South East Asia and North East Asia actually remained very strong while global FDI was contracting (UNCTAD 2003). A further impact of the global contraction in FDI was that many countries actually increased the pace of their liberalization policies in order to compete for shrinking stocks of FDI (UNCTAD 2003). The result of these observations is that there may well be short to medium term falls in FDI in many parts of the global economy, reflecting the short to medium concerns about economic growth in the aftermath of the global financial crisis of 2008 and the subsequent Euro crisis, and such falls are likely to be particularly noticeable in M&As and in high yield foreign investments.

In the long term the institutional and organizational developments outlined so far show no real signs of slowing down, and even though global business confidence is currently suffering, the long-term trajectory towards increasing international investment appears to be inevitable (UNCTAD 2012; Ghemawat and Altman 2011, 2012). Yet, one of the outcomes of these globalizing changes – as already highlighted above – is that, while the number of bilateral investment treaties has accelerated rapidly, as has the number of multinational firms in the global economy, the patterns of international investment agreements suggests that the individual regional trading blocs are actually becoming more important in developing the institutional frameworks for facilitating FDI (UNCTAD 2003).

This is clearly mirrored in the geographical configuration of one of the critical organizational innovations of current globalization processes. Global production networks (GPNs), as we have already seen in Chapter 4, depict the structural and relational nature of the current global organization of production, distribution and consumption of goods and services, and capture specifically the nexus between 'global' and 'local' (Coe et al. 2008). Although GPNs obviously have an intrinsically dynamic nature, and there is little or no comprehensive data on their evolution, they have become especially evident as forms of global production organization since the 1980s, when the modularization of production processes and the outsourcing of specialized functions, combined with ICT technologies, became a viable alternative to vertical integration (Ernst 1997, 2005; Guy 2009). The increasing relevance of the GPN phenomenon has shown that radical changes occurred in the last few decades in the organizational and governance relationships within and across industries at the global level, and, in turn, these relationships have affected the development and upgrading of opportunities for the localities and the firms involved (Coe et al. 2008).

As we saw in Chapter 4, the value-chains and production networks of vertical and horizontal linkages between MNEs and local small and medium enterprises at the global scale 'act as mechanisms to weave together various specialized industrial clusters, giving rise to a *network of clusters*' (Sturgeon 2000). In line with the 'heterarchical MNE' discussed in Chapter 3, the literature on flexible production has long pointed to the network organizational form as being superior in terms of efficiency, market response and innovation to the vertically integrated form of international production (Piore and Sabel 1984; Scott 1988; Walker 1988; Storper and Walker 1989). In such organizational structures rapid cross-border dispersion coexists with agglomeration and current globalization often occurs as an extension of regional clusters across national borders. This implies, firstly, that some stages of the value chain are internationally dispersed while others remain concentrated, and secondly, that even the globally dispersed activities typically congregate in a limited number of sub-national regions and clusters (Ernst et al. 2001).

Such profound organizational changes reflect increasing pressures to exploit complementarities that result from the systemic nature of knowledge. GPNs thus encompass both intra-firm and inter-firm linkages, create a diversity of network participants, link together multiple locations, and cover a variety of value chain stages, including higher-end and more knowledge-intensive activities (Ernst et al. 2001, 2002, 2010). The number of networks and the intensity of competition vary across sectors, reflecting their different stages of development and their idiosyncratic industry structures (Ernst and Guerrieri 1998; Ernst et al. 2001). As reflected in the figures reported above in this section, what has mostly characterized the last two decades is the fact that outsourcing and offshoring do not concern anymore only manufacturing, but also a variety of high-end, knowledge-intensive support services that are intrinsically linked with production. Even if these activities do not involve formal R&D themselves, they may give rise to considerable learning and innovation, such as trial production, tooling and equipment, productivity benchmarking, testing, process adaptation, product customization, and supply chain coordination.

The scope of such GPNs, however, is still clearly macro-regional, and the operations of most MNEs are concentrated in super-regions (Dunning et al. 2007; Osegowitsch and Sammartino 2008; Rugman and Verbeke, 2004b, 2007, 2008a, b; Cantwell 2009). The GPNs managed by large MNEs were highly facilitated by the reduction, or even the almost complete removal in super-regions such as the EU, of trade barriers (Guy 2009). For example, the foreign operations and assets of Japanese MNEs are overwhelmingly located in the East Asia region. Global firms in the automotive industry, such as Ford, General Motors, Volkswagen and Nissan-Renault, organize most of their manufacturing activities on a macro-regional basis to serve the respective 'local' or rather super/macro-regional markets. And, although there are Mexican *maquiladoras* associated with Japanese or Korean MNEs (Kenney and Florida 1994), far more are in fact connected to US corporations (Guy 2010). What once took place within national borders now takes place within super-regions such as the EU, NAFTA, Mercosur, and ASEAN – or within China or India, each with a larger population than any of such regional areas (Guy 2010).

The clearest example of the power of modern multinationals in driving economic growth comes from recent data on US multinationals. Although the 2270 current US multinational enterprises account for less than 1 per cent of US firms, they account for 23 per cent of US GDP. More importantly, they account for 31 per cent of total US economic growth between 1990 and 2007, and 11 per cent of total US employment growth during this period (MGI 2010a). The impact of multinationals on modern aggregate economic growth is enormous.

As discussed in the various chapters of this book, the dynamic performance of multinationals is reflected in various dimensions. For example, US multinational firms displayed a 41 per cent increase in labour productivity between 1990 and 2007 (which is more than twice that of US non-multinational domestic firms over the same period), and an annual growth in value added which was 45 per cent higher than that of domestic firms (MGI 2010a). As already argued in this book, the productivity performance of US multinationals was largely due to the knowledge investments made by these firms, whose R&D expenditure accounts for 74 per cent of all US private sector R&D (MGI 2010a). These huge knowledge investments allow these firms to operate and successfully compete in global markets. US multinational firms accounted for 37 per cent of all US imports and 48 per cent of all US exports in 2007 (MGI 2010a), and they therefore display a much more favourable trade balance than other types of firms in the US economy.

Obviously, following our previous arguments, the international success of these multinational firms has powerful and immediate implications for the US domestic economy in that they have both a decisive role in driving the knowledge economy evolution of the US, and also generate significant domestic demand effects due to their export success. Collectively some 60 per cent of their sales, two-thirds of their workforce, three-quarters of their wage payments and 60 per cent of their assets are accounted for within the US (MGI 2010a). Domestically, in terms of their demands for input factors and their payments of factor rewards, in 2007 US multinationals accounted for 19 per cent of US private sector employment, 25 per cent of US wages paid, 25 per cent of US profits earned (MGI 2010a). As such, these global companies are still heavily embedded in the US economy as well as the global economy, and any benefits they gain from globalization also directly and indirectly impact positively on the US economy. Indeed, this last point is very important, in that as well as being key drivers of the domestic US knowledge-economy due to their out-performing of domestic firms on many knowledge and technological dimensions, the contribution of US multinationals to overall US economic growth also benefits of enormous domestic transmission effects. Indeed, these firms account for 90 per cent of all intermediate inputs sourced by firms within the US. As such, US multinationals have huge indirect multiplier effects on other domestic non-multinational US firms (MGI 2010a), amounting to an extra 14 per cent in terms of their contribution to US GDP (such that their total impact is 37 per cent) and an extra 9 per cent of US private sector employment (such that their total impact is 28 per cent).

Not surprisingly, the domestic economic contribution of US multinationals tends to be very much pro-cyclical, with proportionately much larger positive growth effects than domestic firms during periods of global expansion, and proportionately greater negative effects during global recessions (MGI 2010a). This raises questions regarding their likely role in the US economic recovery during the decade following the Global Financial Crisis of 2008. Moreover, these issues are made additionally complicated by the fact that the US, along with other G7 countries, has also declined in certain aspects of its attractiveness as an investment location (MGI 2010a). In particular, the US has slipped significantly in terms of global institutional and infrastructure rankings: whereas in 2000 36 per cent of the Fortune 500 companies had their headquarters in the US and 16 per cent were located outside of the G7 countries, by 2007 these numbers had fallen to 28 per cent and 33 per cent (MGI 2010a).

Notwithstanding these specific issues in the case of the US, the critical point here is that US multinationals alone account for one third of all aggregate US economic growth in the modern era of globalization. The US is the world's largest economy and the global technological leader, and multinationals are the decisive element in the ongoing evolution of this leadership because the technological, institutional and organizational changes driving modern globalization are all profoundly intertwined. What therefore emerges from the discussion here is that such changes, and the benefits that in principle they should trigger, are disproportionately going to particular places and regions, either at the level of supra-national macro-regions or at the level of sub-national regions, depending on the types and distributions of the MNEs' assets. As such, MNEs play an ever increasing role in determining the local distributions of the costs and benefits of globalization across these different levels of geographies, and these distributions depend critically on the spatial and organizational structures of the MNEs.

Since the early 1990s, the imperative to build on knowledge assets has meant that multinational investment strategies have been dominated by investing in developed countries plus the mega-cities of the developing world (MGI 2011a). These locations have been the major beneficiaries of multinational R&D, technology and innovation-driving investments. However, while MNEs have been crucial in shaping the economic geography of growth in these regions, we still have to consider exactly which types of places will MNEs invest in over the coming decades. Obviously, city-regions are still likely to be major candidates for MNE investments. Today, half of the world's population currently lives in cities, and this urban population accounts for 80 per cent of global GDP. According to McKinsey Global Institute (MGI 2011a), of this urban population, the 600 largest cities in the world, calculated in terms of their city GDP, alone account for some 60 per cent of global GDP, or three quarters of all global urban output (MGI 2011a). At present, 22 per cent (approximately 1.5 billion people and 485 million households) of the global population currently live in these 600 largest GDP cities. The largest 100 GDP cities accounted for 38 per cent of 2007 global GDP, while the 23 mega-cities of over 10 million people alone generated some 14 per cent of global GDP in 2007 (MGI 2011a).

By 2025, the share of population accounted for by the largest 600 cities is expected to have risen to 25 per cent (more than 2 billion people and 735 million households), and accounting for nearly 60 per cent of global output (MGI 2011a). Of the cities which are currently in the top 600 ranking, the 380 cities which are located in the developed world are estimated to account for 50 per cent of global GDP in 2007, with more than 20 per cent of global GDP alone coming from 190 US cities within this group (MGI 2011a, 2012). The other 220 of these largest GDP cities, all of which are located in the developing world, account for another 10 per cent of global output, and this is comprised of 4 per cent of global GDP accounted for by Chinese cities and another 4 per cent accounted for by Latin American cities. Obviously, rapid urban growth rates in developing countries in particular mean that in 2025 the top 600 GDP cities will not be the same cities as those listed today. Many cities from the developed world which are currently in the top 600 rankings will fall out of the rankings and by 2025 will be replaced almost entirely by cities from the developing world (MGI 2011a). Interestingly, however, the increasing role played by cities, and in particular by cities in developing countries, in global economic growth, is not expected to come from the very largest cities. The contribution of the world's largest 100 cities (defined in terms of economic output) to global output is actually expected to fall from 38 per cent in 2007 to 35 per cent in 2025. Indeed, and rather contrary to many popular assumptions, over recent years most mega-cities of over ten million inhabitants have not grown faster than their own economies, and many appear to be experiencing diseconomies of scale. Their share of global output growth in 2025 is forecast to be some 10 per cent, such that the contribution of the mega-cities to global output is expected to fall from 14 per cent in 2007 to 13 per cent by 2025. As such, while major cities and city-regions are obviously going to continue to be major investment locations for MNEs in the coming decades, the emerging patterns of multinational investment both in developed and developing countries may be rather different from what we have observed over recent decades. These issues are discussed in more detail in the following chapters.

6.6 CONCLUSIONS

As we have seen above, globalization is a process which has been taking place over several centuries, and in this sense the current phenomena of globalization represents nothing particularly new in comparison to the previous long run historical trends (Steger 2003). On the other hand, however, there is now enormous evidence that the current era of globalization, which has lasted for approximately two decades, is in many ways qualitatively very different from the previous ones, ensuring that the magnitude of international interdependence is now far greater than in earlier times (Steger 2003; MacGillivray 2006). Yet, the reasons why this is the case is because of the particular set of structural changes which were technological, institutional and organizational in nature, and which took place at more or less the same time during the period spanning the end of the 1980s and through to the mid-1990s. In all respects, trade, cross-border investment, and international transactions appear to be becoming relatively more geographically localized, at the level of both super-regions and also at the level of sub-national city-regions.

These localization observations appear to be fundamentally at odds with the 'flat world' argument, the assumptions of which suggest that the institutional and organizational changes ought to allow businesses in general, and MNEs in particular, to increasingly outsource and offshore activities to other lower wage countries. According to the 'flat world' thesis, the resulting changes in the geographical patterns of foreign investment should lead to a greater geographical dispersion, not a greater localization, of MNE operations. However, the reason that we do not see such a simple or straightforward process of dispersion is that, as we saw in Chapter 5, the nature of the technological changes which have taken place over recent years have often been misunderstood. In economic geography terms, our discussions in Chapters 4 and 5 on both the highly uneven distribution of key knowledge-related inputs for the production of wealth across space, combined with our observations of rising spatial transactions costs for knowledge-related activities, lead to the emergence of new types of linkages between the role of specific city-regions as sources of knowledge, and MNE location behaviour. It is these relationships between certain types of city-regions and the location behaviour of modern multinational firms that we now turn to in the next chapter.

NOTES

1. European international trade in current values grew at 4.1 per cent a year in the period 1870–1913, as compared to 16.1 per cent a year in the period 1830–1870 (Bairoch 1976, p. 77, cited in Daudin et al. 2008).
2. Similarly, the ratio of world trade to GDP had fallen to 9 per cent by 1938, then rising to 27 per cent by 1992 (Taylor and Aranya 2008).
3. As pointed out by Dicken et al. (1997), the study of Japanese investment in the Western economies, and particularly in Europe, has suffered two major shortcomings: first, Japanese investment abroad has had a much longer history than that usually recognized by the literature; second, the bulk of the studies are concentrated on manufacturing FDI, with a serious underestimation of Japanese FDI in various service industries.
4. Data refer to July 2008. See http://www.wto.org/.
5. The APEC Region Trade and Investment 2008.
6. The sensitivity of international M&As to global business confidence may be increased further by the fact that private equity funds are now playing a growing role in international M&As, accounting for just under 18 per cent of the total by value (UNCTAD 2007). Traditionally, private equity investments have tended to resemble shorter time-span portfolio investments with a focus on higher risk, higher yield opportunities (UNCTAD 2006). The long-term effects of these changes in international FDI are hard to gauge. If the increasing role of private equity leads to shorter foreign investments time spans, this may increase the sensitivity and volatility of FDI to changes in global business confidence. On the other hand, private equity may allow for a more diversified range of foreign investments to be developed, thereby providing new growth opportunities for some markets.

7. Multinationals, connectivity and global cities

7.1 INTRODUCTION

Following the arguments in the previous chapters, the reasons why multinational firms should play an increasingly important role in shaping globalization are not dependent specifically on the structural changes – technological, institutional and organizational – which have strengthened and accelerated globalization processes. Rather, it is because of the technological capabilities, skills and knowledge assets of MNEs themselves. In the global competition of the twenty-first century, knowledge, innovation and technology become increasingly critical assets, and the structural changes which have facilitated globalization have also increased the potential returns to such intangible resources. As we have seen in Chapters 4 and 5, MNEs compete primarily on the basis of their internalized knowledge assets and also how these are employed to engage with the knowledge assets of the geographical localities in which they operate. As such, their competitive advantages arise from a variety of sources both internal and external to the firm. However, the crucial competitive advantages MNEs have are in terms of their ability to exploit their organizational capabilities and experience of operating in international markets in order to leverage the benefits from the variety of knowledge sources they are able to access. As such, multinational firms are best placed to reap the potential global rewards from globalization, and at the same time to act both as conduits and facilitators of global knowledge engagement across space.

In terms of economic geography, one of the dominant features of the current wave of globalization is the emergence of a new set of relationships between the location of knowledge assets, the role of particular city-regions as knowledge sources, and the location behaviour of MNEs. The emergence of these relationships has been taking place gradually over the last two decades, during exactly the same period in which the structural changes facilitating the modern era of globalization have occurred. However, over the most recent years, the nature and ordering of these relationships has become more apparent, as their manifested trends appear to be accelerating. In particular, if we consider the economic geography of these relationships, the distinction between 'winners' and 'losers', or rather between those who are either beneficiaries or marginalized by the emerging order, are becoming more marked. In terms of the issue of where multinational firms ought to locate their operations, we will see that this is obviously very complex because it is not simply a matter of locating MNE facilities in the largest cities or urban regions. Other factors are just as important for economic growth as is urban scale, and in particular places which are open centres of knowledge and knowledge exchanges will offer advantages for many firms. Meanwhile, from our knowledge taxonomy analyses in Chapter 5, we also know that other locations which in terms of knowledge flows are more closed will offer different advantages for other types of multinational firms, depending on their specific objectives. In particular firms wishing to avoid unintended knowledge outflows will tend to prefer more closed locations. The optimal location decisions for MNEs are not simply biased towards large, diverse or open regions, as it also depends on the how the ownership O and internalization I strategies of the firm are related to their location L behaviour.

In order to understand these factors and the role played by city-regions in globalization processes, we start our discussion by considering the impact of technological change – and in particular the advent of the information and communication technology paradigm – on the costs of controlling and coordinating economic and innovative activities in multiple cross-borders locations, that is spatial transactions costs.

7.2 SPATIAL TRANSACTIONS COSTS AND TECHNOLOGICAL CHANGE

The costs associated with engaging in and coordinating activities across space can be termed spatial transactions costs. In order to understand the ways in which such costs have changed over recent years we need to split them up into three different types. The first type of spatial transactions costs are those associated with moving products across geographical space: these are transportation costs. The second type of spatial transactions costs are those associated with moving knowledge and information across geographical space. Both the first and second types of spatial transactions costs are explicitly geographical in their construction in that the costs incurred always depend on the distance covered. The third type of spatial transactions costs are those associated with crossing national borders. These costs are institutional costs and the tariffs associated with a particular border crossing. They are not defined geographically, in the sense that tariffs do not vary systematically with the distance travelled before or after arriving at a particular institutional border. As such, from the perspective of economic geography we can consider these costs to be fundamentally aspatial in construction, although clearly geographical in terms of their implementation (McCann 2005). In what follows we will consider only the first and second types of spatial transactions costs, which are explicitly dependent on geographical distance.

7.2.1 Falling Spatial Shipment Costs

A particular subset of spatial transactions costs are those directly related to the costs of moving goods or information across space. These costs are dependent on communication and transportation technologies and, as a combined group, we can refer to these as spatial *shipment* costs.

Since the 1980s we have seen dramatic improvements in the ability of decision-makers and planners to coordinate activities across space. The primary reason for these improvements has been the enormous technological developments in information and communications technologies (ICTs). Information technologies employing satellite and fibre-optical technology allow for greater quantities of information to be transmitted at much lower costs than was previously possible (McCann 2007). These developments have increased market access for individual firms and also meant that complex operations across diverse locations can now be managed both more efficiently and effectively than in the past. For industries trading specifically in information, such as finance, advertising, marketing and tourism, modern ICT technologies provide new possibilities for the supply of information-based services across a global market space. Market access has therefore increased dramatically for huge numbers of firms trading in information-based services (McCann 2007, 2008). At the same time, such technological progresses also allow decision-makers to undertake the coordination of spatial arrangements of activities which were previously unfeasible and, as mentioned in Chapter 6, this is most noticeable in the case of the increasing offshoring and outsourcing of many types of service industry activities. This is evident in examples such as international accounting, where New York banks transfer their bookkeeping requirements overnight to firms in Ireland or India, in order to have them updated in time for the opening of the money markets the next day. Other examples include Silicon Valley firms which subcontract software development activities to producers in Bangalore, while still maintaining daily contact and control of the Indian software development process from California. These observations all imply that information shipment costs must have fallen dramatically over recent decades.

Similarly, there is also evidence that many of the sectors which have benefited from the possibilities of geographic dispersal associated with these technologies are often sectors which are not specifically trading in information. For example, advanced communications and control technologies have been widely applied to the management of supply chains, to production and inventory scheduling control systems, and to logistics and distribution operations. The types of firms which particularly benefit from these technologies are those requiring the precise coordination of complex networks of production and distribution operations across large geographical distances, such as the 'flagship' MNEs (Ernst 2002a, b) controlling GPNs.

Comprehensive evidence across both countries and time pointing towards falling spatial transactions costs associated with the increased usage of information technology usage comes from Ioannides et al. (2007). They find that the increasing use of information and communications technologies over time generally allows for an increased dispersion of activities across space, thereby making city sizes more uniform. It appears therefore that the impacts of falling information shipment costs have been widely felt across a broad range of sectors.

Evidence for falling spatial shipment costs is also mirrored in the case of firms involved in the movement of physical goods and commodities across space. Transportation technologies have improved dramatically over recent years. Examples of this include the rapid growth in roll-on roll-off trucking technology, sophisticated gains in containerization technology and capacity, rapid-turnaround shipping, and the increased speed and efficiency of air transport technologies (McCann 2008). Trade in parts and components now accounts for around 30 per cent of world manufacturing trade, and the share of imports to total inputs for US goods producing sectors has doubled to 18 per cent over the last two decades (Venables 2006). All of these technological developments would imply that the cost of moving goods and commodities across space has fallen significantly over recent years. Indeed Glaeser and Kohlhase (2004) suggest that the costs of transportation for goods fell by as much as 95 per cent during the twentieth century. Currently, over 80 per cent of US shipments occur in industries where transport costs are less than 4 per cent of the total value (Glaeser and Kohlhase 2004). However, whether this is primarily due to technology or markets is not entirely clear. Evidence from France (Combes and Lafourcade 2005) suggests that most of the transport cost reductions of the last three decades appear to be primarily due to deregulation within the transport industries, rather than to technological changes.

These various observations all point to falls in the costs of both transmitting information and moving goods across space. According to the 'flat world' logic seen in Chapter 6, falling spatial shipment costs lead to convergence across space. The reason is that reductions in the real costs of shipping information and goods across space implies that geographical peripherality is becoming relatively less of a handicap to accessing global markets. Indeed, there is much evidence to suggest that such convergence is already taking place in many parts of the global economy such as Europe (Fingleton 2003a) and the US (Higgins et al. 2006). However, there is also a great deal of evidence which points in the opposite direction (e.g., Fagerberg and Verspagen 1996, 2002; Fagerberg et al. 1997; Dunford and Smith 2000; Fagerberg et al. 2007; Brakman and van Marrewijk 2008; Fagerberg and Srholec 2008). Although there are difficult empirical issues associated with measuring convergence (Higgins et al. 2006), the rate of convergence in many arenas of the global economy appears to be either surprisingly slow, non-existent, or even negative.[1] Thus, convergence is far from being an established tendency despite the development of communications and transportation technologies. The fact that spatial *shipment* costs have fallen does not imply that spatial *transactions* costs have fallen. Indeed, as we discuss in the next two sections, there are both theoretical arguments and broad empirical evidence which suggest that while the first have declined, the latter have actually risen.

7.2.2 Rising Spatial Transactions Costs

The theoretical argument that spatial shipment costs have fallen while spatial transactions costs have risen is that improvements in ICT technologies themselves increase the quantity, variety and complexity of the knowledge and information produced and handled. This in turn increases the costs associated with acquiring and then diffusing such knowledge and information across space. These rising costs become apparent when one replaces the 'widespread knowledge' assumption with the alternative assumption regarding how knowledge may be generated, reproduced and exchanged if it is not treated as equivalent to information (Steinmueller 2005).[2] The process by which knowledge evolves and spreads through the economy involves changing its nature between tacit and codified forms (Cowan and Foray 1997). While modern ICT technologies 'have rendered codification easier and more valuable in certain situations, they have not changed the fundamental structure of costs and benefits' (Cowan and Foray 1997, p. 620). In the same line, Fagerberg and Verspagen point out that technological 'diffusion in some sense seems to have become more "difficult" and demanding over time' (Fagerberg and Verspagen 2002, p. 1303). As was discussed at length in Chapters 4 and 5, much of the information which is transmitted easier and faster through ICTs will originally have emerged from knowledge of a non-standardized tacit nature, which will never be totally convertible in codified and standardized information. In other words, as many scholars have argued, the distribution of knowledge between tacit and codified knowledge has not really changed. It is true that tacit knowledge is still needed to use codified and standardized information (e.g., Lundvall and Johnson 1994; Senker 1995; Steinmueller 1995a; Dosi 1996; Cowan and Foray 1997). If the new technologies are adding up to the codified knowledge base, there must also be an addition to the tacit knowledge base by which economic actors are able to use the new codified knowledge (Cowan and Foray 1997). It follows that the acquisition and transmission of larger, more various and more complex bits of knowledge require greater levels of absorption capacity, capabilities and skills.

As such, for many knowledge-intensive activities, the required frequency of face-to-face interaction has increased over recent years (Gaspar and Glaeser 1998; Storper and Venables 2004; McCann 2007, 2008), because the time (opportunity) costs associated with not having face-to-face contact have increased with the quantity, variety and complexity of the knowledge and information produced. The outcome is that the optimized frequency of interaction across space will have increased (Rietveld and Vickerman 2004) for many knowledge-intensive sectors, thereby raising the level of spatial transactions costs for any given distance over which communication takes place (McCann 2007b). The increased importance of face-to-face contact is also manifested in terms of the development of customized products or services. Lower transport costs can be shown to imply that firms increasingly switch to the production of higher quality customized goods (Duranton and Storper 2008), whose sensitivity to distance is greater than for standardized products. This is because the cost of providing a given level of service quality becomes higher with distance (Duranton and Storper 2008). Spatial transactions costs as a whole will therefore have increased because of the rising importance of transacting knowledge via face-to-face contact, even though the spatial shipment costs of information may have fallen. This argument is also consistent with the point made by Glaeser and Kohlhase (2004) that while the costs of moving goods have fallen dramatically, the costs of moving people have not.

7.3 PROXIMITY, CONNECTIVITY AND URBAN GROWTH

The arguments above imply that, from a theoretical point of view, even though spatial shipment costs have fallen, the relative advantages of proximity and accessibility, as well as the disadvantages associated with geographical peripherality, have increased. Indeed, as we contend in Chapter 5, following the recent challenges raised by some scholars to the centrality of spatial proximity itself as a facilitator of knowledge flows (e.g., Malmberg and Maskell 2002; Howells 2002; Gertler 2003; Boschma 2005; Torre and Rallet 2005; D'Este et al. 2012), geographical proximity may not only have a direct impact on knowledge flows, but it could also play an *instrumental* role in spurring knowledge linkages and spillovers. Yet, the literature on proximity emphasizes the largely indirect role of the spatial dimension in fostering knowledge creation and interactive learning processes by bridging and strengthening other forms of propinquity such as cognitive, organizational, social, relational, and institutional proximity (e.g., Kirat and Lung 1999; Nooteboom 1999; Torre and Gilly 2000; Boschma 2005; Bouba-Olga and Grossetti 2005; Torre and Rallet 2005; Moodysson and Jonsson 2007; Ponds et al. 2007; Vicente et al. 2007). Such arguments are also consistent with three broad sets of empirical evidence that we will discuss here below.

The first broad set of empirical evidence which suggests that spatial transactions costs have increased despite the fall in spatial shipment costs, pointing to a complementary role of geographical proximity, comes from the usage patterns of ICTs. Gaspar and Glaeser (1998) find that the closer people are geographically to each other, the more they interact using information technologies. Moreover, the extent of this interaction is also associated with the local density of the urban area, such that large dense urban areas exhibit the greatest internal communications per head (Gaspar and Glaeser 1998). Additional anecdotal evidence in support of this argument comes from the fact that many of the industries which are most dependent on information technologies, such as the advanced semiconductor-electronics industry (Arita and McCann 2000; 2006) and the international financial services industry, are themselves amongst the most geographically concentrated industries in the world. Even Internet transactions exhibit this localization behaviour (Blum and Goldfarb 2006). While for some activities ICTs and face-to-face contact are likely to be substitutes, there are also many activities or roles for which information and communications technologies and face-to-face contact appear to be primarily complements (McCann 2008). This seems to be the case also for large MNEs operating in service industries, for which face-to-face interaction continues to have a critical role in shaping success or failure in the context of globalization (Jones 2007a). Foreign direct investment in high knowledge-intensive activities appears to be complementary with face-to-face interaction whereas low knowledge-intensive activities use information technology as substitutes for direct personal interaction (Nachum and Zaheer 2005). Even IT-services markets themselves exhibit this dual nature, with some activities being closely related to the need for face-to-face contact, whereas others are largely independent of such a need (Arora and Forman 2007). Moreover, the activities for which face-to-face contact and ICTs are complements appear to be those located in urban areas with extensive transportation infrastructure (Haynes et al. 2006). Further evidence which is suggestive of this comes from the fact that frequency of airline business travel between major cities has increased more or less in line with the growth in telecommunications usage between them (Gaspar and Glaeser 1998). The fact that cities are increasingly being dominated by high human capital individuals (Berry and Glaeser 2005) implies that the importance of engaging in face-to-face relationships is positively associated with knowledge levels.

The second set of empirical evidence which suggests that spatial transactions costs have increased comes from trade modelling and studies on logistic costs. Empirical research on distance costs finds that distance effects are not only persistent (Disdier and Head 2008), but also that such a persistence cannot be explained simply by observing the behaviour of shipping costs. Using a meta-analysis, Disdier and Head (2008) find that bilateral trade exhibits an average elasticity of 0.9 with respect to distance, which implies that on average bilateral trade is nearly inversely proportionate to distance. Given that a 1 per cent rise in the share of GDP accounted for by exports is associated with a per capita income increase of up to 1 per cent (Frankel and Romer 1999), this average elasticity of trade with respect to distance suggests that there are very large proximity-productivity effects. Moreover, Disdier and Head (2008) find that although distance effects declined slightly between 1870 and 1950, analyses employing recent data suggest that after 1950 they began to rise again.

The most likely explanation for these persistent distance effects is associated with the issue of time. Hummels (2001) and Deardorff (2003) suggest that the influence of time on trade is increasing. Time in transit is costly, being up to as much as 0.5% of the value of the goods shipped per day (Hummels 2001). The high cost of time in transit comes partly from the costs of carrying stock and also from the likelihood that long transit times reduce the reliability and predictability of deliveries (McCann 1998; Harrigan and Venables 2006; Venables 2006). It also makes firms slower to respond to changing demand conditions or costs levels (Venables 2006). Since the early 1980s, the opportunity costs of time appear to have increased for both household and industrial consumers (Piore and Sabel 1984; Best 1990). Consumer demand requirements are becoming ever more sophisticated and exhibit an increasing preference for retail services characterized by reliability, timeliness and quality. Modern household and industrial consumers now require a level of service customization and delivery speed which is only possible by employing more frequent shipments of goods (McCann 1998). This accounts for the almost universal trend towards Just-in-Time (JIT) type systems of shipments, which allow for Total Quality Management (TQM) principles to be applied on the basis of minimum inventory supply chains (Schonberger 1996). As we saw in Chapter 6, over the last three decades JIT logistics and distribution systems have spread progressively from the Japanese automotive industry into almost all modes of global manufacturing, retailing and distribution. Obviously, the increasing sophistication of both consumer preferences and also the advanced logistics systems responding to them are mediated by the advances in communications and transportation technologies.

Further empirical evidence suggesting the spatial transactions costs involved in shipping of goods have increased over the last two decades comes from the analysis of logistics costs as a whole, rather than simply observations of transport costs. Logistics costs are the combined costs of all the transportation, storage, and inventory-handling costs which are associated with moving goods across geographical space, and these are not only much greater than transport costs, but are also related to the costs of both time and space (McCann 1998). Firstly, logistics activities accounted for 16 per cent of global GDP in 2000 and 18 per cent of European GDP (Leinbach and Capineri 2007): during the 1990s, the growth rate for the logistics sector as a whole was of the order of 6 per cent per annum (Leinbach and Capineri 2007). Between 1987 and 1995 there was a 60 per cent increase in outsourced dedicated third-party logistics operations (Chatterji and Tsai 2006). Secondly, relative to the value of output (Schonberger 1996), the average inventory levels for almost all manufacturing and distribution sectors in the developed world have fallen dramatically since the 1980s. This implies, as was mentioned above, that the average lead times of goods-shipments have fallen over recent years, with a concomitant increase in goods-shipment frequencies. Thirdly, by carefully disentangling the various components of transport costs Hummels (1999) demonstrates that, for many sectors, the proportion of global output which is accounted for by the combination of logistics and transportation activities has not fallen over recent decades. More specifically, while the transportation cost component of bulk materials has indeed generally decreased, in the case of manufactured goods there is evidence that this proportion has actually increased over the recent decades, in spite of the improvement in transportation and logistics technologies (Hummels 1999, 2001). Fourthly, and consistently with the organizational changes shown by the regionalization of global production networks and global value chains discussed in Chapter 6, industries which are very dependent on JIT shipments have tended to reorganize their trade patterns in favour of geographically close suppliers and customers. Moreover, this behaviour is even evident in industries in which the product value-weight ratios are extremely high and for which transport costs typically account for less than 1 per cent of value (McCann and Fingleton 1996). In other words, increasing localization behaviour is present even in the very industries which proponents of the 'flat world' thesis would have deemed it to be entirely unnecessary (McCann 2007, 2008).

The third set of indirect empirical evidence which suggests that spatial transactions costs have in fact increased, again highlighting the complementarity of various forms of proximity, comes from observations of urban growth. The role of urban scale and centrality appear to be becoming ever more critical as sources of economic growth. Over the last three decades, the increasing importance played by cities as engines of national, regional and global economic growth is demonstrated by the fact that the proportion of people living in urban areas has increased in all parts of the global economy (Heitor 2004; Richardson and Bae 2005). The number of cities in the world with a population of more than one million went from 115 in 1960 to 416 in 2000; for cities of more than 4 million the increase was from 18 to 53, and for cities of more than 12 million it was from 1 to 11 (Venables 2006). By 2006, for the first time, more people worldwide lived in urban areas than in non-urban areas (Mastercard 2007). This unprecedented urban growth suggests that it is becoming increasingly important for firms and people to be clustered together, and in particular, for high skills and high knowledge workers and firms to be co-located. Recent evidence suggests that not only is there an increasing share of university educated human capital living and working in cities (Berry and Glaeser 2005), but that this proportion of highly educated workers is also correlated with the existing human capital stock (Berry and Glaeser 2005), and both are correlated with the growth of the city (Glaeser et al. 1995; Glaeser and Shapiro 2003; Berry and Glaeser 2005; Shapiro 2006). Cities are increasingly being dominated by high human capital individuals as more and more mobile workers respond to the rising wage premia associated with high value-added knowledge work in cities. After conditioning on individual characteristics it is clear that wages are indeed higher in high human capital cities (Shapiro 2006). Furthermore, US cities are found to be becoming more dissimilar in terms of their human capital composition such that regional divergence appears to have superseded previous decades of regional convergence (Berry and Glaeser 2005; Lazonick 2008).

Further evidence in support of the argument that access to cities is becoming increasingly important comes from US counties (Partridge et al. 2007) and European sub-national regions (Caniëls and Verspagen 2003). In both cases, local growth is found to be directly related to an area's proximity to major urban centres. Meanwhile, within Europe, investment capital and information have become more concentrated in capital cities and large urban centres during the very decades in which advances in technology and deregulation trends have rendered both factors more mobile (Rodríguez-Pose 1998a). As discussed above, the reasons for this appear to be that urban areas are increasingly seen to be sources of productivity growth (Ciccone and Hall 1996; Fingleton 2003b) due to their role in facilitating the production of knowledge, human capital interaction (Berry and Glaeser 2005) and the generation of innovations (Acs 2002; Carlino et al. 2007). A doubling of city size is associated with a productivity increase of some 3–8 per cent (Rosenthal and Strange 2004), such that moving from a city of 50,000 to one of 5 million is predicted to increase productivity by more than 50 per cent (Venables 2006). As such, being in close proximity to one-another appears to becoming ever-more important for an increasing number of people, and in particular, for people whose activities contain a high degree of knowledge or skills.

The result of all of these changes is that while between-country inequality has been falling over the last three or four decades (e.g., Crafts 2004), within-country inequality has actually been increasing since the 1980s (e.g., Fagerberg and Verspagen 1996; Fagerberg et al. 1997; Caniëls 2000; Brakman and van Marrewijk 2008; Rodríguez-Pose and Crescenzi 2008). Allied with the fact that the rate of convergence among advanced economies has slowed down since the 1980s (e.g., Greunz 2003; Cappelen et al. 2003), the outcome of the increasing importance of urban areas is that economic convergence at a continental scale coexists in many cases with increasing divergence at sub-national local and regional scales. The three broad sets of evidence briefly presented here all imply that in many contexts spatial transactions costs have increased and that spatial proximity has been complemented by various forms of proximity channelled by different networks: thus, in many cases the world seems to have become *steeper*, rather than flatter.

7.4 THE CURVED, SPIKY, AND UNEVEN WORLD

A careful assessment of the variety of evidence available to us supports the argument that the outcome of two opposing trends will actually be different in different situations. Different types of changes in transactions costs have tended to take place in different types of sectors, activities and contexts, and there are regularities to the pattern of these changes.

As discussed above, taking a broad view of spatial transactions costs it appears that the latter have fallen for most goods and services, allowing ever greater market areas to be supplied from a small number of locations. In particular, most of the evidence for falling international and geographical transactions costs relates to trade in relatively standardized types of activities and goods, for which the nature and frequency of the spatial transactions undertaken have not changed fundamentally over time. This is typically the case in industries producing semi-finished or finished manufactured products at mature stages within the product cycle, or service industries characterized by relatively routine activities in which the nature of the information being transacted is itself rather standardized. That is not to say that these are necessarily low-skilled jobs, as in many cases they are actually medium-skilled (Manning 2003), but the important point is that many of the routines undertaken in such jobs can increasingly be mediated or facilitated by the use of ICTs, as well as other forms of electronic mechanization. In such cases, geographical peripherality appears to be less of a disadvantage than it might have been previously, and the world appears to be getting flatter. These are exactly the types of standardized activities which currently dominate the offshoring trends to countries such as China, India, Vietnam, Philippines and South Africa (COL 2005b), and comprise much of Leamer's (2007) 'manly man' type of manufacturing work, as well as many types of tele-services and accounting activities.

In contrast, in sectors where demand lead-times have fallen dramatically, or in industries in which there has been an increasing variety and complexity of knowledge inputs associated with the customization of products and services, spatial transactions costs appear to have risen. In particular, the spatial transactions costs associated with the high knowledge inputs required for high value-added outputs, and particularly those embodied in human capital, have grown. In these sectors, which correspond to Leamer's (2007) 'geek work', the world therefore appears to be getting steeper, the result of which is that these types of activities tend to be increasingly localized. As such, the offshoring or outsourcing of many of these types of activities is actually less possible nowadays than might previously have been the case (COL 2005b). Overall, the general pattern we currently observe is therefore one of increasing globalization of both high value-added outputs and low value-added inputs, along with an increasing localization of high value-added inputs (McCann 2005). Given these changes in spatial transactions costs, a pattern emerges in which high value-added outputs are only produced in a limited number of locations and then sold all over the world, while low value-added outputs can be produced all over the world. This is also consistent with the observation that R&D and production may not necessarily be in the same place or even the same country. The real value-added component embodied in the output good is innovation and R&D, and the location of these will generally be tied to knowledge centres that may be far from the production sites. The global economy therefore appears to be simultaneously characterized both by global flattening and local steepening, the outcome of which is to make the world more curved, more uneven and more spiky (McCann 2005, 2008).

The evidence in favour of the contemporary role played by agglomeration economies in shaping economic geography is now so overwhelming that it is more or less beyond question (Venables 2006). Yet, much of the reason why agglomeration has once again become so important is related to the increasing spatial transactions costs associated with engaging in high-value knowledge activities. At the same time, for many output activities or low-knowledge input activities, spatial transactions costs have fallen. In order to understand the impacts of these simultaneous changes it is useful to consider the arguments of Krugman and Venables (1995), who analyse the case in which economies produce two sets of outputs, one under conditions of constant returns to scale and the other under increasing returns to scale. Krugman and Venables (1995) demonstrate that with high transport (spatial transactions) costs all countries have similar production patterns, because high transport costs act as a trade barrier, thereby encouraging local production. However, as transport (spatial transactions) costs begin to fall an increasing centre-periphery divergence emerges and regions in the periphery suffer decline. Finally, as transport costs fall to very low levels, convergence starts to occur. The outcome of these changes is that the relationship between degree of spatial concentration (agglomeration) and the level of transport (spatial transactions) costs exhibits an ∩-shaped form.

The question therefore arises as to where in the Krugman-Venables framework does the global economy currently stand? While the 'flat world' thesis implicitly assumes that we are at the third stage in such a framework, the above arguments suggest that this is not the case. The evidence indicates that high knowledge-intensive activities are increasingly produced in increasing returns to scale (agglomeration) environments, while low knowledge-intensive activities are increasingly produced in environments of constant returns to scale not requiring a high degree of proximity. At the same time, we recall that although there is evidence that spatial transactions costs in general have fallen due to decreasing shipment costs, there is also sign that they have probably not fallen by very much, if not even arisen overall. The combination of these various observations implies that we are currently in the second stage of the Krugman-Venables transition process. In order to explain this we will employ a diagrammatic approach (Figure 7.1).

Figure 7.1 depicts a one-dimensional economic geography model of the global economy which spans the distance *AB*. In this particular spatial

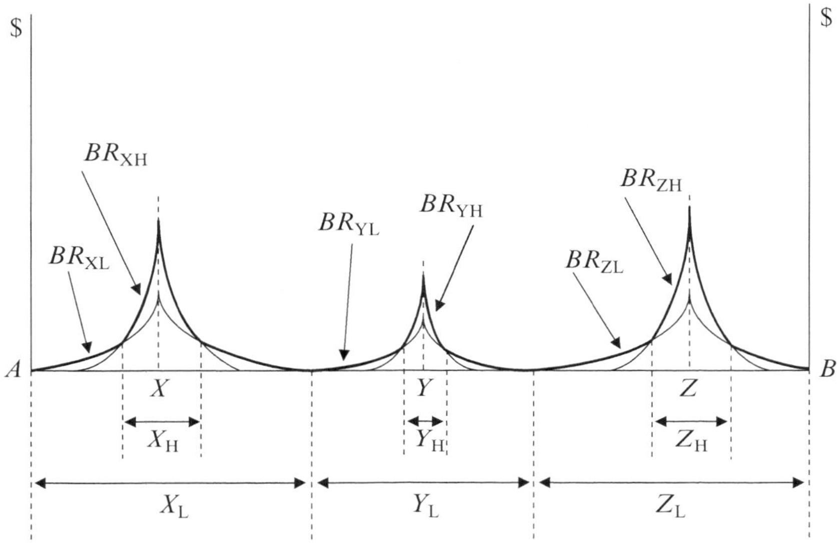

Source: McCann (2008).

Figure 7.1 A three city one-dimensional economic geography

economy there are three cities, *X*, *Y* and *Z*. Cities *X* and *Z* are larger than city *Y*, and all three cities exhibit two types of production, namely high value goods and services *H*, and low value goods and services *L*. The high value goods and services *H* are produced by those sectors for which the spatial transactions costs have risen, while the low value goods and services *L* are produced by those sectors for which spatial transactions costs have fallen. For simplicity we also assume that the high value *H* sectors are those which exhibit the agglomeration economies, precisely because of the increasing spatial transactions costs of acquiring their inputs, whereas the low value *L* sectors do not require any agglomeration effects precisely because of the falling spatial transactions costs of acquiring their inputs.

Following urban economic theory we construct what are known as bid-rent curves, whereby we plot what land prices firms are willing to pay per square metre of land in order to be located at a particular location, such that their profits are equalized across space. The curvature of these bid-rent curves is described as being convex, and the reason for this is either due to factor substitution (McCann 2001) or to the frequency of interaction (McCann 2007). Firms with high spatial transactions costs will have steeply-sloping bid-rent curves, whereas firms with low spatial transactions costs will have shallow-sloped bid-rent curves. This is because with high spatial transactions costs the opportunity costs of distance away from the market location are high, and therefore in order to make equivalent profits a firm must be able to pay much lower land prices. In contrast, with low spatial transactions costs the opportunity costs of distance are low, and therefore a firm is able to make equivalent profits a significant distance away from the market, while paying land prices which are not a great deal lower than at the market itself.

In Figure 7.1, the associated bid-rent curves for the production of each respective goods produced by each city are denoted as BR_{XH}, BR_{YH} and BR_{ZH} for the high value goods produced in city *X*, *Y*, and *Z* respectively, and BR_{XL}, BR_{YL} and BR_{ZL} for the low value goods produced in city *X*, *Y*, and *Z* respectively. Assuming that there is competition for land at each location, the land is allocated to the activity which is willing to pay the highest land price at that location, and this sector can be identified simply by comparing the bid-rent curves of each sector at each location. The land prices paid and the sector producing at each location in the global economy *AB* are given by what is known as the 'envelope' of the individual local bid-rent curves, and this is depicted in Figure 7.1 in bold.

As cities *X* and *Z* are larger than city *Y*, this implies that there is both a larger local market and also more competition for each good in *X* and *Z* than in *Y*. The result is that at the central city market locations of *X* and *Z*, land prices are higher for the production of each good *H* and *L* than at *Y*, and the bid-rent curves for both goods *H* and *L* produced in *X* and *Z* extend further outwards from their city centres than the equivalent bid-rent curves associated with city *Y*. The economic geography of the global economy *AB* is such that the spatial production area for good *L* in city *Y*, denoted as Y_L, accounts for less than one-third of the total economic geography of production of the low value good *L*. Meanwhile, cities *X* and *Z* each account for more than one-third of the economic geography of production of the low value good *L*, with the respective production areas denoted as X_L and Z_L. In addition, city *Y* also has a very small local production area Y_H for good *H*, while cities *X* and *Z* have much larger local production areas, denoted as X_H and Z_H, for the high value good *H*.

Figure 7.2 depicts the situation in response to the types of globalizing and localizing trends discussed above. As we have seen, falling spatial shipment costs apply primarily to low value goods. This means that the equilibrium bid-rent curves for the production of these goods becomes shallower. In Figure 7.2, the shallower sloping bid-rent functions for the production of low value goods and services *L* imply that such activities can be profitably undertaken in all locations. This is the spatial economics application of the 'flat world' thesis. However, this is not the end of the

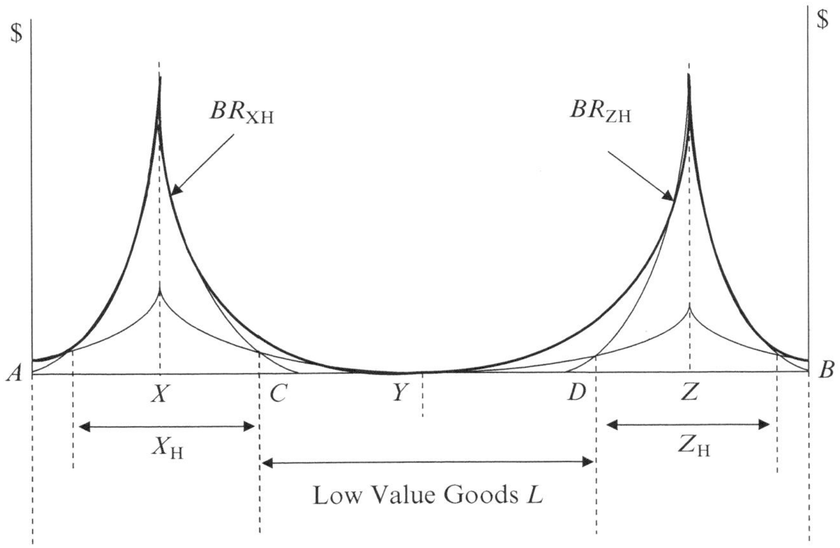

Source: McCann (2008).

Figure 7.2 Globalization, localization and economic geography

story. The reason is that the actual global distribution of production and activities depends on the interaction between this flattening effect and the other opposing localization trends discussed above. If we consider the localizing tendencies regarding the production of high value goods and services *H*, the arguments above imply that the production of these goods and services will benefit from location-specific agglomeration economies. The existence of agglomeration economy advantages as well as increasing spatial transactions costs in the production of high value goods and services *H* means that the land in the major urban areas *X* and *Z* is increasingly dominated by the production of these high value goods. Cities *X* and *Z* expand outwards, the production of high value goods is concentrated in just these two locations *X* and *Z*, and the bid-rent functions associated with the production of these high value goods become steeper. The production location *Y*, which previously had a local dominance in both the production of low value *L* and high value *H* goods for its own hinterland area, now disappears as an independent production centre. Instead, the whole region between *C* and *D* now continues to function simply as a location for producing low value goods and services *L*. The interval CD corresponds to the 'flat terrain' of Leamer (2007), while *X* and *Z* correspond to his 'hills and mountains' (Leamer 2007). The global economy is becoming spikier. As Fujita et al. (1999) demonstrate, the reason for this is that when agglomeration economies are allied with falling spatial transport costs, the existence of such agglomeration economies can lead to situations in which intermediate production locations such as *Y* disappear altogether. This is partly because the previous economic raison d'être of the former intermediate production location now vanishes, and also because the low distance costs no longer provide any possibility for 'protection' from the major urban centres and the maintenance of limited local monopoly power. In the presence of agglomeration economies the same result can also be generated in a situation where border tariffs are reduced or removed. Therefore, the combination of both falling transport costs and falling tariffs reinforces the whole process.

The equilibrium land prices at each location in the global economy *AB* are given by the envelope of the individual bid-rent curves, and as in Figure 7.1, these are depicted in Figure 7.2 in bold. As we see, the envelope rental curves are now far more curved and convex than previously. Assuming that land and labour prices are highly correlated, because of the need to maintain local real wages at competitive levels, we can assume that the envelope land price curves also closely resemble the spatial variations in both local incomes and productivity levels. The differences between the major centres and the relatively smaller centres will tend to increase. This is not to say that all major centres will increase relative to smaller centres, as it will also depend on the range of technologies and industries present in particular cities and regions: as seen in Chapters 4 and 5, different industries and technologies rise and fall over time. However, the arguments outlined earlier still do imply that globalization will lead to increasing differences between the fortunes of regions and cities even within the same country. This is because particular major urban centres will benefit from the increasing scale advantages associated with being nodes in global exchange networks.

We can also consider the effects of these various changes in spatial transactions costs on the distribution of income and productivity. In Figure 7.1, if the areas denoted as X_L, Y_L and Z_L also define the borders of the individual national or regional market areas within the global economy *AB*, the transition from the environment depicted in Figure 7.1 to that depicted in Figure 7.2 would imply that the differences between the locations had increased, resulting in income or productivity divergence. On the other hand, if there were only two countries or regions, whereby one country or region contained only city *X* and the other country or region contained both cities *Y* and *Z*, the transition from the environment depicted in Figure 7.1 to that depicted in Figure 7.2 would imply that the differences between the two countries or regions had decreased, resulting in income and productivity convergence. As such, the outcomes depend in part on the definition of the spatial units, and this is known to economic geographers as the modifiable unit area problem (Openshaw and Taylor 1979). However, observation of the increasing convexity of the envelope curves demonstrates that the transition from the environment in Figure 7.1 to that in Figure 7.2 is associated with increasing inequality across space. Therefore, from the perspective of economic geography, as we move to a world of falling spatial shipment costs for low value goods and routine activities while at the same time increasing spatial transactions costs for high-value activities and functions, the world will become more unequal and more spiky, as new rounds of unevenness are superimposed on the previous spatial structure (Warf 1995; Harvey 2006).

As Leamer (2007) points out, rather than the world shrinking or becoming flatter, economic activity is dispersing, but from the perspective of economic geography dispersal and flattening are very different phenomena. The arguments presented here suggest that, at the global scale, the various institutional, technological and organizational changes which have taken place since the late 1980s mean that we are currently in the second phase of the Krugman-Venables framework. This phase is characterized by an increasing centre-periphery divergence in which regions and countries in the periphery tend to suffer, whereas those which are centrally located benefit from agglomeration effects. More specifically, the divergence between central and peripheral areas is manifested by an increasing distinction between the locations of high versus low knowledge-intensive activities. Obviously, these general arguments and observations are not meant to imply that all central regions will benefit and all peripheral regions will suffer from globalization. The actual observed outcomes will depend on the nature and scale of the economic activities already located within the countries and regions (Fujita et al. 1999). However, regions which include large and diverse agglomerations will tend to benefit from falling distance-transport costs, as the advantages of scale give rise to what Krugman (2007) terms 'home-market' effects, whereby the scale of the local region and its hinterland allows local firms to develop economies of scale in production, which then allow them to service much wider market areas. For firms which are mobile – such as, by definition, MNEs – locating in a major city or region therefore not only provides for a large domestic market, but also for a high export capability. In contrast, regions without such agglomerations will find it increasingly difficult to attract investment from MNEs wishing to export or to access wider market areas. Indeed, even centrally located regions without such large and diverse agglomerations within them, as is the case of region *Y* in Figure 7.1 above, will tend to suffer. Regarding the increasing role of areas of economic integration in which the potential 'home market effects' increase, these arguments also imply that the major cities and regions within such integrated areas will become even more important, as they are at the same time repositories of knowledge, capabilities and networks. Thus, in the presence of agglomeration economies, the creation of transnational areas of economic integration within which trade barriers are reduced further increases the unevenness and spikiness of the spatial economy.

Similar arguments also hold for labour. For people who are mobile, migration to such centres increases employment opportunities. In particular, the technological, institutional and organizational changes associated with the recent phase of globalization have actually increased the ability of highly skilled individuals to move between locations in order to reap the rewards of their human capital. Most of these human capital movements are directed towards the dominant cities or regions (COL 2007c), which benefit disproportionately from such inflows relative to other localities. Therefore, a rapidly widening income gap between high- and low-skilled individuals has already emerged within advanced economies (Scheve and Slaughter 2007), and this is also reflected in greater interregional disparities within countries (Fagerberg and Verspagen 1996, 2002; Fagerberg et al. 1997; Dunford and Smith 2000; Brakman and van Marrewijk 2008). Similarly, what is taking place within countries is also taking place between them. It is well documented that the gap between the wealthiest and poorest countries has steadily increased over the last centuries: 250 years ago the difference in GDP per head between the richest and the poorest countries in the world was about 5:1, while today it is approximately 400:1 (Landes 1998). Most of these inequality increases have taken place during the periods in which shipping costs have fallen at the fastest rates and global trade has increased at the fastest rates (Venables 2006). Moreover, inequality between richer and poorer countries has actually increased at an even faster rate over the last two decades (Leamer 2007), as international migration increases. Once again, individual earning power and increasing migration is associated with the increasing unevenness and spikiness of the global economy.

These complex transactions costs relationships have direct analogies in terms of the spatial-industrial typologies in the knowledge taxonomies discussed in Chapter 5. The transactions costs relationships best captured by the industrial complex model most closely reflect the standardized and routinized activities discussed in this chapter. Where activities or transactions can be decomposed into any form of routine, system, framework, blueprint or template, then these activities can most easily be defined and standardized, and therefore also most easily safeguarded and controlled via legal and contractual means. As we have just seen, over recent years the transactions costs of undertaking these standardized and routinized activities across geographical space are often observed to be falling, and these falls spur even greater dispersion tendencies on the part of the industrial complex spatial-organizational forms. As we also discussed in Chapter 5, in terms of the transactions costs characteristics of the inter-establishment relationships within the confines of the parent corporate boundaries, MPDEs, and particularly MNEs, are the clearest example of the industrial complex spatial-organizational structure. The relationships between the MNEs establishments located in different regions are standardized, defined, predictable and tightly controlled via an industrial complex type of logic. As such, falls in these transactions costs between MNE establishments accounts for much of the enormous spatial dispersion of MNE establishments over recent decades.

If the local knowledge and relational characteristics of the MNE subsidiary or affiliates are of the industrial complex type, then the combination of these along with an industrial complex spatial-organization structure within the overall MNE corporate hierarchy implies that increasing numbers of individual MNE affiliates will be located in the 'plains' in Figure 7.2 above, where production costs are low. This is exactly in accordance with the logic of the Weber (1909) and logistics-costs models (McCann 1993, 1997) discussed in Chapter 3. In contrast, if the spatial transactions costs associated with many knowledge activities are increasing, and as such more subsidiary or affiliate decision-making autonomy becomes advantageous, the relationships between multinational affiliates and their individual immediate localities will increasingly exhibit relational characteristics which are more typical of either the pure agglomeration or the social network models than the industrial complex model. Given that the MNE corporate hierarchy will still exhibit the spatial-organizational features of an industrial complex, in terms of Figure 7.2 above, this implies that MNE subsidiaries and affiliates undertaking knowledge activities will increasingly be located in the spikes, or the peaks. This largely accounts for the increasing sub-national regional specificity of the location behaviour of many MNE affiliates.

Many modern multinational have affiliates and subsidiaries which undertake quite different activities and roles, whereby some affiliates deal with standardized activities whose relationships can best be described by the transactions costs features of an industrial complex model, while others undertake knowledge-intensive activities whose relations are best captured by the agglomeration or the various social network models. Given that the spatial transactions costs exhibited by the industrial complex model have systematically fallen in many cases, thereby permitting ever-tighter MNE corporate control systems, in terms of Figure 7.2, the result is that MNEs are increasingly able to locate the former type of affiliate in the 'plains' and the latter type of affiliates in the peaks. Today's MNEs are therefore able to better match the production and knowledge characteristics of the individual subsidiary establishment to the knowledge and cost features of the specific location than at any time in the past. As such, modern MNEs have been able to take advantage of globalization not only via a general increase in the possibilities of corporate dispersal across countries in order to exploit factor price differences (resource seeking), but also via more accurate and appropriate subsidiary locational behaviour within individual cities and regions (specific asset and strategic asset seeking).

Increased geographical dispersal possibilities may also be associated with some of the old social networks which are based on very long-standing relationships. Examples here might include the outsourcing and offshoring of Italian leather goods, textiles, and ceramics industry activities to countries such as Romania. In this case, the outsourcing is often to particular regions with long-standing cultural connections to the origin region. New clusters of subsidiaries emerge in these destination regions facilitated by spatial transactions between the parent region and the destination region which are characterized primarily by the industrial complex types of cross-border control systems typical of multinationals. Whether the in-migration of MNE subsidiaries and affiliates into a developing region triggers the long run development of a local spatial industrial system more typical of an agglomeration, a complex, a new social network or a new social network or an old social network, depends on the various issues discussed in this book, Indeed, this issue regarding the broader structural and behavioural impacts on local host regions of the technology transfer possibilities afforded by MNE relocation, outsourcing and offshoring, is a much broader issue in general for developing economies, and these issues will be examined in Chapter 8. However, as already explained, the transactions costs features of the cross-border intra-MNE relationships will always operate under the very tight organizational O and knowledge-internalization I arrangements typical of the industrial complex, irrespective of the specific arrangements between the affiliate and its locality.

7.5 THE ROLE OF GLOBAL CITIES

The arguments above imply that falls in particular types of spatial transactions costs provide MNEs with greater possibilities for sophisticated cross-border control systems than at any time in the past. As we have seen, the corporate organizational logic of these cross-border control systems will largely mirror the spatial industrial logic of the industrial complex. These improved control possibilities in turn better allow MNEs to make more accurate subsidiary or affiliate location or acquisition choices, whose knowledge-transactions characteristics with the local region will be most appropriately tailored to the local context so as to maximize the knowledge-related benefits. On the one hand, the integration processes taking place between larger groups of both rich and also emerging economies clearly offer greater rewards to MNEs than ever before from exploiting these possibilities for better affiliate-regional location matching as well as coordination between dispersed subsidiaries. At the same time, the higher modern demands for timeliness, the greater requirements for higher frequency transactions, and the increased preferences for customization and variety, all tend to increase the distance costs associated with knowledge-related transactions (McCann 2007,) and thereby increase the opportunity costs of sub-optimal MNE affiliate locations.

These observations all lead to an additional set of questions regarding what exactly is the role of scale in economic geography and also in the design of MNE locational decisions. As we saw in Chapter 2 international business arguments typically phrased location L decisions at the level of a country, whereas as we saw in Chapter 3 economic geography arguments phrased location decisions at specific places within countries. More recent new economic geography thinking emphasizes the role of home market effects, irrespective of whether or not they cross national borders. However, the knowledge, technology and innovation arguments in this book imply that geographical specificity is becoming more important than ever for MNEs in terms of their corporate investment strategies. However, exactly what we mean by locational or geographical specificity is as yet not clear. In terms of MNE strategies, is the scale of a city-region more important than the scale of a country, or is the scale of a supra-national area of integration more important than the scale of a country or city-region, or is the scale of a country's firms actually more important than the scale of the market? In order to answer these questions it is necessary to combine different strands of arguments. Chapters 4 and 5 in this book have discussed in detail different aspects of the importance of knowledge for location behaviour; Chapter 5 in particular examined the knowledge, technology and innovation capabilities of firms and MNEs and used these insights to develop the knowledge taxonomy of firm-geography interactions. Combining these arguments with the insights of the present chapter suggests that certain super-regions and the city-regions contained therein ought to play an increasing role in the global economy. This is the issue which we turn to here.

7.5.1 City Size and Productivity

While clear trends towards global regionalism, rather than simply globalization, are emerging, traditional economic geography theory based on the Marshallian tradition often emphasizes the importance of local and urban scale as a driver of productivity advantages, for all of the reasons discussed in Chapter 5. In terms of the location strategies of multinational firms, we therefore have to ask where will MNEs locate in the areas in which they are regionalizing?

Large cities provide increased opportunities for a wide range of sharing, sorting and matching mechanisms to operate, which increase the overall efficiency of the local activities. Indeed, estimations using US data suggest that a doubling of city size is associated with something of the order of a 5–7 per cent increase in productivity. As such, there is still clearly a very important role for large cities in the industrialized world in terms of driving productivity (Rosenthal and Strange 2004). However, when we consider issues of globalization and the potential advantages of different locations for MNEs, the relationship between city size and productivity is nowadays much less straightforward, and this is particularly so for most advanced countries.

In order to see this we can first compare the largest fifteen cities in the world at the turn of the millennium and compare these with the largest fifteen cities in the early part of the twentieth century. As we see in Table 7.1, all of the world's largest cities in 1925 were located in the world's richest countries. Cities clearly played the role as the internal 'engines' of the economy, in which urban scale and productivity were intrinsically interconnected, exactly as Marshall (1890) suspected. Yet, today, the largest cities in the world are all in developing economies, except for those located in USA, Japan and Korea. As shown in Table 7.2, apart from the case of these three countries, none of the world's largest cities is located in an advanced economy. As such, during the twentieth century something appears to have changed.

It is still very much the case that the world's most productive cities are currently located in the world's most productive economies. As we see in Table 7.3, 14 of the world's 15 highest per capita productivity cities – so defined in terms of GDP per capita – are located in the USA; London, which is ranked 13th, is the only non-US city in the top 15. Indeed, the 23 largest urban areas of the USA all rank in the top 27 most productive cities in the world. The next 55 of the world's highest productivity cities are all located in the high income OECD countries (OECD 2007), plus Taiwan, Israel, Hong Kong and Singapore.

As we see in Table 7.3, the majority of the world's highest productivity

Table 7.1 The world's largest cities in 1925

1925	City population 000s (% change 1900–1925)	Country population 000s (% change 1900–1925)	GDP $000s (% change 1900–1925)	GDP per capita $ (% change 1900–1925)
New York	7774 (83.2)	116,284 (52.2)	730,545 (233)	6282 (53.5)
London	7742 (19.5)	45,059 (9.48)	231,806 (25.4)	5144 (14.5)
Tokyo	5300 (354)	59,522 (86.0)	112,209 (216)	1885 (59.7)
Paris	4800 (44.1)	40,610 (11.7)	169,197 (44.9)	4166 (44.8)
Berlin	4013 (48.2)	63,166 (87.2)	223,082 (37.4)	3532 (18.3)
Chicago	3564 (208)	116,284 (52.2)	730,545 (233)	6282 (53.5)
Ruhr	3400 (443)	63,166 (87.2)	223,082 (37.4)	3532 (18.3)
Buenos Aires	2410 (299)	10,358 (221)	40,597 (233)	3919 (53.5)
Osaka	2219 (228)	59,522 (86.0)	112,209 (314)	1885 (18.3)
Philadelphia	2085 (47)	116,284 (52.2)	730,545 (216)	6282 (53.5)
Vienna	1865 (9.8)	6582 (10.2)	22,161 (233)	3367 (204)
Boston	1764 (64.1)	116,284 (52.2)	730,545 (28.7)	6282 (53.5)
Moscow	1764 (57.5)	158,983 (27.2) (USSR)	231,886 [1928] (50.5)	1370 [1928] (10.0)
Manchester	1725 (20.2)	45,05 (9.48)	231,806 (25.4)	5144 (14.5)
Birmingham	1700 (36.2)	45,059 (9.48)	231,806 (25.4)	5144 (14.5)

Sources: City Population Data (Chandler 1987); Country Population, GDP and GDP per Capita Data (Maddison 2006); McCann and Acs (2011).

cities are not what the OECD (2006) classifies as 'mega-cities' of over 7 million inhabitants. In fact, of the world's 75 highest productivity cities, 29 are what the OECD (2006) classifies as 'small metro areas' of less than 3 million inhabitants; 32 are classified as 'medium to large metro areas' of between 3 and 6.99 million inhabitants; and only 14 are 'mega-cities' of at least 7 million inhabitants. Excluding the US, and therefore focusing just on the rest of the world, of the world's most productive 52 non-US cities, 21 are 'small metro areas'; 20 are 'medium to large metro areas'; and only 11 are 'mega-cities' of at least 7 million inhabitants. For OECD cities of over 1.25 million inhabitants, there is only a very weak cross-sectional link between city per capita productivity and population, which if anything, is slightly negative (OECD 2007).

Table 7.4 lists the top 15 per capita productivity US cities and also the top 15 non-US OECD cities, and indicates the city per capita productivity relative to its respective national average. Obviously the US rankings remain unchanged, but the non-US rankings differ significantly when

Table 7.2 The world's largest cities in 2000

2000	City population[3] 000s (% change 1975–2000)	Country population 000s (% change 1975–2000)	GDP $000s (% change 1975–2000)	GDP per capita $ (% change 1975–2000)
Tokyo	29,896 (30.0)	126,737 (13.6)	2,589,320 (204)	20,431 (80.0)
New York	24,719 (44.1)	270,561 (25.2)	7,394,598 (210)	27,331 (67.8)
Seoul	20,674 (275)	46,898 (30.7)	624,582 (559)	13,317 (421)
Mexico City	19,081 (68.3)	98,553 (62.0)	655,910 (209)	6665 (29.5)
Sao Paulo	17,396 (73.2)	169,897 (56.0)	926,918 (203)	5459 (30.2)
Manila	16,740 (310)	79,376 (78.5)	181,886 (201)	2291 (12.9)
Los Angeles	15,807 (76.4)	270,561 (25.2)	7,394,598 (210)	27,331 (67.8)
Mumbai	15,769 (223)	991,691 (63.3)	1,803,172 (3.31)	1818 (202)
Djakarta	15,086 (284)	207,429 (58.9)	628,753 (3.2)	3031 (201)
Osaka	15,039 (−3.0)	126,737 (13.6)	2,589,320 (204)	20,431 (80.0)
Delhi	13,592 (309)	991,691 (63.3)	1,803,172 (3.31)	1818 (202)
Kolkata	12,619 (60.2)	991,691 (63.3)	1,803,172 (3.31)	1818 (202)
Buenos Aires	12,297 (44.7)	36,235 (39.2)	334,314 (57.8)	9219 (13.2)
Shanghai	11,960 (49.5)	1,252,704 (36.6)	4,082,513 (509)	3259 (372)
Cairo	11,633 (38.4)	66,050 (78.7)	140,546 (339)	2128 (89.8)
World [1998]		5,907,680 (45.3)	33,725,631 (202)	5709 (39.4)

Sources: City Population Data (Chandler 1987; Le Gales 2002); Country Population, GDP and GDP per Capita Data (Maddison 2006; McCann and Acs 2011).

measured relative to their national productivity levels. In Table 7.5 we repeat this exercise after excluding all US cities. The first three columns of Table 7.5 also exclude cities located in the former transition economies, as well as in Mexico and Turkey. The second three columns include all OECD countries. If we consider the first three columns of Table 7.5 we see that 12 of the cities are the same as those in the second three columns of Table 7.4. However, the top fifteen cities in terms of relative per capita productivity in the first three columns of Table 7.5 are smaller on average than the top 15 absolute productivity non-US cities. If we now consider the second three columns of Table 7.5, however, we see that the rankings change dramatically, and are now dominated by cities in the poorer nations of the OECD, some of which are very large cities indeed. Ten out of the top 15 relative productivity cities are located in either former transition economies or developing countries members of the OECD. As such, the productivity advantages of very large cities appear to be relatively more important for lower income and transition economies than for the most advanced ones.

Table 7.3 The world's most productive cities 2002–2004

US cities	City population[4] (million)	City per capita productivity (US $000 PPP)	Non US OECD cities	City population (million)	City per capita productivity (US $000 PPP)
San Francisco	4.2	62.3	London	7.4	46.2
Washington DC	5.1	61.6	Paris	11.2	42.7
Boston	4.4	58.0	Dublin	1.6	38.9
Seattle	3.2	54.4	Vienna	2.2	37.6
Minneapolis	3.1	53.0	Stockholm	2.2	36.7
New York	18.7	52.8	Stuttgart	2.7	36.4
Denver	2.3	50.8	Milan	7.4	35.6
Philadelphia	5.8	50.5	Lyon	1.6	35.2
Dallas	5.7	50.1	Munich	6.1	35.2
Atlanta	4.7	47.8	Oslo	1.7	35.0
Houston	5.2	47.4	Sydney	4.2	35.0
San Diego	2.9	46.8	Brussels	3.8	35.0
Chicago	9.4	45.6	Toronto	4.7	34.9
Los Angeles	12.9	45.3	Helsinki	1.8	34.0
Detroit	4.5	44.0	Frankfurt	5.6	33.6

Sources: OECD (2007, pp. 38–40); World Bank (2006, 2008b)[5]; McCann and Acs (2011).

For the rich countries, apart from the cases of USA, Japan and Korea, city size is not an indicator of city productivity. Indeed, within the OECD, for cities of over 6 million inhabitants, there is actually a weak but statistically significant negative cross-sectional relationship between city per capita productivity and city population (OECD 2006). Obviously, if the largest cities from the developing countries listed in Table 7.2 were also to be included in such cross-sectional estimations, then the effect of population scale will be very much more negative amongst mega-cities than is the case with just the OECD cities. On the other hand, for cities of less than 10 million, the positive relationship between city per capita productivity and OECD city population is both statistically significant and strong (OECD 2007). Further evidence in favour of this argument comes from the fact that amongst the rich countries, 12 out of the 15 most entrepreneurial cities are small to medium sized cities (Acs et al. 2008), while eleven out of the world's fifteen most competitive cities are small to medium sized (COL 2008). As such, amongst OECD cities there appears to be something of an ∩-shaped relationship between city per capita productivity and population scale. Moreover, over recent decades, more cities in the developed world

Table 7.4 The highest relative productivity cities in the OECD

US cities	City Population[6] (million)	Relative productivity	Non US OECD cities	City population (million)	Relative productivity
San Francisco	4.2	1.72	London	7.4	1.59
Washington	5.1	1.70	Paris	11.2	1.53
Boston	4.4	1.60	Dublin	1.6	1.18
Seattle	3.2	1.50	Vienna	2.2	1.27
Minneapolis	3.1	1.46	Stockholm	2.2	1.29
New York	18.7	1.45	Stuttgart	2.7	1.34
Denver	2.3	1.40	Milan	7.4	1.29
Philadelphia	5.8	1.39	Lyon	1.6	1.26
Dallas	5.7	1.38	Munich	6.1	1.30
Atlanta	4.7	1.32	Oslo	1.7	0.95
Houston	5.2	1.31	Sydney	4.2	1.07
San Diego	2.9	1.29	Brussels	3.8	1.19
Chicago	9.4	1.26	Toronto	4.7	1.08
Los Angeles	12.9	1.25	Helsinki	1.8	1.19
Detroit	4.5	1.21	Frankfurt	5.6	1.24

Sources: Calculations based on OECD (2006 pp. 38–40); OECD (2008); World Bank (2008b); McCann and Acs (2011).

have actually shrunk in size than the number of cities that have grown (UN-HABITAT 2008). As such, the ∩-shaped relationship between city per capita productivity and population scale may have actually shifted to the left slightly, as well as changed shape, as labour out-migration from cities specialized in declining industries gives rise to the growth of other cities specialized in growing sectors.

7.5.2 City-regions and Connectivity

Part of the problem here is that, while estimations on the relationship between city size and productivity within individual countries make some sense, estimations across countries are far more complex because of the spatial, structural and institutional heterogeneity of countries, and pooled cross-sectional samples of cities from different countries will provide different overall pictures. As well as the ∩–shaped relationship between city size and productivity, across the OECD it is also generally the case that the higher income cities are actually outgrowing lower income cities, irrespective of population scale (OECD 2006). Taken together, all of these various observations strongly suggest that other characteristics of cities

Table 7.5 The highest non-US relative productivity cities in the OECD

Non US OECD cities excluding former transition economies, Mexico and Turkey	City population (million)	Relative productivity	Non US OECD cities (all OECD countries)	City population (million)	Relative productivity
London	7.4	1.59	Warsaw	3.0	1.99
Paris	11.2	1.53	Monterrey	3.2	1.98
Lisbon	2.7	1.39	Istanbul	11.4	1.60
Auckland	1.2	1.34	London	7.4	1.59
Stuttgart	2.7	1.34	Budapest	2.8	1.59
Milan	7.4	1.31	Paris	11.2	1.53
Munich	6.1	1.30	Prague	2.3	1.51
Stockholm	2.2	1.29	Mexico City	18.4	1.49
Vienna	2.2	1.27	Izmir	3.4	1.46
Lyon	1.6	1.26	Ankara	4.0	1.41
Frankfurt	5.6	1.24	Guadalajara	3.5	1.39
Madrid	5.6	1.24	Lisbon	2.7	1.39
Rome	3.7	1.21	Puebla	2.1	1.36
Brussels	3.8	1.19	Auckland	1.2	1.34
Helsinki	1.8	1.19	Stuttgart	2.7	1.34

Sources: Calculations based on OECD (2006 pp. 38–40); OECD (2008); World Bank (2008b); McCann and Acs (2011).

are also just as important for productivity and performance as scale, and the arguments in Chapters 4 to 5 point to the crucial role that cities or urban regions can play as centres of knowledge (Caniëls 2000), creativity (Florida 2005a,) and innovation (Acs 2002). Indeed, comparing Tables 7.3 and 7.4 with Tables 7.2 and 7.5, and also considering the arguments developed in previous chapters, would indicate that for advanced economies today issues of knowledge, technology and innovation are likely to be far more important for fostering productivity than simply scale.

All these findings therefore suggest that during the twentieth century there have been qualitative changes in the role of cities in the industrialized world which favour the competitive advantages associated with cities being centres of knowledge. The clues as to why particular cities are highly productive lie in the types of firms which are located there. In order to generate the required returns to their knowledge investments, many of these knowledge-based firms must capture markets which extend well beyond the borders of their own country. Traditionally, these returns were generated by exports. However, as we have already seen, one of the key features of the current phase of globalization is that there is now an increasing premium associated with face-to-face contact. This implies that the global engagement facilitated by international investment is becoming far more important than exporting. As such, the relationship between cities, countries, and globalization is changing, and the arguments in this book all point to the increasingly important role of multinational firms as conduits and facilitators of such global engagement.

We therefore must return to our location theory principles discussed in Chapter 3 in order to ask why multinational firms would want to locate in particular cities in the first place. Larger cities are more expensive investment locations and, as we have seen here, we cannot simply assume that agglomeration effects are the answer, because city productivity performance is not simply a matter of scale. Yet, combining these insights with those of Chapters 4 and 5 – and in the light of the Hymer's legacy discussed in Chapter 2 – the location behaviour of MNEs cannot be understood without recourse to detailed considerations of organization, technology and innovation and, most importantly of all, to the firm's ability to access and exploit internal, local and global knowledge sources. Indeed, the arguments in this book actually go further than this in that they point to the critical role which MNEs play in making cities highly productive by their global connections which go well beyond the scale aspects of the local city or urban region.

As such, following all of our previous arguments, the location behaviour of MNEs would therefore appear to be critical in explaining why particular types of cities are knowledge centres. Indeed, in this vein, building on interdisciplinary seminal studies on cities and economic growth (e.g., Hall 1966; Jacobs 1969; Hymer 1972; Friedmann and Wolff 1982; Friedmann 1986; Sassen 1991; Henderson et al. 1995), there is now substantial evidence from the literature on 'global cities' that economic growth at the international scale is being increasingly dominated by networks of particular major urban centres and regions (e.g. Borja et al. 1997; Sassen 2001; Scott 2001; Scott et al. 2001; Derudder et al. 2003; Button et al. 2006). Similarly, there is also much evidence which examines the role played by these global cities as the principal location bases for globally connected firms (e.g., Sassen 1994, 1995, 2002; Knox and Taylor 1995; Enright 2000a, b: Taylor 2004; Bel and Fageda 2008; Verbeke and Goerzen 2009). The analysis of global cities suggests that in the current phase of globalization, the links between a city and other parts of the global economy are a key determinant of the city-region's performance.

In the global cities literature, which draws heavily on sociological approaches, the importance and influence of a city in the global economic

Table 7.6 Global city rankings

Global city	2004 population (000s)[7]	2008 Global City Index	Global financial centre	Population (000s)	2008 Global Financial Centre Index
London	7400	79.17	London	7400	795
New York	18,700	72.77	New York	18,700	786
Tokyo	34,200	66.60	Hong Kong	7000	695
Singapore	4000	66.16	Singapore	4000	675
Chicago	9400	65.24	Zurich	2500	665
Hong Kong	7000	63.94	Frankfurt	5600	642
Paris	11,200	63.87	Geneva	450	640
Frankfurt	5600	62.34	Tokyo	34,200	628
Seoul	23,500	61.83	Sydney	4200	621
Amsterdam	7500	60.06	Boston	4400	618
Madrid	5600	58.34	San Francisco	4200	614
Sydney	4200	58.33	Dublin	1600	613
Toronto	4700	58.16	Paris	11,200	612
Copenhagen	2400	57.99	Toronto	4700	610
Zurich	2500	56.86	Washington	5100	597

Sources: Global City Index (Mastercard 2008); Global Financial Centre Index (COL 2008); City Populations (OECD 2006); World Bank (2008b);[8] McCann and Acs (2011).

system is discussed in terms of the extent of its global 'connectivity' (Sassen 2002), whereby 'connectivity' refers not only to the various aspects of the knowledge and information exchanges which take place between particular locations, but also to the discretionary decision-making power to act on those exchanges. Incidentally, this argument can be extended to different geographical levels by combining the innovation systems' perspective with that of system openness, discussed in Chapters 4 and 5 respectively. As such, global connectivity may be manifested via a variety of different mechanisms such as corporate headquarter functions, foreign subsidiaries, decision-making links, human capital mobility patterns, research collaboration, trade exchanges, transport networks, financial linkages, and asset management roles (e.g., Reed 1981; Friedmann 1995; Beaverstock et al. 1999; Godfrey and Zhou 1999; Taylor 2004; Sassen 2006; Taylor and Aranya 2008; Taylor et al. 2011).

In order to provide an index of this much broader concept of connectivity it is possible to apply weighting measures and algorithms to data on these various connectivity characteristics (Taylor 2004; Taylor et al. 2008) and use them to rank cities according to their degrees of global connectivity. Both the global-city centres of commerce rankings (Mastercard 2008) and also the rankings of global financial centres (COL 2008) are calculated in this way, and they therefore provide composite indicators of the level of global connectivity of different cities viewed from different perspectives. Not surprisingly, there is a very close correspondence between the level of global connectivity of the cities and their GDP per capita. Of the top 50 most productive cities in the world, 35 of them are in the top 50 of the global-city centres of commerce rankings (Mastercard 2008). If we were also to include the highest productivity non-OECD cities of Tel Aviv, Taipei, Singapore and Hong Kong,[9] then the number would rise to 39 out of the top 50; in other words, the most productive cities in the world are also in the top 50 globally connected cities. In terms of the global financial centres, ignoring small tax havens,[10] 30 out of the top 37 global financial centres are among the world's 50 most productive cities. Of the ten largest cities from the newly-industrializing world which from Table 7.2 are all amongst the 15 largest cities in the world, nine are amongst the top 70 worldwide centres of commerce (Mastercard 2008), and two are also amongst the top 37 global centres of finance (Mastercard 2008).

This striking correspondence between the worldwide centres of commerce rankings (Mastercard 2008), the global financial centre rankings (COL 2008), and the world city-productivity rankings (OECD 2006), is consistent with the argument that the cities with the highest levels of global connectivity are also largely the world's most productive cities. Moreover, these observations are consistent with the previous arguments that the performance of these global cities is largely related to the scale of the global engagement of the companies located there, and in particular, the globally competitive MNEs. There are good reasons for this. The combination of localized agglomeration advantages plus a range of specific network effects which are characteristic of key nodal locations, allied with major economies of transportation and communication, appears to maximize a firm's global market potential, exactly as the theoretical chapters in this book suggest. These 'global' urban centres are therefore locations which not only exhibit significant agglomeration advantages, but which also primarily interact with other similar globally-oriented cities in other countries, rather than with other smaller urban centres within their own countries, which tend to be oriented more towards the provision of local goods for local markets. In many sectors such as financial services (COL 2007a, b; Mastercard 2007) there is already much evidence that global markets are increasingly dominated by networks of global cities such as London, Paris, Tokyo, Sydney and New York (Sassen 2002). The increasing relative dominance of these global cities appears to be associated with the density of knowledge and information technology assets in the city (e.g., Sassen 2002; Taylor et al. 2002a, b; Derudder et al. 2003; Simmie

2004; Button et al. 2006; Taylor et al. 2011). Moreover, empirical evidence suggests that the importance of major urban nodes (Limtanakool et al. 2007) within such networks is also reinforced by the existence of hubs within the global air (COL 2002; Burghouwt 2005), rail, and marine transportation systems (Leinbach and Capineri 2007). In addition, there is also clear indication that the performance of these global-cities is also playing an increasingly important role even with respect to their own hinterland local, national and continental economies (e.g., Glaeser 2005; COL 2005a; 2006b; 2007c; HMT-DTI 2001, BTRE 2004).

For example, if we take one particular aspect of this notion of connectivity, namely that of the relationship between the location of major corporate headquarter functions and the spatial structure of global intercontinental airline linkages, recent evidence from European regions suggests that, controlling for endogeneity bias, the supply of direct intercontinental flights is found to be a major determinant of corporate headquarter location decisions (Bel and Fageda 2008). While proximity to large markets and specialist suppliers is also important as expected, the size of the city has little if any explanatory power, and the size of the city relative to the country is not at all significant. As such, in the modern European context, urban scale and national scale alone appear to be much less important as location determinants for key corporate coordination and control functions than the structure of global airline networks (Bel and Fageda 2008). These observations are also supported by the findings of Ni and Kresl (2010) who found that the most important element in the competitiveness rankings for global cities is connectivity, rather than urban size or structure.

Additional empirical evidence in support of these findings comes from Button et al. (1999) and Wickham and Vecchi (2008). Button et al. (1999) examined the relationship between US high technology employment and the location of hub airports. They found that proximity to a hub airport increases local high-technology employment, and the Granger causality test implies that the former drives the latter. Following the case study research of Wickham and Vecchi (2008) the reason for this is that proximity to hub airports allows companies, and small companies in particular, to easily access a much wider market, thereby reducing the constraints associated with a lack of scale. Moreover, the importance of the access to the hub airport appears to dominate any role played by local cluster institutions, a finding which is consistent with the Bel and Fageda (2008) finding that the size of the city is not significant.

Following the connectivity arguments, the reasons for these findings are that the spatial network structure of global airline system (Grubesic et al. 2008, 2009) determines the geographical patterns of knowledge flows embodied in particular high human capital individuals with the discretionary power to act on that knowledge. More specifically, the spatial network structure of the global airline system determines the ease and frequency (McCann 2007) with which business and corporate decision-makers are able to engage in direct face-to-face contact with similar decision-makers in other locations (McCann 2008; Aguilera 2008). As such, the implication of these arguments is that global hub airport functions are critical for facilitating the types of higher-order knowledge flows which result in investment decisions being made. Greater proximity to such infrastructure should increase both the likelihood of investment being forthcoming in the nearby regions and also the resulting level of innovations generated by the regions.

There are two final points that it is important to note about the correspondence between global cities-regions and the location behaviour of MNEs. Firstly, these global cities are not the world's largest cities. Indeed, many of the cities listed in Table 7.2 would not be classed in any way as global cities in this particular literature, and the reason is that the definition of 'global cities' is not related to the city size, but rather to its degree of 'connectivity', 'openness' and 'accessibility'. As yet, and despite the emergence of global hubs in some emerging economies, as we will see in Chapter 8 most of the world's largest cities which are located in developing countries still do not exhibit the knowledge, technology, trade, financial, decision-making, transportation and human capital flows and networks with the rest of the world which a genuinely global city exhibits.

Secondly, the emphasis on the role played by the presence of MNEs and their network structures in making cities 'global cities' should not, in any way, undermine the role that small and medium sized enterprises (SMEs) and agglomeration play for local economic growth. On the contrary, and as also emphasized in Chapter 5, small entrepreneurial start-ups are critical for innovation and growth, and agglomeration is crucial for promoting their success well beyond the urban configuration of externalities (Acs 2002). However, our arguments also imply that in the current era of globalization the probability of success for small and medium sized firms and SME-based clusters will be higher in the very city-regions which are the most globally-connected and 'open' to new ideas and people. The reason is that the suggestion from sociological studies that 'networks . . . may act as a substitute for agglomerations . . . providing some or all of the utility gains and productivity increases derived from agglomeration' (Johansson and Quigley 2004, p. 166) does not contradict the claim that networks and agglomeration are in fact *complements*, as suggested by the proximity literature. Such a claim has found strong support in the empirical evidence coming from the study of global productions networks and value chain as MNEs and SMEs production systems (e.g., Guerrieri et al.

2001; Belussi et al. 2010; Cusmano et al. 2010; Ernst 2010). The types of local spillovers which are likely to operate mean that the export (of products and knowledge) potential of uni-national (multiplant or not) firms will be higher, *ceteris paribus*, in regions which are characterized by substantial MNE presence and embeddedness. Multinational firms are indeed the primary conduits via which global knowledge flows operate and the natural channels via which domestic firms can distribute their products. As such, MNEs facilitate and provide for regional connectivity. This is exactly what was found by Aitken et al. (1997) who demonstrated that it is proximity to multinationals, and not proximity to other exporters, which increases export propensities. Moreover, their finding was robust to the inclusion of other measures such as the overall industrial activity of the region, proximity to the capital city and border regions, and price and costs variables. More recent evidence from Sweden shows that the export propensities of non-MNEs is not only higher for those firms located in larger and denser cities, but in addition is also correlated with the degree of MNE activity in that city (Andersson 2009; Johansson and Loof 2009). As such, as has been emphasized by sociologists and geographers working on world city networks, it is clear that the urban scale is only part of the story: global connectivity is also a critical part of the story, and one which has been rather overlooked by urban economists, institutional economists, and trade modellers.

7.6 CONCLUSIONS

The arguments developed in this chapter and in Chapter 6 suggest that in terms of economic geography the institutional, technological and organizational features of modern globalization are together making the world more uneven and spikier, rather than flatter and more uniform. Yet, these arguments do not entail that all developing nations and less advanced regions are being left behind by globalization processes, but that convergence and 'flatness' are far from being achieved. A first implication of our discussion is that the economic growth of both developing economies and lagging behind regions very much depends on the cities which already exist in these countries and regions. Indeed, the emergence of many of the major newly-industrializing countries is driven by the dominant cities, as witnessed by the rising international importance of centres such as Shanghai, Mumbai, Sao Paulo and Moscow. However, our arguments here also imply that the spread effects to other parts of the developing world will continue to be relatively restricted (Crafts 2004). The main point is that the rapidly-growing economies in the newly-industrializing regions of the world already all contain major cities (Collier 2007), whereas poor countries without any such major cities have great difficulty in achieving rapid growth.

The arguments and evidence presented in this chapter also indicate that globally dominant firms will be those which are strongly embedded in the global city knowledge networks. Similarly, the leading global city-regions located in both developed and developing economies will be those which have the strongest presence of such multinational companies. As we saw in Chapter 2, this is exactly the correspondence between MNEs and cities that Stephen Hymer first predicted. The various examples discussed here therefore serve to highlight the difference between simple 'connectedness', defined in terms of the architecture of transport and communications infrastructure, and the much broader concept of 'connectivity'. Connectivity is a behavioural concept incorporating the capability of individuals, firms, organizations and institutions to interact, engage, take initiatives and make decisions across different locations and within networks, and is a concept showing a substantial degree of two-way openness: openness to 'ideas' (new knowledge), to 'business' (new investment) and to 'people' (new diversity) (Fagerberg and Srholec 2008). As we have argued in this book, global networks and local agglomeration act as complementary forces strengthening each other in determining the 'spikes' of the world economy.

Following our discussions in Part I of this book, understanding connectivity is therefore essential in order to make sense of the locational strategies of MNEs. In the international business and international management literatures the importance of such connectivity mechanisms is more or less taken for granted (Dunning 2000), whereas amongst economists and even geographers there are still many who give little credence to these issues. The arguments contained in this book are intended to change some of these perceptions.

NOTES

1. It seems useful to remind here that catching up and convergence, although partially overlapping, are distinct concepts. The first relates to the ability of a country, or a group of countries, to narrow the gap in productivity and income with the leader economy. Convergence refers to a trend towards a reduction of the overall productivity and income differences at the worldwide level. If all countries below the frontier catch up, convergence will necessarily follow; otherwise, if just some countries catch up while others fall behind, the outcome with respect to convergence is far from clear. The huge empirical evidence available shows that, at best, convergence is restricted to some groups or 'convergence clubs' of national or regional economies (Lall 2004; Fagerberg and Godinho 2005).

2. See Chapters 4 and 5 in this book on the distinction between knowledge and information.
3. The figures reported here are for conurbations, and come from GEOPOLIS database, rather than just city administrative boundaries. Therefore, New York City (five boroughs) has only 7.549 million, the greater New York area has 18.7 million (OECD 2006), and the total New York conurbation (including Philadelphia) has over 24 million inhabitants. Similarly, depending on the source used and methods of defining urban areas, London city has only 7.4m inhabitants (OECD 2006) but the London conurbation has a population of 8.5 million according to the *World Urbanization Prospects* (2008) [http://esa.un.org/unup/p2k0data.asp], a population of over 9.66m according to GEOPOLIS, or a population of 11.22 million according to the *World Gazetteer* (www.world-gazetteer.com). See Venables (2007).
4. The city population figures here are from the OECD metropolitan database (OECD 2006) and differ slightly from the urban definitions employed in Table 7.2.
5. Combining the OECD (2006) metropolitan productivity data with PPP national productivity data at current prices (World Bank 2008) would also rank Singapore as the 30th highest productive city in the world, below Stuttgart and above Milan, and Hong Kong as the 48th highest productivity city in the world, below Auckland and above Hamburg.
6. The figures here are from the OECD metropolitan database (OECD 2006 pp.38–40) and differ slightly from the urban definitions employed in Table 7.2. Moreover, the OECD (2012) has just very recently produced new city-size estimates based on particular standardised definitions of population densities, population contiguity and commuting patterns. However, combining these new estimates with the latest rankings (COL 2009; Long Finance 2011) still leaves the overall arguments outlined here unaffected.
7. Population data comes from the OECD metropolitan database (OECD 2006 pp. 38–40). The data for Amsterdam is that of the Randstad, and the data for London is only the area within the London city boundaries and does not include the whole London commuting area (OECD 2006).
8. Combining the OECD (2006) metropolitan productivity data with PPP national productivity data at current prices (World Bank (2008b) would also rank Singapore as the 30th highest productive city in the world, below Stuttgart and above Milan, and Hong Kong as the 48th highest productivity city in the world, below Auckland and above Hamburg.
9. For our purposes we can regard Singapore and Hong Kong as city-states.
10. Of which there are 13 in the top 50 world financial centres.

8. Multinationals, emerging economies and the changing economic geography

8.1 INTRODUCTION

As we have already seen in Chapter 6, in the current phase of globalization the importance of the role played by trade blocs and areas of integration in the world economy has increased. At the same time, the importance of multinationals has increased (Rugman 2000, 2005). For example, if we take the case of the three super-regions of US, EU and East Asia, we find that the average same-region sales share of the world's top 500 MNEs is over 70 per cent (Rugman 2005). Moreover, as discussed in Chapters 4 and 5, MNEs play a crucial role in facilitating knowledge flows across space, and this is particularly evident within super-regions. However, along with FDI flows, the overwhelming dominance of developed economies in global R&D effort is still very marked. Of the 2600 foreign affiliates in the global economy whose primary role is R&D, almost 70 per cent are still located within the triad of US, EU and Japan (UNCTAD 2005). In terms of global shares, the geographical origins of global R&D reflect almost exactly the geographical origins of FDI outflows and, if anything, they are slightly even more skewed in favour of most developed macro-regional blocks.

As such, developed economies not only dominate global output, global R&D and global trade, but also are dominated by the trading relationships of the MNEs located in the same region. Why is this the case? From the arguments in the previous chapters this is related mainly to the ability of the multinational firm to access key knowledge sources and assets in the local economy of origin, as well as from other locations in the global economy. Indeed, evidence from patent citation data suggests that firms typically learn 80 per cent of their knowledge within the local region and 89 per cent within the same country (Peri 2005). Location in knowledge centres is therefore critical for high value activities, and the arguments in Chapters 3 to 5 imply that this spatial knowledge concentration will be even stronger for service industries dependent on trust relations and interpersonal networks.

However, new trends are clearly emerging with respect to the combined importance of MNEs and developing and transition economies. The increasing role in R&D played by the foreign affiliates of MNEs, on the one hand, and the growing attractiveness of developing countries as locations for MNE foreign affiliates, on the other, have sharply accelerated their pace in the last decade. These trends undoubtedly represent favourable opportunities for developing economies in order to improve and upgrade their competencies and capabilities, and their capacity to accumulate knowledge and technological assets. However, the distinction between winner and losers at the global level is likely to be further exacerbated by these new drifts (e.g., Verspagen and Schoenmakers 2004; Athreye and Cantwell 2007; Zhao 2006; Cantwell 2009). Following the arguments provided in previous chapters on the growing interactions between knowledge and geography, the economies within the developing world who are most likely to be winners and see a remarkable increase in their knowledge assets are those where agglomeration economies favour the formation and growth of cities-regions which are well connected and integrated with the rest of the global economy. The global firms located in these core cities-regions will increasingly reap the economic rents associated with their knowledge assets via genuinely global production, communication and financial networks (Coleman 1996; Cohen 1998; Zook 2005), thereby generating further connectivity and knowledge spillovers. The location behaviour of MNEs firms is therefore critical, and the local actors – firms and other organizations – which are already embedded in leading centres and regions of the developing and emerging world are those which will achieve the major returns from globalization along with the MNEs.

8.2 DEVELOPING ECONOMIES AND KNOWLEDGE-RELATED MNE INVESTMENT

In terms of the emerging economic geography of the twenty-first century, we see in this book that there is now a growing polarization between different types of activities. Knowledge-intensive activities are increasingly being located in high agglomeration city-regions. Low-skill, routine, and low value-added activities are being increasingly spatially dispersed across the world, and this dispersion is particularly noticeable in terms of shifts away from developed economies to developing and transition economies. On the other hand, as we have already seen through the examples discussed in Chapters 4 and 5, many knowledge-intensive activities are also becoming increasingly localized and concentrated in particular types of localities (Revilla Diez and Berger 2005). These localities can be labelled as knowledge-regions, as they are characterized by high levels of knowledge assets, such as human capital, skills and R&D. Following the arguments in this book, the types of localities which are increasingly attractive for knowledge-intensive activities tend to be those economically diverse regions with a high potential for the exploitation of technological complementarities, and which are also highly connected in the global transportation and communication networks. Knowledge assets, global connectivity, and knowledge-related foreign investment are increasingly co-determined, and some of the emerging economies from the developing and transition countries are also now becoming beneficiaries of these processes. At a global scale, the long term result of these trends ought to be that emerging countries in general will play a much more important role in the global economy of the twenty-first century than they did in the previous one (Lucas 2000). At the same time, however, these arguments also imply that not all emerging countries will benefit from these globalization trends and that there will be increasing differences between the winners and losers amongst the developing and transition economies (e.g., Athreye and Cantwell 2007; Leamer 2007; Fagerberg and Srholec 2008; McCann 2008; Ernst 2010; Fifarek and Veloso 2010). Those which are most likely to benefit from globalization will be the countries containing the regions which most closely resemble the global city-regions. These new emerging locations will increase their knowledge assets markedly vis-à-vis both other more peripheral regions within the same national borders and also relative to other developing economies without such features. Regions lacking these knowledge assets will increasingly be left behind to specialize only in the production of low value-added activities.

In order to understand the current and future role of MNEs in globalization, the arguments in this book point to an ever-increasing importance of the interactions between knowledge and geography. The global cities literature along with the literatures on global production networks and value chains discussed in the previous chapter are ways of identifying these interactions, but other forms of evidence can also be brought to bear on these issues. For example, we can use R&D expenditure as a proxy for the process of knowledge creation. As was highlighted in Chapter 4, although R&D functions are now mainly seen as only part of a wider innovation process, it still remains true that they gain in importance as technological progress becomes more complex. As MNEs on average show relatively higher levels of accumulated competencies, skills and organizational expertise, they tend to be much more research-intensive than other firms (Cantwell 1991). Thus, R&D is a useful tool for uncovering the role of multinationals in the global development of knowledge.

At the same time, as is also shown by the figures reported in Chapter 1, although the advanced economies overwhelmingly dominate R&D expenditure, there is evidence that the small role played by the developing countries in innovation and technology is actually increasing. In fact, in 1991 the developed world as a whole accounted for 97 per cent of total global R&D expenditure (UNCTAD 2005), whereas by 2002 this share had fallen to 91 per cent. In 1996, the top six developing and transition economies (excluding Korea, Singapore, Taiwan and Hong Kong) accounted for just 4.1 per cent of total global R&D expenditure, whereas by 2002 this had risen to over 5 per cent; in terms of private sector R&D, their share rose from 1.2 per cent in 1996 to 3.5 per cent in 2002 (UNCTAD 2005). At the same time, there has also been a substantial rise in innovation outputs on the part of MNEs from developing or transition economies. Between the two periods of 1991–1993 and 2001–2003 the share of foreign patenting in the US from developing and transition economies increased from 7 per cent to 17 per cent (UNCTAD 2005).

While global R&D is increasing, there have also been some important structural changes taking place over recent decades in the nature of multinational R&D, exactly as highlighted by the literature discussed in Chapters 4 and 5. In particular, the knowledge creation role played by MNE foreign affiliates – and their growing specific- and strategic-asset seeking nature – is increasing for almost all locations of origin of MNEs and almost all types of host economies (UNCTAD 2005). There are three aspects to underline here in this respect.

First, while in general the large scale internationalization of R&D by multinationals began in the 1980s and accelerated in the 1990s (UNCTAD 2005), the role played by foreign affiliate establishments in the global R&D effort has increased even faster. As a result, between 1993 and 2002 the total R&D expenditure of all MNE foreign affiliates increased from \$30bn to \$67bn, raising their share of global business R&D expenditure from 10 per cent in 1993 to 15 per cent in 2002 (UNCTAD 2005, 2007). The rate of such an increase was more than twice that of MNEs as a whole, clearly demonstrating that the global R&D effort of MNEs is becoming more geographically dispersed (UNCTAD 2005).

Second, the increasing role of foreign multinational affiliates in the global R&D effort is overwhelmingly a story of advanced economies. The share of global business R&D accounted for by the foreign affiliates of MNEs located in developed countries rose during this period from 10 per cent to 13.9 per cent (UNCTAD 2005). If we focus only on multinational R&D, the share of all multinational R&D undertaken by MNE foreign affiliates located in developed countries was 93.1 per cent (UNCTAD 2005).

Third, however, the developing countries are starting to play an increasing role in these processes. The developing world as a whole increased its share of MNE foreign affiliate R&D activity from 1.7 per cent in 1996 to 6.9 per cent in 2002, and this comprises of 0.8 per cent and 6.2 per cent by developing countries, and 0.4 per cent and 0.7 per cent by transition economies, respectively for the two years considered (UNCTAD 2005). As we discussed in Chapter 5, the major motivation for MNEs to locate their R&D in knowledge-regions or centres of excellence in developed economies is to benefit from localized knowledge spillovers and to tap into and monitor technological and innovative activities of local firms. The location of MNE R&D functions in developing and transition economies is instead motivated by one, or a combination of, the following factors: to support foreign production by adaptations to the specific conditions of the market in which products are sold; the availability of R&D personnel or other resources required for R&D activities at a relatively lower cost than in the home location; in order to benefit from the R&D subsidies provided by host governments (Kumar 1996, 2001).

Overall, the increasing role in R&D played by the foreign affiliates of MNEs and the growing attractiveness of developing countries – and of specific regions within their boundaries – as locations for their foreign affiliates are clearly emerging trends which are likely to continue in spite of the global financial crisis. As such, although the relative scale is still much smaller, the role played by developing economies in the global multinational R&D effort is increasing faster than that of the advanced economies (UNCTAD 2005). For example, in terms of the foreign R&D expenditure of US MNEs, the share accounted for by developed countries (including Korea, Taiwan, Hong Kong and Singapore) decreased from 95.1 per cent in 1994 to 88.3 per cent in 2002, while the share of developing and transition economies more than doubled from 4.9 per cent in 1994 to 11.7 per cent in 2002 (UNCTAD 2005; see also Revilla Diez and Kiese 2006).

Understanding these trends of geographical dispersion by multinationals, not just in terms of the levels of foreign investment, but also in terms of the changes in their foreign affiliates' role in R&D generation and in innovation processes more generally, is critical because these types of investments open up the host economy to the latest technology, knowledge and learning, all of which are crucial growth engines and competitive assets. As we have seen in Chapters 4 and 5, MNEs can act as critical conduits for technology transfer. In FDI-led manufacturing and service industries – especially, but not exclusively, those with relatively high technology or knowledge content – in less advanced countries, MNEs are critical actors in the creation of technological capabilities (Revilla Diez and Berger 2004; Padilla-Perez 2006; Iammarino et al. 2008). Their interactions with, and indirect impact on the other actors of the local recipient system are crucial. MNEs have been a very important source of technology transfer –

whereby technology transfer is understood as the reception and utilization by one country of technology developed in another (Graham 1982). In particular, MNEs have been a major source of technology transfer to developing countries, since these countries commonly face structural constraints in terms of finance, technology, marketing and management (Radosevic 1999c). MNEs influence host economies through a wide array of formal (e.g., R&D generation) and informal mechanisms such as technical assistance to local companies, knowledge and skills acquisition by local personnel working for the MNE, imitation of new technologies by domestic firms, etc. (e.g., Caves 1974; Grossman and Helpman 1991; Dunning 1994; Blomström and Kokko 1998; Radosevic 1999; Pietrobelli 2000; Padilla-Perez 2006; Iammarino et al. 2008; Saliola and Zanfei 2009). On the other hand, in the short- and even long-term in many cases, such technological and knowledge resources cannot be internally generated by a developing economy. As seen in Chapters 4 and 5, the absorption, adaptation and improvement on foreign technologies are not automatic and costless processes. Firms and local organizations must engage in deliberate and integrated efforts and devote substantial resources to start up and sustain a gradual process of knowledge accumulation which is conducive to technological capability building (e.g., Pietrobelli 1994; Young et al. 1994; Hobday 1995; O'Donnell and Blumentritt 1999; Narula and Dunning, 2000; Narula 2004.). Therefore, identifying the pattern and nature of MNE investment flows is crucial in determining the long-run growth trajectories of emerging economies in the modern global economy.

8.3 THE BRIICS COUNTRIES

As both their levels of global inter-connectedness and knowledge-related activities grows, at least partly in response to MNE investments, the group of emerging economies known as the BRIICS economies is experiencing an ever-increasing role in the global economy over the twenty-first century. The six BRIICS countries of Brazil, Russia, India, Indonesia, China and South Africa appear to be a rather diverse group of nation states, although to some extent they share several common economic and structural features. Firstly, these are the world's six largest developing economies. Leaving aside the oil-rich state of Saudi Arabia and also the two large and emerging OECD economies of Mexico and Turkey, the BRIICS countries are by far the largest economies in the developing and transition world, and the only developing or transition countries with gross national incomes of over $200bn per annum (World Bank 2007). For comparison purposes, the smallest of the BRIICS countries, namely South

Africa, has an economy which is just slightly larger than that of Greece. Second, all six countries have undergone fundamental structural upheavals and transformations over the last two decades. Third, the role played by multinational investment in all six countries has increased dramatically over recent years. Fourth, as we will see below, the role played by agglomeration processes has also significantly increased in all six countries in the latest periods. The economic geography of restructuring is very apparent in each case.

If we consider the growth performance of the BRIICS countries during the current phase of globalization we see that China and India are currently the two fastest growing large economies in the world, with 2005 growth rates of 10.2 per cent and 9.2 per cent respectively (World Bank 2007). The third fastest growing large economy is Russia, with a growth rate of 6.4 per cent. In 2005, Indonesia's growth rate has picked up again to 5.6 per cent, South Africa's growth was 4.9 per cent, while that of Brazil was 2.3 per cent (World Bank 2007). If we consider the scale of the individual BRIICS economies, we also see that the relative rankings are rather different. China's gross national income in 2005 was $2269.7bn, which ranked it as the world's fifth largest economy, almost identical in size to that of the UK; India's economy was some $804.4bn, and ranked as the tenth largest in the world, just slightly larger than that of Korea; the economy of Russia, at $638.1bn, was ranked 16th in the world, just slightly smaller than the economy of the Netherlands (World Bank 2007). Brazil, with a 2005 gross national income of $662bn, was ranked as 14th; Indonesia, with an economy of $282.2bn, was in the 23rd position, and South Africa, with a gross national income of $223.5bn, was ranked as the 27th largest economy in the world. On the other hand, if we consider per capita income rather than gross income, the rankings are very different again. South Africa, with a 2005 per capita income of $4770, was ranked 85th in the world; Russia, with $4460, was 90th, Brazil, with $3550, was 96th; China, with $1740 per capita, was ranked 128th; Indonesia, whose per capita income was $1280, occupied the 139th position; while India, with an income per capita of just $730, was ranked 158th in the world (World Bank 2007).

From observation of the growth rates of these countries, we see that the growth performance of the three largest BRIICS economies of China, India and Russia has been both higher and also more consistent over a longer period than for the other three BRIICS countries of Indonesia, South Africa and Brazil. These observed differences in both scale and growth rates suggest that it is the three larger BRIICS countries that are currently benefiting the most from the changes in the world's economic relations taking place under the current phase of globalization. These growth and scale observations appear to support the Krugman (2007) "home market effect" theory, whereby firms locate in large markets the activities which are subject to increasing returns to scale, and the efficiency gains associated with investing in these localities then allow firms to export these goods. As we will see in what follows, this argument appears to be particularly appropriate when we consider FDI investments in China.

Two common features of all of the BRIICS economies are that they are very large in both area and population and also contain large cities with extended regional hinterlands. In terms of economic geography, this allows these countries to exhibit both large home market effects and also large agglomeration effects. Amongst the BRIICS countries, by far the most important is China, followed by India, and South and East Asia is, as previously discussed, the most dynamic macro-region for multinational activity.

The dynamic growth of East Asia predates the dramatic institutional and technological changes which took place since the end of the 1980s. As we have already seen in Chapter 6, the 1970s and 1980s witnessed the rapid rise of the Asian multinational firms. Initially, the growth in East Asia during this period was driven primarily by the global rise of the Japanese multinationals. During the 1980s and 1990s this was followed by the Korean and Taiwanese manufacturing MNEs, and the Hong Kong and Singaporean service MNEs. These different waves of multinational firms which emerged from Asian economies have changed the nature of economic relations within the East Asia region, as well as those between East Asia and the rest of the industrialized world. While the other super-regions of NAFTA and the EU are maintaining their roles, Asia is continuing to play an increasingly important part in the global economic system. Indeed, whereas the US and the EU have grown at similar rates since 1970, East and South Asia has shown over twice the rate of both of these regions (Fujita 2007a). In the last two decades East Asia has continued on its dynamic trajectory and by 2005 its growth rate was 8.9 per cent (in South Asia 8.7 per cent), some 2.7 times the equivalent growth rate for the US and 6.7 times that of the Eurozone (World Bank 2007). On the whole, this rapid growth in the East and South Asia region has been the result of several forces, among which the economic liberalization of China in the early 1990s, the increasing mutual openness of the ASEAN countries, and the density and efficiency of global production networks localized in this macro-region. The various waves of outward FDI from Asian economies, as well as increasing inflows of FDI from North America and Europe, have also been facilitated by the enormous growth of bilateral investment and taxation treaties and by the overall process of intra-area economic integration.

While it is true that over time Asian economies will generate more of their own multinational outputs, the knowledge centres controlling their technological operations are still primarily located in other geographical arenas, such as the EU and the US. For example, at the end of the 1990s the bulk of overseas R&D activities of Japanese MNEs – over 90 per cent – was concentrated in Europe and the US, with a share of around 9 per cent accounted for by other newly industrialized Asian economies such as Korea and Taiwan (Kumar 2001). Although over the past decade or so Japanese MNEs have also located their R&D operations in developing countries, particularly those in East and South Asia such as China, Indonesia, Malaysia and Thailand, the bulk of their operations in these countries consists of adaptive R&D, that is adjustment of existing technologies to the local market conditions (Shimizutani and Todo 2007). Allied with the fact that the EU and US economies still dwarf the combined economies of both China and India, these observations would suggest that views of Asia as an example of 'flattening' trends need to be treated with caution. Moreover, the colossal population shifts from rural to urban areas and from small towns to major cities in both China and India demonstrates the increasing convexity (McCann 2008) argument perfectly. China expects its urban population to double to more than one billion by 2030 (Heitor 2004; Venables 2006).

In response to the increasing globalizing trends, and also to the need to take advantage of both the competitive opportunities and technology transfer possibilities associated with foreign investment, many developing and transition countries have moved not only towards a progressive liberalization of their international investment policy regimes, but also to promote themselves as host locations for inward FDI. In general, the policies initiated to achieve these goals focus on issues such as reducing corporate taxes, structural liberalization and deregulation, or the expansion of promotional efforts, as has been extensively undertaken already in India and Brazil (UNCTAD 2007). In contrast, there is a relatively small number of cases where governments have moved to restrict foreign ownership, such as in the case of Russia, where the strategic sectors of the defence and extraction industries only permit minority foreign ownership (UNCTAD 2007).

From the perspective of investment managers and decision-makers, many of the countries with the highest potential for investment returns also tend to be relatively risky locations. Therefore, international investment decisions have to consider and weigh the risk-return profiles of lower labour costs versus greater communication, coordination and transportation costs. Taking account of all the structural stability and institutional issues affecting FDI yields, USA is ranked as the number 1 country in terms of its potential as a location for inward FDI, and all the top 25 locations are occupied by developed economies, including Singapore as number 5, and Hong Kong as number 15 (UNCTAD 2005).

Yet, as we have seen, one of the major outcomes of the general trends towards market liberalization by developing and transition economies is that these countries are increasingly sought out as locations for FDI from developed economies. Nonetheless, inward FDI exhibits different levels of relative importance in different host economies. Amongst developing and transition countries, the trans-nationality index of openness, which indicates the scale of inward multinational investment in terms of FDI inflows, stocks, value-added and employment relative to total GDP, ranks South Africa 12th, Brazil 22nd, Russia 31st, China 32nd, India 36th, and Indonesia 38th (UNCTAD 2007). In general, across all developing and transition economies, the relative trans-nationality openness to FDI tends to be higher in small countries and lower in the larger economies. This is also broadly true for the BRIICS countries, although Indonesia is relatively more closed than its scale might suggest, while China is relatively more open than its scale would suggest. On the other hand, in terms of the national FDI performance rankings for developing and transition economies, China is ranked 45th, Brazil 62nd, Russia 88th, India 112th, South Africa 126th and Indonesia 136th (UNCTAD 2005). Yet, FDI performance rankings are affected by investment yields, and these tend to favour countries dominated by high risk primary, extraction and raw material industries.

Overall, taking account of all of the relevant risk, structural and potential performance factors, prior to the global financial crisis of 2008, UNCTAD survey evidence of multinational executives' perceptions regarding the most attractive locations for FDI over the coming years found that China was ranked as the number 1 country in the world, followed by India in second position, Russia with number 4, and Brazil with number 5 (UNCTAD 2005, 2007). The sluggish post-crisis performance of advanced countries suggests that perceptions are likely to have been bolstered over recent years. Consequently, these perceptions would suggest that outward FDI – and many related MNE functions and operations – will continue to flow in very large quantities from developed countries into these developing and transition countries over the foreseeable future. As such, these countries are likely to become increasingly open and integrated into the global economic system. However, while the flows of foreign investment into particular developing and transition economies are expected to increase over the next few years, for the foreseeable future the nature of these flows is still likely to remain markedly different from the inflows into developed economies. There are two qualifying aspects to such differences, and these relate to the mode of FDI and the relative importance of R&D-related foreign investment in these countries.

8.4 MODES OF FDI IN THE BRIICS COUNTRIES

Observation of the increasing levels of FDI in developing and transition economies demonstrates that the dominant mode of FDI entry in these countries is quite different to that in developed economies. In particular, if we consider mergers and acquisitions (M&As), developed economies are by far the largest destinations for FDI. As we have noted in Chapter 2, this corresponds to MNEs investing in advanced locations with a blending of different aims, of which the most dominant are likely to be related to the search for specific- and strategic-assets. However, the picture which emerges is very different if we consider the so-called 'greenfield' projects. Greenfield FDI projects are the foreign affiliate investments whereby a brand new establishment is constructed on a new site. The number of such greenfield FDI increased globally by 13 per cent to some 11,800 projects in 2005. Manufacturing accounted for 54 per cent of these projects, with the service sector accounting for 42 per cent and primary industries for 4 per cent (UNCTAD 2007). In terms of broad regions, South, East and South East Asia accounted for 3515, or around 30 per cent of all greenfield projects. China alone accounted for 1378 greenfield FDI projects, or 11.6 per cent of the global total, while India received 981, representing 8.3 per cent of the overall total. For the other BRIICS countries the numbers of greenfield FDI projects are very much lower: 386 for Russia, 145 for Brazil, 93 for Indonesia and 74 for South Africa (UNCTAD 2007). In order to give a sense of the relative scale of these numbers, as we see in Figure 8.1, in the same year the number of greenfield inward FDI projects in the US was 723, in the UK 669, and in France 582 (UNCTAD 2007). Asian economies, and in particular China and India, are by far the most important locations for greenfield FDI projects.

China has now by far the largest number of domestically located MNE foreign establishments, with 42,753 foreign affiliates in 2004 employing some 24 million employees (UNCTAD 2007). These affiliate establishments are heavily concentrated in manufacturing industries. The 24 million employed in China in foreign affiliates represents one-third of the global total workforce currently employed in foreign affiliates: this number has increased fivefold from less than 5 million in 1991 (UNCTAD 2007).

If we consider sectors other than manufacturing, the importance of greenfield FDI is seen to be rather different for some other developing economies. Brazil, for example, hosts the ninth largest number of financial foreign affiliates in the world, as well as hosting by far the largest number of such affiliates amongst developing and transition economies,

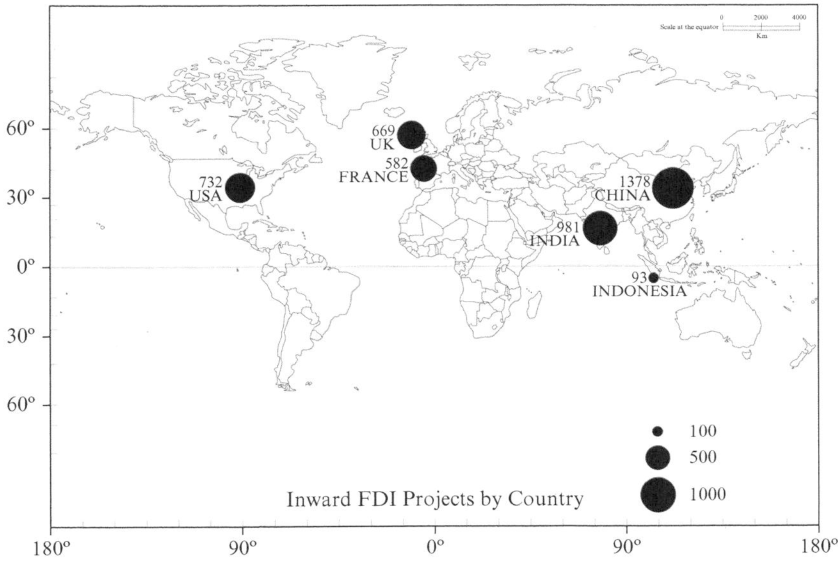

Source: Arita et al. 2011.

Figure 8.1 The number of greenfield inward FDI projects by country in 2005

and even more than Hong Kong and Singapore or China (UNCTAD 2007).

These observations all point to the current dominance of greenfield modes of FDI in developing and transition economies. In these countries inward FDI is therefore qualitatively quite different in nature to the dominant modes of FDI in developed countries. The reasons for such qualitative differences are that mergers and acquisitions tend to be successful where there are potentially significant two-way knowledge flows between the acquiring and the acquired organizations. Therefore, M&As tend to be the preferred mode of foreign investment by strategic- and specific-asset seeking MNEs. On the other hand, in the case of FDI into developing or transition economies, the potential reverse flows of knowledge and technology from the host location to that of origin are in general relatively low. The flows of knowledge and technology transfer therefore tend to be unidirectional. As such, MNEs from developed economies prefer to build new greenfield establishments in developing or transition economies, as this best allows them to organize, configure and control all aspects of the production or service delivery process.

8.5 MULTINATIONAL R&D IN THE BRIICS COUNTRIES

Following these arguments, it is clear that the modes of FDI in developing and transition economies ought to evolve over time from greenfield investments to increasing numbers of mergers and acquisitions, as the knowledge base of these countries steadily expands. The possibilities for upgrading R&D and innovation capabilities of many BRIICS countries over the long term are bound to grow. For example, in 2001 China, India and Russia together accounted for one third of the world total number of tertiary educated technical people (UNCTAD 2005), while the Bangalore high technology industries alone employ 35,000 US educated or trained people (UNCTAD 2005). Yet, of all developing economies, China's growth in its R&D capacity has been the most remarkable. Between 1996 and 2003 China increased its domestic R&D expenditure by over 3.8 times, such that by 2002 China became the only developing and transition economy in the world's top ten R&D expenditure countries. It was ranked 6th in the world in terms of total R&D expenditure, and 7th in terms of business R&D expenditure (UNCTAD 2005). Amongst developing and transition economies, China is now ranked number 1 for both total and business R&D expenditure (UNCTAD 2005), Brazil is ranked number 2, immediately followed by Russia and India. In terms of business R&D Russia is number 2 and Brazil is number 3 (UNCTAD 2005). After the US and UK, China is now the third ranked country in the world in terms of the total number of R&D-related foreign affiliates (UNCTAD 2005).

However, although in absolute figures the scale of China's growth in both domestic R&D and inward FDI-related R&D is so notable, we can get a sense of the relative global and regional contribution of China's knowledge sectors by considering other indicators of innovation. For example, the total 2002 R&D expenditure of developing countries in South, East and South East Asia, including both China and India, is only 14 per cent of the value for Japan, and 6.7 per cent of US R&D expenditure (UNCTAD 2005). Similarly, if we consider the number of US patents and trademarks granted to the residents of particular countries during 2001–2003, we see that China had 1543, India 1022, Russia 956, Brazil 524, South Africa 428, and Indonesia 108. As we see in Figure 8.2, for comparison the respective figures for Taiwan and Korea are 20,414 and 12,195 (UNCTAD 2005). Similarly, in terms of US patents and trademarks granted to firms or organizations during 2001–2003, we see that India had 558, China 475, Brazil 254, South Africa 167, Russia 163, and Indonesia 31. Once again for comparative purposes, as we see in Figure 8.2, the equivalent figures for Taiwan and Korea are 12,686 and

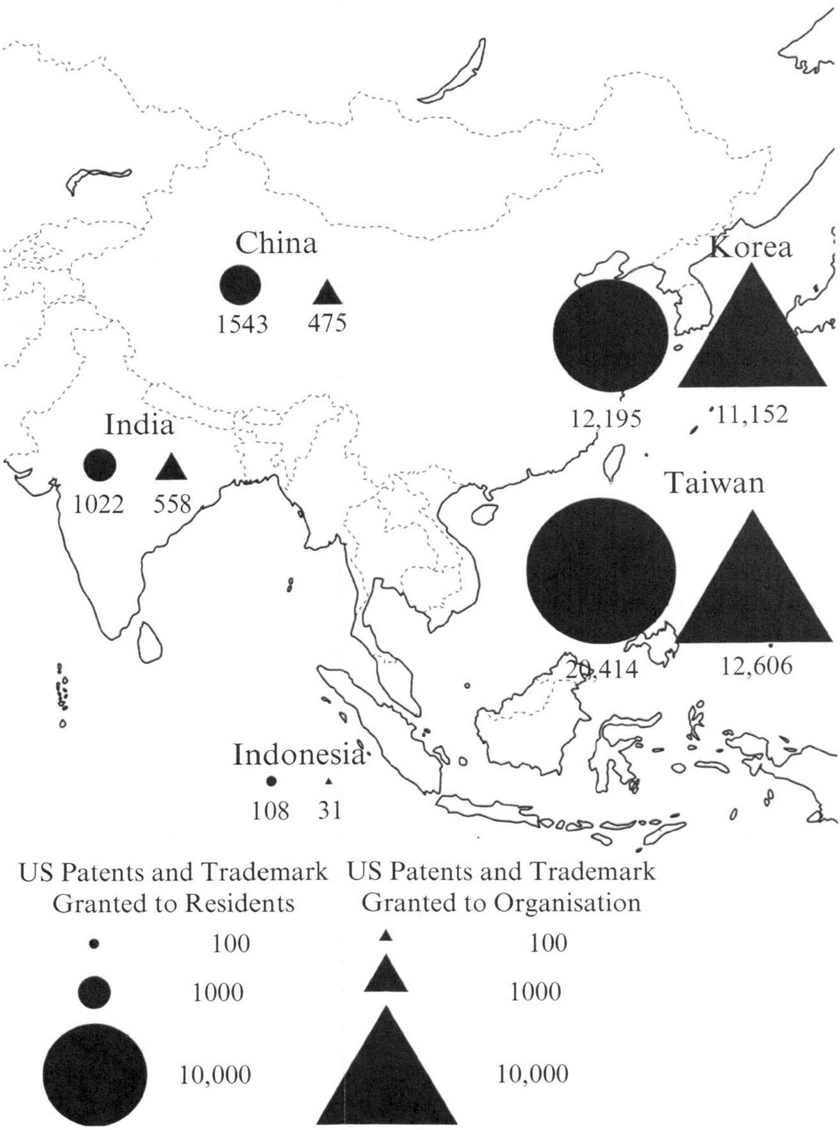

Source: Arita et al. 2011.

Figure 8.2 Asian knowledge indicators: patents

11,152, respectively (UNCTAD 2005). Therefore, although amongst the BRIICS countries China and India appear to be the leading knowledge-generators, their relative contributions are far smaller than those of other advanced Asian economies. Meanwhile, the share of global R&D expenditure accounted for by the BRIICS countries combined is only 4 per cent, of which China accounts for more than half of this percentage (UNCTAD 2005). In contrast, the world's developed economies account for 94.7 per cent of global R&D expenditure.

As such, there is still a very long way to go for these countries to catch up with the knowledge base of the developed world. Yet, the outward flows of FDI by the multinational firms from developed economies will facilitate, encourage, and enable this process. In particular, MNEs will play a catalytic role in the knowledge growth of developing and transition countries as long as there is a positive cumulative causation mechanism at work between a sufficient local absorption capacity and the types of FDI increasingly associated with knowledge investments. Indeed, there is already some evidence that this process of knowledge creation and transfer is well underway. The share of total domestic business R&D undertaken in developing countries which is accounted for by the R&D undertaken by multinational affiliates located in these developing countries rose from just 2 per cent in 1996 to 17.7 per cent in 2002 (UNCTAD 2005).

Amongst developing regions, it is the economies in South and East Asia which are the major locations for MNE R&D investment. During 2002–2004, of the 1773 inward FDI projects involving an R&D component, 1095 (62 per cent) were undertaken in developing or transition economies, of which 861 projects (49 per cent) were located in developing Asia alone (UNCTAD 2005). In the case of developing Asia, the share of R&D accounted for by the foreign affiliates of US multinational firms increased from 3 per cent of their total foreign-located R&D in 1994 to 10 per cent in 2002 (UNCTAD 2005). Similar trends are also observable for MNEs from other developed economies which are locating R&D-related investments in Asia.

Within developing Asia itself, it is China in particular which dominates inflows of multinational R&D investment, and the impact on China of these inflows has also been the most marked. Between 1998 and 2002, the share of total domestic business R&D in China accounted for by foreign affiliates increased from 18 per cent to 22 per cent. The R&D expenditure associated with this R&D-related inward FDI now accounts for 13.5 per cent of China's total domestic (public plus private sectors) R&D expenditure (UNCTAD 2005). To get a sense of how important these MNE R&D investments are to China, we can observe that the

42,000 foreign affiliates located in China are currently employing some 24 million Chinese, and this still only represents 3.1 per cent of the total employed workforce in China (UNCTAD 2007). As such, the relative importance of MNE R&D expenditure to China's knowledge-related activities is four times greater than the relative importance of MNEs to China's overall employment.

However, the obvious importance of inward FDI to the growth of China's R&D capacity is not a general observation across either all developing Asian economies or even across just the BRIICS countries. For example, in the case of India, the most recent reliable estimates indicate that in 1999, multinational R&D expenditure accounted for just 3.4 per cent of domestic private sector R&D, whereas in Brazil the same share is 49 per cent (UNCTAD 2005). Yet, this percentage is increasing slowly in China and India, but is more or less constant in the case of Brazil (UNCTAD 2005).

The types of R&D undertaken in different countries and in different regions of the world also appear to be different. In India, over three quarters of R&D expenditure is on services, and primarily on software development, whereas in Brazil the R&D of foreign affiliates tends to be adaptive R&D, not only for local markets but also increasingly for exports (UNCTAD 2005). In China, most multinational R&D focuses on adaptive innovations for the Chinese market, but there is increasing evidence across Asia suggesting that innovative R&D is growing (Kumar 2001; Shimizutani and Todo 2007). As such, there is significant variation across the BRIICS countries in terms of both the relative importance and the role of R&D-related FDI. Only a small number of developing economies are significant hosts for inward R&D-related FDI, and amongst the BRIICS countries China is by far the most important country.

8.6 THE CHANGING ECONOMIC GEOGRAPHY OF THE BRIICS ECONOMIES

It is well-known in development studies that, in terms of the internal economic geography, as countries develop their employment shares in agriculture tends to fall. Urbanized countries with the lowest proportion of employment in agriculture have higher GDP per capita than transforming economies, which also have higher GDP per capita than agriculture-based countries (World Bank 2008c) Historically, the reasons for this obviously relate to structural effects, and many developing economies which have undergone rapid industrialization from agriculture to manufacturing have also experienced increasing urbanization. In countries undergoing this rural to urban transformation, the rate of growth of labour productivity in non-agricultural activities has been on average two-thirds higher than that for agriculture, in addition to the fact that urban employment has been increasing at nearly five times the general rate (World Bank 2008c). This would suggest that agglomeration effects also operate, and urban-rural income disparities have consequently increased in almost all economies undergoing transformation. If we follow the logic of economic geography arguments, it also becomes clear that economic growth will increasingly be associated with divergence and polarization, with rising differences between the sub-national regions which are winners and those which are losers. Moreover, there is already mounting evidence that such trends are well underway. Here we investigate some of this evidence by initially observing the behaviour of by far the largest, the fastest growing, and the most open of BRIICS economies, namely China. We then extend the analysis to the other BRIICS economies.

8.6.1 The Changing Economic Geography of China

Of all developing and transition economies, the most remarkable transformation has been that of China. Between 1980 and 2000 China increased its share of global exports from 0.9 per cent to 6 per cent, its share of global imports from 1.1 per cent to 4.1 per cent, and its share of global GDP from 2.9 per cent to 3.4 per cent (Fujita 2007b). The result of the increasing involvement in international exchanges was that between 1980s and 2000 China's GDP per capita increased by ten-fold (World Bank 2005). However, the initial impetus for China's restructuring and growth came from fairly modest reforms (World Bank 2005). China began with the introduction of a rudimentary system of property rights in order to create incentives, and only recently gave constitutional recognition to private property (World Bank 2005). However, acknowledging the need to access global capital, technology and knowledge assets via inward multinational investment, China has subsequently also liberalized many rules regarding services industry ownership, as well as manufacturing industry. These changes now allow for greater levels of overseas ownership in many advanced sectors, and have been instituted because China is aiming to attract both a broader range and a higher quality of inward FDI (UNCTAD 2005). In particular, the National Economy and Social Development Plan 2005 emphasized the need to improve the quality of FDI by encouraging it in technology-intensive industries, advanced manufacturing, modern services and agriculture and environmental protection. The plan also encourages the establishment of R&D centres, regional headquarters, and bases of advanced manufacturing, and welcomes the role of FDI in the reform of state-owned enterprises (UNCTAD 2005).

The growth of R&D-related FDI investments in China began in 1993 and reached some 700 projects by 2004 amounting to around $4bn of inward FDI. Most projects were implemented after China's accession to the World Trade Organization (WTO) in December 2001 (UNCTAD 2005). These R&D investments are mainly focussed on technology-intensive industries such as ICT, automotive and chemicals. As we see in Figure 8.3, there is clear economic geography logic to these investments, which are heavily concentrated in a small number of locations. In 2004, Beijing had 189 foreign-owned R&D centres (60 per cent of those in ICT), Shanghai had 140 – of which 91 are in Pudong – and the Guangdong and Jiangsu provinces in the south close to Hong Kong are home to a combined number of over 100 foreign R&D centres (UNCTAD 2005).

The fact that multinational R&D centres are being located in Shanghai, Beijing, Guangdong and Jiangsu provinces represents an unambiguous pattern (see also Head and Ries 1996; Tuan and Ng 2003; Lu et al. 2011). These locations are the core knowledge regions of the country, are growing quickly, and form the major attractive centres for all types of international investment. Both Beijing and Shanghai now exhibit many world city characteristics. For example, Shanghai is now listed in the top ten cities in the most recent global financial centre rankings (COL 2009; Long Finance 2011) with both Beijing and Shenzhen (close to Hong Kong) also both listed in the top twenty global financial centres. At the same time, in 2003 the south eastern provinces of Guangdong and Jiangsu individually accounted for 28 per cent and 19 per cent of FDI, respectively (UNCTAD 2005). Not only FDI in general, but knowledge-related FDI in particular, are being increasingly located in these high-growth cities and regions of China.

During the 1970s and 1980s, inequality between provinces, and also between urban and rural areas in China fell consistently (Golley 2007). Until the mid-1980s, the growth in per capita productivity and expenditure was higher in rural than in urban areas, which suggested a slow process of rural-urban convergence (Angang et al. 2005). Between 1978 and 1985, the ratio of per capita disposable income between urban and rural residents had fallen from 2.57 to 1.85, and that of per capita consumption had fallen to just over 2.1 (Angang et al. 2005). However, from the mid-1980s onwards these urban-rural ratios have been reversed: by 1990 the ratio of both per capita disposable income and consumption had risen to above 2.0. Since the economic reforms started, as expected on the basis of the earlier arguments, inequality between provinces in China has risen continuously (Golley 2007). By 1997, the urban-rural ratios of both

Foreign-Owned
R&D Centers
● 100
● 200

Source: Arita et al. 2011.

Figure 8.3 The spatial distribution of China's R&D centres

per capita disposable income and consumption had increased to approximately 2.5, and by 2001 they had reached approximately 3.0:1, while the ratio of per capita income had risen to 2.1:1 (Angang et al. 2005).

As we see in Figure 8.4, the result of this urban rural divergence was

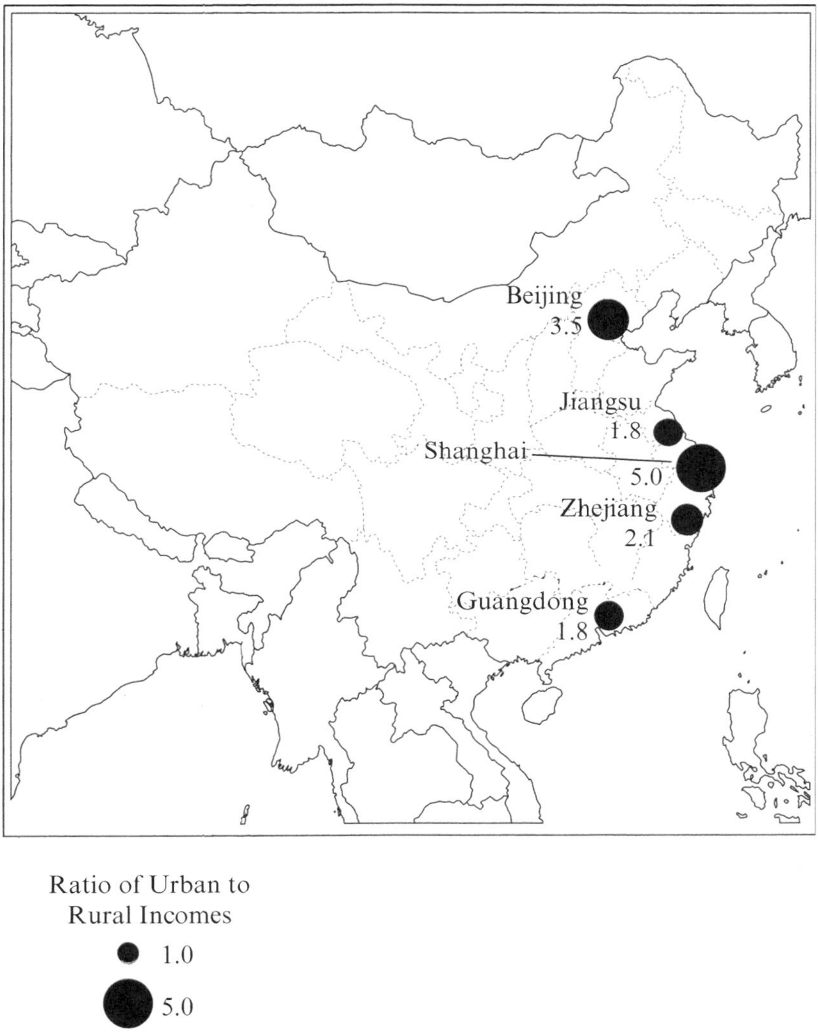

Source: Arita et al. 2011.

Figure 8.4 China's urban to rural income differences

that, by 2004, the largest city-regions in China exhibited the highest per capita incomes, with the ratio of provincial per capita GDP to national per capita GDP being highest for Shanghai (5.0), Beijing (3.5), and Tianjin (2.7). The next highest ratio areas were the regions of Zhejiang (2.1) and

Jiangsu (1.8) – both close to Shanghai – followed by Gangdong (1.8) close to Hong Kong, and then the coastal regions close to the main cities of Fujian (1.6), Liaoning (1.6), and Shandong (1.4) (Golley 2007). These ratios imply that the dominant cities currently exhibit GDP per capita levels which are approximately 13 times those of the lowest regions (Fujita 2007b).

If the three major city-regions of Shanghai, Beijing and Tianjin are left outside, then the increase in regional inequality across China is noticeably reduced, whereas if we group together all of the coastal regions, including the major city-regions, then regional inequality between the coastal and interior regions of China increases even more dramatically (Golley 2007). This demonstrates the role played by particular city-regions in the dramatic growth of China over the last two decades. In 2000, the coastal region between Beijing and Hong Kong as a whole produced 71 per cent of China's total industrial output. This enormous production accounted for more than 60 per cent of national output in all but two sectors, and at least 80 per cent in almost half of the industry sectors, including 97 per cent of China's cultural, educational and sports (Golley 2007). Thus, the growth of China is largely a coastal phenomenon. However, even within the coastal region of China there is a core region, which consists of the South East areas adjacent or close to the major cities and represents broadly a territorial arc bounded by Shanghai and Hong Kong (Golley 2007).

As expected on the basis of our earlier arguments, growing interregional inequality is now a general phenomenon in China. In terms of economic geography, the increasing growth and wealth of certain regions is also strongly associated with the increasing agglomeration of activities in these regions. Once again, this is predicted by economic geography arguments. The core regions of the South East are not only the fastest per capita growth regions, but they are also those of the most rapidly increasing agglomeration. Golley (2007) calculates that, between 1989 and 2000, 26 out of 28 major manufacturing and industrial sectors have become more spatially concentrated, as reflected by increasing spatial Gini coefficients. As such, the general trend towards increasing intra-national inequality apparent across many countries is strikingly manifest also in China.

Regional economic restructuring in China has meant that poverty reduction since the mid-1980s has been most dramatic in the eastern regions, followed by the central regions, with poverty increases in the western regions (Angang et al. 2005). However, this is not just an urban phenomenon. The ratio of per capita farming incomes in the eastern and central regions relative to the West of the country has also increased between 1980 and 2000, from 1.27 and 1.05 to 1.92 and 1.30, respectively (Angang et al. 2005). Unsurprisingly, the areas of highest growth are broadly the regions with highest rates of literacy (Angang et al. 2005). In addition, Chinese income per capita disparities are also exacerbated by a fiscal and transfer system which significantly benefits urban residents (Angang et al. 2005). More generally, however, the competition and wealth effects associated with buoyant regional growth across a range of local sectors tend to spill over to other local sectors, and agriculture in the faster-growing regions will have benefited from such effects.

There is, however, increasing evidence that the spatial concentration effects of economic growth are starting to reverse in China, with economic growth becoming increasingly dispersed across a wide range of locations. MGI forecast that some 350 million will be added to China's urban population by 2025, and that 1 billion people will live in China's cities by 2030 (MGI 2009). This urban population is expected to be comprised of over 220 cities with populations of more than 1 million people, (Europe has 35 such cities), driving an economy which will have expanded by close to 500 per cent between 2010 and 2025 (MGI 2009). This likely trend towards greater geographical dispersion is also being fostered by the institutional decentralization within the Chinese economic system (Xu 2011). These current and emerging trends would suggest that the geography of MNE investment patterns within China will also evolve over the coming decades in ways which will probably be rather different from the inward FDI patterns evident since the late 1980s, which tended to be heavily focussed only on the mega-cities.

8.6.2 The Changing Economic Geography of the Other BRIICS Countries

Many of these same economic phenomena evident in China are increasingly evident also in the case of India. Like China, India also began its economic restructuring with initially modest reforms by reducing trade barriers and distortions to competition within the economy. In 1991 the average tariff was 83 per cent, and only 13 per cent of goods were importable without a license (World Bank 2005). By 1998 tariffs had been reduced to 30 per cent, and the range of goods importable without a license was 57 per cent (World Bank 2005). Since then, Indian GDP per capita has increased by four-fold between 1980 and 2002 (World Bank 2005). As in the case of China, recognizing the need to access global capital, technology and knowledge assets via inward multinational investment, India has also moved to increase its attractiveness as a location for FDI. The Indian Investment Commission aims at attracting foreign as well as domestic investment, while the Foreign Investment Board is intended to act as a one-stop service centre and facilitator for FDI (UNCTAD 2005). In 2004 foreign equity ceilings in Indian aviation services, private banks, non-news print publications, and the petroleum industry were all adjusted upwards in order to attract more international investment (UNCTAD 2005). Yet, even though China and India are often associated in the general globalization discussion, there are actually fundamental differences between the two economies and their responses to globalization. First, one obvious difference is simply in terms of sheer scale. The Chinese economy is almost three times the size of the Indian economy, with per capita incomes of well over twice those of India. Second, there are major trade performance differences between the two countries. In 1950, China's share of global trade was 1 per cent while that of India's was 2.2 per cent, whereas by 2002 the same share had increased to 4.8 per cent for China, while that of India had actually declined to 0.8 per cent (Lardy 2005). In part, these trade performance differences are because China's rapid growth began slightly earlier than India's growth, and also because China's growth has been more dramatic than India's particularly in terms of its increasing trade openness. Third, there are also major differences between the two economies in terms of their industrial structure and sectoral composition.

The structure of GDP by industry in India shows a much greater emphasis on services than is the case in China, where manufacturing is still relatively dominant (Panagariya 2005). In 1980, in China agriculture accounted for 30.1 per cent of GDP, industry for 48.5 per cent (of which manufacturing alone accounted for 40.5 per cent) and services for 21.4 per cent of GDP (Panagariya et al. 2005). On the other hand, in India agriculture accounted for 38.6 per cent of GDP, industry for 24.2 per cent (of which manufacturing accounted for only 16.3 per cent) and services for 37.2 per cent (Panagariya et al. 2005). Even though both countries have undergone enormous changes during the last three decades, the legacy of these inherited industrial structures still remains. In 2002 in China agriculture accounted for 15.9 per cent of GDP, industry for 50.9 per cent (of which manufacturing alone accounted for 34.5 per cent) and services for 33.2 per cent, whereas in India agriculture contributed to 24.9 per cent of GDP, industry to 26.9 per cent (of which manufacturing accounted for only 15.8 per cent) and services to 48.2 per cent (Panagariya et al. 2005).

As a result of its different industrial composition and also its English language advantages, the growth of FDI in India, and particularly the growth of offshoring activities, has been dominated by a range of service industries, rather than manufacturing. This greater emphasis on service industries in India, however, is not yet reflected in terms of the financial services sector. Mumbai, which is currently ranked just within the top 60 global financial centres, is the only Indian city ranked amongst the top 75 financial centres in the world (COL 2009; Long Finance 2011). Yet, in many other aspects the fast-growing industries of India have some similar features to the case of China. First, most of the trade of the Indian and Chinese economies is still in the form of re-exports of finished or semi-finished products or services produced by MNEs which are based in Europe or the US. Second, many of the key growth centres are dominated by links with multinational companies and are often critical nodes in GPNs. In the Indian IT industry, highly concentrated in the Bangalore region, two-thirds of all sales are accounted for by foreign owned MNE affiliates located there (Scheve and Slaughter 2007). Third, as India undergoes continuing regional economic restructuring, firms located in the regions with large home markets earn higher profits (Kambhampati and McCann 2007). This implies that, in terms of economic geography, the large home market effects associated with agglomeration are driving the internal economic growth and restructuring of the Indian economy.

On the other hand, abundant highly trained and cheap human capital, the existence of some public research centres of excellence with international reputation, and the accumulation of sufficient technological capabilities, have attracted to India the location of MNE R&D laboratories and software development centres. Most of these foreign investments have been highly concentrated in space, targeting the regions endowed with the highest technological capabilities and knowledge resources in the vicinity of the centres of excellence (Lall et al. 2003, 2004). Moreover, MNEs have also established collaborations and research contracts with some of India's public R&D institutions such as the National Chemical Laboratory in Pune, the Indian Institute of Chemical Technology in Hyderabad, and the Indian Institute of Science in Bangalore, thereby tapping into the local innovative capabilities and knowledge assets, benefiting at the same time by cheaper costs as compared to the technological centres of excellence of the developed world (Kumar 1996, 1999, 2001). The critical impact of ICT multinational investment on Indian economic development, and its uneven distribution across space within the national boundaries, have been highlighted by a number of studies looking at the rise and evolution of regional cores and clusters located either in medium-sized cities, such as Bangalore, Hyderabad, Chennai and Pune, or within the metropolitan regions of large global cities such as Mumbai and New Delhi (Arora and Athreye 2002; Athreye 2003, 2004; Athreye and Srinivasan 2005; Ramachandran and Ray 2005).

As with the case of China, ongoing economic growth in India is also expected to be associated with newly-emerging cities and regions (MGI 2010b). This should lead to somewhat lower levels of spatial concentration over the coming decades, although to a lesser extent than is the case with China (MGI 2010b). This should also lead to a rethinking of the geography of multinational investment strategies in the Indian subcontinent, which have until now been heavily focussed on a very small number of regions and cities. India's economy is expected to grow by around 500 per cent between 2010 and 2030, by which time almost 600 million people are expected to be living in cities (MGI 2010b). These cities are expected to account for 70 per cent of net new employment, and forecasts suggest that 68 cities will have populations of over 1 million people (MGI 2010b). These cities are also forecast to account for an expected 270 million net increase in the working-age population by 2030, and some 70 million newly-emerging middle class households, thereby increasing the stock of middle-class Indian households from the current level of 22 million to 91 million by 2030 (MGI 2010b).

A similar picture emerges in the case of Indonesia. In terms of internationalization and globalization, the current phase of economic restructuring was driven by the Jakarta Declaration, which outlined the government's vision for infrastructure development, and also its commitment to removing the bureaucracy which was impeding private investment (UNCTAD 2005). Bureaucracy is a major problem in Indonesia: in recent years, the time taken to start a business in Indonesia is 151 days, which is almost four times as long as in China (UNCTAD 2005). The government has also introduced a one-stop investment service for FDI. Other initiatives proposed at the time included abolishing the requirement for foreign affiliates to sell part of their shares to local investors after a certain number of years; and also the abolition of the 30 year limit on business licenses for foreign investors (UNCTAD 2005). These initiatives are all aimed at opening up the country to further foreign investment, with the aim of promoting internal competition and technology transfer. However, increasing trade and investment openness is also likely to be associated with increasing internal disparities between Indonesia's regions, as inward investing MNEs seek out the pecuniary advantages associated with agglomeration. This is evidently so, for example, even in the case of the relatively low knowledge intensity of apparel and textiles industries, the sector in Indonesia which has exhibited the greatest revealed comparative advantage since the 1980s (James 2006). To the extent that technology transfer and spillover effects from inward FDI also operate in these sectors, these are likely to exacerbate the existing advantages of agglomeration in the leading urban regions of Java (Amiti and Cameron 2007).

Like China, GDP per capita in Indonesia is closely related to city size, with the capital city of Jakarta exhibiting GDP per capital levels which are approximately 13 times that of the poorest regions (Fujita 2007b). As in the case of India, in Indonesia it is only the largest city of Jakarta which is listed in the world's top 75 financial centres (COL 2009; Long Finance

2011). Indeed, rather than finance, this region is dominated by manufacturing. Java – particularly the Jabotabek region (i.e. Jakarta, Bogor, Tangerang and Bekasi) in West Java, and Surabaya in the East part of the island – has by far the highest concentration of manufacturing industries of the country (Deichmann et al. 2005). Since the rapid industrialization of the 1980s and 1990s, the bulk of foreign direct investment has been located in the Greater Jakarta region. A wide array of regional development policies – ranging from investment incentives for infrastructure and human capital formation, to tax exemptions and transport networks projects – have been put in place to tackle the huge disparities across the country, and to address the serious concerns about remote regions such as Eastern Indonesia (Hill 1991; Deichmann et al. 2005). Although these peripheral regions in the Indonesian archipelago are endowed with rich natural resources, their level of socio-economic development and attractiveness remains very poor due to several factors: low population density, remoteness of some locations and communities, inadequate general infrastructure, high transport costs, cultural and language barriers (Deichmann et al. 2005). Empirical estimations (Deichmann et al. 2005) seem to suggest that further public investment in infrastructure and transport networks to increase the attractiveness of lagging regions to MNEs operating in standard industrial activities is likely to prove ineffective, as the strong urban and regional agglomerations, particularly of capabilities and skills observed in the Java region, will continue to exert centripetal forces towards the activities of both domestic and foreign firms.

In the case of Latin America, Brazil is by far the largest recipient of inward FDI in the region, receiving approximately $19bn of annual inward FDI, which represents over 40 per cent of FDI in South America (UNCTAD 2007). However, although inward FDI in South America grew by 18.5 per cent between 2004 and 2006, annual inward FDI inflows into Brazil remained fairly static during this period (UNCTAD 2007). This is in contrast with the trend registered by Brazil's FDI outflows which increased dramatically to the largest in the region in 2006, amounting to the record level of $28bn. For the first time, Brazil's FDI outflows surpassed its FDI inflows: though this was mainly due to one mega-deal in which a Brazilian mining company purchased a foreign company (UNCTAD 2007). Apart from this one case, however, over the medium term it is still likely that FDI inflows into Brazil will continue to significantly outweigh the FDI outflows, which previously were typically between $3bn and $9bn per annum (UNCTAD 2007).

Both inward and outward FDI flows from Brazil tend to be concentrated in extractive industries, resource-based manufacturing industries, and also in the infrastructure and telecommunications sector (UNCTAD 2007).

On the other hand, foreign investments in various key sectors of the Brazilian economy, such as the automotive industry, sugar refining, and the steel industry, have increased over recent years. During the 1990s and early 2000s, foreign MNEs made \$20–\$25bn of investments in the MERCOSUR automotive industries, 80 per cent of which were in Brazil (UNCTAD 2005). These investments were primarily of a market-seeking nature, focussed on the large domestic markets. More recently, FDI in the automotive industry of MERCOSUR, which is dominated by Brazil, is increasingly focussed on export markets such as Mexico, rather than on domestic consumption, and this has been helped in part by major currency devaluations (UNCTAD 2005). Nowadays, the two largest major cities of Brazil, namely Sao Paulo and Rio de Janeiro, are both ranked just within the top 50 global financial centres, but more significantly, they are the two principal financial centres in Latin America (COL 2009; Long Finance 2011).

As in the case of Indonesia, these developments are likely to strengthen further the very significant regional inequalities of Brazil. Brazil is a country which exhibits very sharp centre-periphery features which have proved to be particularly resistant to the wide range of policy instruments applied over time to reduce sub-national imbalances. The Northeast has historically been the poorest region in the country, with per capita incomes about one-half those of the rich Southeast areas (Lall and Shalizi 2003). Moreover, differences in regional per capita incomes have remained steady over long periods of time. Per capita income in the Southeast was 2.9 times that of the Northeast in 1939, and 2.8 times in 1992 (World Bank 1998; Lall et al. 2004). Only over recent decades has evidence emerged of something of an interregional process of convergence taking place, leading to a narrowing of regional productivity differentials (OECD 2011). However, the ongoing dominance of the large metropolitan regions around São Paulo and Rio de Janeiro, along with the cities of the south and east coastal regions (Monteiro Monasterio 2008), means that the economic fortunes of the peripheral regions in the North will still be largely dependent on the behaviour of the core regions, whose own performance is largely independent of that of the periphery (Perobelli et al. 2008).

Across Latin America as a whole, by 2025, some 315 million people are expected to be living in almost 200 large cities in Latin America, up from the current numbers of some 260 million people today (MGI 2011b). This urban population growth of 65 million people is expected to comprise some 50 million new workers entering the labour force. Differently to the case of China, for example, the urban population growth in Latin America is expected to occur largely within the existing system of large cities, rather than via the rapid growth of new large cities (MGI 2011b). If such urbanization patterns do operate as expected, the changes in the overall economic geography of inward FDI are likely to be rather less dramatic than the expected changes foreseen in the cases of China and India (MGI 2011b).

In the case of Russia, regional transformation has been extremely rapid. Apart from the major natural resource extraction regions, economic growth has been increasingly localized in a small number of regions, dominated by the major cities of Moscow and St Petersburg (Hanson and Bradshaw 2000), both of which are now listed just within the top 70 financial centres in the world (COL 2009; Long Finance 2011). FDI inflows to the Russian federation doubled between 2005 and 2006 to $28.7bn, and are dominated by natural resource-based activities (UNCTAD 2007). However, the Russian government has moved to restrict foreign ownership in the strategic sectors of the defence and extraction industries, only permitting minority levels of foreign ownership for firms in these industries (UNCTAD 2007). Meanwhile FDI outflows from Russia have continued to increase between 2005 and 2006 by 41 per cent, reaching $18bn. This was particularly in resource-based firms attempting to globalize, as well as in banking FDI outflows into other former Commonwealth of Independent States (CIS) countries (UNCTAD 2007).

Recent empirical evidence, however, indicates some peculiar trends in the economic geography of Russia. Firstly, and most predictably, the dramatic increase, particularly during the late 1990s, in the weight of the capital city, Moscow: the GDP per capita in Moscow in 1995 was 75 per cent above the country average, while in 2000 it was more than 250 per cent (Benini and Czyzewski 2007). Secondly, there has been a strengthening of the natural resources and energy endowed regions, in line with the emphasis put by the government on a development strategy labelled as "natural resources led growth economy" (Benini and Czyzewski 2007, p. 131). Thus, on the one hand, the dominant trend toward agglomeration in the Western cities with relatively high quality infrastructures and services point to "new polarization effects, away from the old type of spatial concentrations, irrespective of distance and costs considerations" (Benini and Czyzewski 2007, p. 132). However, on the other hand, the economic and demographic decline experienced by peripheral regions rich in natural resources, such as Siberia and the Far East, appears to be somewhat at odds with the main economic natural resources-based strategy of the government. In particular, such peripheral regions are in fact those most penalized by the constraints imposed on MNE investment which, at least as far as the large and most remote periphery is concerned, offer the greatest development opportunities. The FDI restrictions therefore appear to clearly prevent a more balanced regional development in the country.

South Africa opened up its economy to the global market in 1994. Trade liberalization had a very significant impact, with exports plus imports rising from 47 per cent in 1996 to approximately 60 per cent of GDP in 2004. In terms of inward investment, South Africa in 2005 accounted for 21 per cent of all FDI inflows into the African continent. In 2005 the country actually witnessed a fall in FDI inflows due to the sale of a foreign equity stake in a domestic gold-mining firm to a domestic firm, although South Africa remains Africa's major location for inward FDI, and the continent's major source of outward FDI (UNCTAD 2007). As well as the traditional investors from the UK and USA, there is now also increasing interest in foreign investment in South Africa on the part of investors from Asia, a trend which reflects the overall increasing interest of Asian MNEs in Africa in general (UNCTAD 2007).

As South Africa's trade, and in particular its manufacturing exports, have increased dramatically over the last two decades, the relationship between trade and economic geography has become much sharper. The dominant commercial centre Johannesburg is currently just outside the top 50 financial centres in the world (COL 2009; Long Finance 2011), and is the only city in the whole of Africa within the world's top 75 global financial centres (COL 2009; Long Finance 2011). More generally, however, in terms of the internal economic geography, economic growth in South Africa is being increasingly dominated by the urban centres, and particularly those with major transportation infrastructures (Naude and Krugell 2003). In 2000, some 84 per cent of South Africa's manufacturing exports were accounted for by only 6 per cent of the magisterial districts (Naude and Krugell 2003). As already noted in the case of the other BRIICS economies, considering agglomeration effects enforces constraints in the typical policy instruments for regional development (Nel and Rogerson 2009). For example, increasing productivity by attracting FDI in high value added and technology intensive manufacturing activities will tend to benefit those South African urban centres where such activities are already strongly agglomerated (Flaherty 1995). In agriculture, successful export crops are also disproportionately cultivated in the richer regions (Flaherty 1995). Moreover, the trade-off between growth and equity is particularly difficult to tackle in the case of South Africa, as interregional disparities are also strongly correlated with interclass and interracial differences (Flaherty 1995). However, empirical studies have pointed out that the reduction of such inequalities may indeed depend on the economic growth of South Africa's largest cities, and on the way that their further spatial agglomeration processes interact with smaller cities and the wider urban hierarchy (Naude and Krugell 2003).

8.7 CONCLUSIONS

In this chapter we have seen that the role of multinational affiliates is increasing globally. One aspect of this is that the R&D and innovation generation role of MNE affiliates is also increasing, in both the developed and developing economies, exactly as expected from the arguments presented in Chapters 4 and 5. In terms of economic geography, global FDI flows appear to have certain characteristics to them. First, such FDI flows are still overwhelmingly dominated by the developed economies, both as destinations for inward FDI and as origins of outward investments. Moreover, such a leadership has been maintained over recent years during the current phase of globalization. Second, and consistent with the investment development path thesis (IDP) discussed in Chapter 2 (Dunning 1981; Dunning and Narula 1996), global FDI inflows into developing and transition economies are far more important than the global outflows from such economies, in spite of their increasing share in outward FDI flows in the most recent years (e.g., Amighini et al. 2009; Athreye and Kapur 2009). Third, the major MNEs are almost entirely from the developed areas of the world. Fourth, although the outward flows of FDI from developing and transition countries are increasing markedly, in absolute terms these flows are still very small in relative terms, and their destinations are still mainly orientated towards the most developed economies (Revilla Diez and Kiese 2006). Fifth, greenfield FDI plays a much more important role in developing countries than M&As, and this is particularly noticeable in the case of the Asian economies. This is in contrast with the main modes of foreign investment among the developed economies, for which mergers and acquisitions represent the major form of multinational expansion. This supports the fact that the types of MNEs examined in previous chapters – even when they belong simultaneously to all four broad categories described in Chapter 2 – differentiate their strategies across different levels of economic development and geography.

These various observations suggest that as the scale of the developing economies, and the BRIICS economies in particular, increases over the coming decades, the role which these countries will play in multinational activity is bound to increase at all levels. Outflows of multinational investment will increase from these countries into both other developing economies and also into the developed countries, as MNEs seek new knowledge assets as well as new resources and markets. Thus, the forms of FDI outflows from these emerging countries are also likely to increasingly take the form of mergers and acquisitions in the developed economies, and of greenfield projects in the lower wage developing countries. Within these general trends, the growing multinational expansion of firms from the largest and fast-growing BRIICS countries, such as China and India, is already particularly noteworthy and reflects all the above expectations (Athreye and Kapur 2009). The internationalization of MNEs from developing countries is motivated by the search for strategic assets, such as new technologies, capabilities and brands, for access to raw materials and distribution networks. As succinctly put by Athreye and Kapur (2009, p. 213), rather than exploiting existing assets, MNEs from developing and transition countries are seeking to acquire or enhance their assets and to upgrade their technological capabilities.

At the same time, agglomeration phenomena within the BRIICS economies, and more generally in the developing and transition areas of the world, as we have seen, are occurring at an impressive speed. According to the UN report on the State of the World Population 2007, the current twenty-first century could also be labelled as the "Urban Millennium" (UNFPA 2007). It is estimated that in the next decades up to 2030, 93 per cent of urban growth will occur in developing countries, with 80 per cent of urban growth occurring in Asia and Africa (UNFPA 2007). Thus, major cities in these countries are expected to play an ever more prominent role in the global city rankings, and their surrounding regions will increasingly acquire the status of core regional locations within the global economy.

The patterns evident in terms of the geography of growth in the major emerging economies are not, however, static or predetermined. Indeed, recent evidence suggests that even these patterns are again evolving. The mega-cities of the developing world, with populations of over 10 million people, on average account for per capita GDP levels which are some 80 per cent higher than their national averages (MGI 2011a). As such, over the last two decades, corporate and global investment strategies have tended to focus primarily on investing in the developed economies, along with the mega-cities of the large emerging countries, because together the developed economies plus the emerging economy mega-cities account for more than 70 per cent of the global economy (MGI 2011a). This pattern, however, is likely to change significantly over the coming decades and there are two reasons for this. Firstly, developing countries as a whole are expected to contribute 74 per cent of economic growth up to 2025 (MGI 2011a), three times the total global economic growth contribution of developed economies. Secondly, as has already been mentioned, the increasing role in global output growth played by cities in the developing countries is not expected to come from the very largest cities. As we have seen, the contribution of the world's largest 100 cities (defined in terms of economic output) to global output is expected to fall from 38 per cent in 2007 to 35 per cent in 2025 (MGI 2011a). This reflects an ongoing trend, whereby over recent years most mega-cities have not grown faster than their own economies, and indeed many appear to be experiencing diseconomies of scale (OECD 2011; MGI 2011a). The share of global output growth to 2025 accounted for by mega-cities of over ten million people is forecast to be some 10 per cent, such that the overall contribution of the mega-cities to total global output is expected to fall from 14 per cent in 2007 to 13 per cent by 2025. The changing geography of economic growth away from both developed economies and also from the mega-cities of the developing countries is likely to have profound impacts on the economic geography of MNEs and FDI, as these firms increasingly adapt their location strategies in order to achieve the first-mover advantages necessary to capitalize on the new global growth patterns.

We can get a good sense of the likely orders of magnitude and the emerging spatial patterns of global economic growth from a range of indicators. In terms of the scale of the emerging markets, taken together, the urban populations of China, India and Latin America are expected to increase by some 700 million people by 2030, a figure which is of a similar order of magnitude to the current combined populations of the USA and the European Union (MGI 2009, 2010b, 2011b). In the developing world as a whole, 235 million urban households are expected to have household incomes of $20,000 or more (MGI 2011a). In China alone, some 50,000 skyscrapers are expected to be built (equivalent to ten times as many as New York City), with 40 billion square metres of floor space in five million buildings (MGI 2009); in terms of infrastructure this could amount to 170 new mass transit systems being built, along with some 5 billion square metres of new roads (MGI 2009). In the case of India, some $1.2 trillion of capital investment are projected to be required to meet the population and demand increases of India's cities by 2030 (MGI 2010b). This is expected to include 700–900 million square feet of commercial and residential space needs to be built every year, which is equivalent to the current scale of Chicago. In terms of infrastructure, some 2.5 billion square metres of new roads need to be built (MGI 2010b), along with 7400 kilometres of new metros and subway systems (equivalent to 19 London Undergrounds, 22 New York Metros, 35 Paris Metros, or 36 Tokyo Metros).

These remarkable growth figures might appear to point to the geography of economic growth in developing countries being still concentrated in the mega-cities. However, as we have already seen, while developed countries are expected to contribute only 26 per cent of economic growth up until 2025, the mega-city regions of the developing world are only expected to contribute another 8 per cent of economic growth As such, almost two-thirds of global economic growth to 2025 are expected to come from the *non*-mega-city regions of the developing world (MGI 2011a). Of this two thirds of global growth, 29 per cent is forecast to come from rural regions with small cities of less than 150,000 people, and 14 per cent is expected to come from the small to mid-sized urban regions with city populations of between 150,000 to 2 million people; 13 per cent from regions with city populations of 2–5 million people; and 10 per cent from regions with city populations of 5–10 million people (MGI 2011a). Therefore, 43 per cent of global economic growth to 2025 is forecast to come from developing regions with city populations of less than 2 million people, and some 56 per cent of total global economic growth is expected to come from developing regions with city populations of less than 5 million people (MGI 2011a). The fastest rates of growth are forecast for the middle-sized cities of between 150,000 and 10 million people, a grouping of cities which will account for half of global economic growth to 2025. Around 150 of these middle-sized cities are in developed economies while over 420 of these are in developing economies (MGI 2011a).

At this stage, what is not clear is exactly how long these trends will take to unfold and how many years it will be until the developed world will rely on these newly emerging economies as critical sources of new knowledge, capabilities and technology. What it is clear, however, is that for the foreseeable future the developing world is bound to be far 'spikier' than 'flatter', although the relationships between the different geographical spikes and MNEs will again evolve. MNEs play a critical role in contributing to the spread of knowledge and production activities, and the actual economic geography of FDI, R&D and innovation in developing countries will depend both on how MNEs choose to spearhead or respond to the evolving economic geography of economic growth.

9. Conclusions: review of the arguments and implications for future research

In this book we have argued that understanding economic geography is increasingly important for understanding multinational enterprises (MNEs), and in turn MNEs are nowadays playing a critical role in shaping economic geography at various spatial levels, from the local to the global. In particular, our aim has been to demonstrate that, although these relationships between geography and firms are in principle valid for any kind of business or firm, they are *particularly* valid in the case of MNEs, and also more than ever are *particularly* important now for MNEs. Moreover, this book has argued that the pivot on which all of these relationship turn is the creation, diffusion and management of new knowledge and technology. Each of the chapters of the book, and also the sequencing of the topics discussed in these chapters, are designed to elaborate these arguments, bringing the reader right up to date with the latest debates.

The book began by arguing that multinational activities and multinational expansions were the key dominant characteristic of the modern era of globalization. Chapter 1 presented a variety of evidence to demonstrate that understanding the investment and decision-making behaviour of modern multinationals is key to understanding the modern global economy. More specifically, our contention in this book is that the economic geography of globalization is much less well understood than ought to be the case, given the scale of the transformations we are currently witnessing, and much of the blame for this is attributable to the fact that the locational behaviour of multinational firms has traditionally been heavily under-researched. The reasons for this are variously related to academic fashions, perceived political-economy realities, and issues of data availability. Yet, our reason for writing this book is to demonstrate that actually we know far more than we typically realize about these matters, but that previously this has not been understood. This is because the various analytical and empirical jigsaw pieces which need to be assembled in order to present a clearer image of the relationships between multinationals and economic geography, have in the past been far too scattered across different literatures and methodologies to be presented as a coherent whole. As such, this book has attempted neither to offer any new theories nor any new data on these matters. Rather the objective of the book has been to assemble the currently-available arguments and evidence in a manner which better allows for the image to appear. In order to do this we have adopted a particular analytical lens through which we observe and interpret these issues, namely that of knowledge and technology, and this approach runs throughout the various chapters of the book. This is not an arbitrary choice, but reflects what we contend is the best approach for unifying all of the currently-available arguments and evidence regarding the two-way relationships between multinational corporations and economic geography.

Chapter 2 reviewed the various international business theories which have examined the rationale for multinationalism – in other words the *why multinational?* question. The answers offered to this question are typically based on Dunning's (1977) eclectic OLI paradigm which explains the why multinationalism arises in terms of the ownership O advantages which firms wish to exploit. Similarly, the answers to the questions regarding the mode of multinationalism – in other words the *how multinational?* question – are generally found in the various knowledge-internalization I explanations evident in the international business literature and in neoclassical trade theories. Moreover, the links between O and I are also examined in managerial and sociological approaches such as the resource-based view and the evolutionary theories of technological change. O and I are seen to frequently co-evolve, such that *being* multinational can be regarded as a stage within the evolutionary development of a firm.

If we consider the three pillars of the OLI paradigm, however, it is clear that the major shortcoming of the literature on MNEs so far is the issue of location L – in other words the *where multinational?* question. In comparison to our understanding of O and I, and also of the interactions between O and I, the progress made so far in understanding the economic geography of the MNE has been very limited indeed. In part, the limited progress on the L-location pillar in comparison to the O and I pillars was related to the perceived lack of importance of any needed fine-grained locational analyses during earlier decades when countries were still largely closed entities. Now that countries are opening up to one another and multinational firms make ever-more specific locational decisions, there is now an increased urgency to understand these issues. Chapter 3 presented some orthodox location arguments in order to foster discussion around these issues, and what becomes evident is that very simple and classical location models provide powerful insights into the behaviour of both multiplant domestic enterprises (MPDEs) as well as multinational enterprises (MNEs). In particular, these models underline the critical importance of both knowledge and transactions costs in determining the locational behaviour of these firms.

The orthodox firm location models discussed in Chapter 3 also highlight the fact that the MNE is simply a special case of the MPDE, and many aspects of the locational analysis of both MNEs and MPDEs therefore share common elements. However, to the extent that the MNE is a special case of the MPDE, it is special in the sense that the variety and magnitude of knowledge sources that MNEs access and the variety of transactions costs challenges that MNEs must overcome, are very much greater for MNEs than for MPDEs. However, it is not simply a matter of scale and magnitude. As we saw in Chapter 4, the much greater complexity of the environment in which MNEs operate means that there are fundamental differences between MPDEs and MNEs, the differences of which centre on the importance of issues of knowledge generation, knowledge acquisition, knowledge transfer, capabilities, learning and innovation. The innovation systems literatures discussed in Chapter 4 emphasizes that the linkages between a firm and its environment are central to the modern economics of technological change and evolutionary views, and early location theory also builds on these arguments. However, the way in which these arguments must be applied in order to analyse the geography of MNEs is very much more sophisticated than the ways in which the arguments can be applied to understanding the geography of MPDEs.

On the basis of all of the foregoing arguments Chapter 5 offered a framework for positioning MNEs within the context of economic geography by means of a knowledge-transactions costs taxonomy of different types of spatial-industrial structures and geographies. This allows us to identify the key linkages operating between particular types of firms and particular types of regions, and to distinguish between which types of regional structural conditions will be favourable to which types of MNE firms. We argue here that this approach provides much richer insights than dichotomous models based on distinctions between horizontal integration versus vertical integration, home country versus foreign country, or concentration versus dispersion. This taxonometric approach demonstrates that multiple types of firm-cluster interactions are discernible, all of which can involve multinational investment, and all of which can evolve in various ways. The crucial issue which the taxonomy uncovers is that the extent to which the rationale and characteristics of the MNE 'fit' with those of the local economy is critical for understanding the relationship between location and innovation. What becomes evident is that the degree of 'fit' between the knowledge and transactions costs characteristics of the MNE and its local environment determines the extent of knowledge spillovers and locally-embedded innovative linkages.

One of the major contentions of this book is that in order to understand the current local linkages evident between multinationals and their local environments it is necessary to understand the historical evolution of the local spatial-industrial configuration in which the MNE establishments, affiliates and subsidiaries operate. Similarly, in order to understand the current global context in which multinationals invest and locate facilities in general, it is necessary to understand the evolution of the global economic order in which MNEs currently operate.

As we saw in Chapter 6, the modern era of globalization was ushered in by the global transformations which took place between 1988 and 1994. This period saw dramatic changes in technology and institutions at a global scale, which forever changed the global economic context in which commerce operates. However, globalization was a process which had been taking place for over 400 years. In contrast, much of the twentieth century was a period of anti-globalization, and as such it was a period which ran counter to long-run historical trends, and from which the global economy emerged in the latter decades of the twentieth century. By observing these historical patterns it became clear in Chapter 6 that the modern era of globalization contains some elements in common with previous eras of globalization, in terms of high degrees of trade and investment openness. However, the principal distinguishing feature of the modern era of globalization is characterized by enormous and rapid movements of knowledge, and in particular knowledge embodied in multinational firms operating new production, management, distribution and decision-making systems, in finance, services, and manufacturing. Global cross-border investment is nowadays dominated by two-way financial flows between adjacent rich countries and in particular within services industries. In contrast, the first three centuries of globalization were dominated by investment flows from rich countries to distant and poorer regions within an empire context and primarily these flows were in manufacturing and extraction sectors. The result is that mergers and acquisitions nowadays dominate greenfield investments, the dominant mode of FDI in earlier eras of globalization. As such, FDI and globalization are very much dependent on international business confidence, and while recent trends may be partially disrupted in the aftermath of the 2008 global financial crisis, the opening up of the BRIICS countries will continue to drive globalization processes.

These issues of the 'fit' between a multinational establishment and its local environment raised in Chapter 5 have therefore been thrown into sharp relief by the dramatic changes in the global economic context described in Chapter 6 which have taken place since the advent of the modern era of globalization commencing in the late 1980s and early 1990s. As we saw in Chapter 7, changes in transport, information and communications technologies along with transformations of the global institutional context all mean that commerce today operates within a fundamentally different set of relationships between firms, knowledge, technology and geography than in previous periods. The modern relationships between knowledge acquisition, knowledge transfer and the transactions costs associated with generating and transferring knowledge across geographical space, imply that the global economy is in many ways becoming rather more uneven and more 'spiky', and not flatter and more even as many popular commentators contend. Following the taxonometric arguments developed in Chapter 5, in Chapter 7 we saw that in today's global economy, these spikes are the most appropriate regions for many MNE establishments to be located, and in particular for MNE establishments dealing with high value knowledge-related activities. In other words, MNE establishments focussed on seeking of specific assets or strategic assets will be located in these 'spiky' places. In contrast, MNE resources-seeking establishments will increasingly be located outside of these spiky expensive locations and in lower cost regions. In terms of their knowledge, technology and innovation features, the relationships between an MNE establishment and its immediate locality can reflect either of the four spatial-industrial typologies discussed in Chapter 5, and the modern MNE explicitly aims to design these relationships to maximize the value of the subsidiary's activities to the whole corporation. As such, not only do MNEs exploit globalization opportunities by increasing the cross-border geographical dispersal of their establishments, subsidiaries and affiliates, but also they exploit globalization opportunities by ensuring better matches between the features of the individual establishments and the potential resources and/or assets which can be derived, developed and exploited from their localities. This combined ability to exploit both global and local dimensions is exactly why MNEs are the key drivers of modern globalization. Their ability to play this dual role is contingent upon developing a cross-border corporate control and organizational system, the inter-establishment logic of which in knowledge-management terms is akin to the industrial complex model, but which can be tailored to the knowledge-relations specifics of the local context.

What these arguments point to is the critical importance nowadays of clear definition of exactly what we mean by the location space L which MNEs consider in their geographical investment decisions. As we saw in Chapter 2 in most of international business arguments the location choice space L has traditionally been interpreted simply in terms of a country, whereas in today's increasingly open global economy, this concept of geography is totally inappropriate. MNE headquarter decision-makers obviously consider countries as locational options. However, nowadays they also explicitly consider different cities and city-regions both within countries and between countries as key locational spaces for their locational choice options. As such, the importance of L very much operates at the sub-national regional level. At the same time the importance of L is also very important at the super-national cross-border level. For all types of MNE establishments, the empirical evidence suggests that the spatial patterns MNE establishments are generally designed to be in specific sub-national regions, places which serve super-regional cross-border markets rather than either national or global markets per se. These super-regional markets comprise groups of adjacent countries which are undergoing economic integration, and the locations from which these integrating super-regional markets are served are nowadays often described as being 'global cities' or global city-regions. However, as we saw in Chapter 7 the key features of these global cities are not simply scale-agglomeration advantages of a form familiar to the classical authors writing over a century ago on the economics of cities. Rather the key feature of these global cities is that of the city's connectivity in terms of its global knowledge inflows and outflows. Indeed, in rich countries the most productive cities are not the largest cities but those which display the highest levels of connectivity in the global economy, and particularly knowledge flows mediated by multinational firms. The geography of multinational firms increasingly mirrors the global urban hierarchy, and vice versa. As such, the 'correspondence principle' arguments of Hymer (1972) have been proved to be prophetic, while the original urban economic arguments of Vernon (1960) are also to some extent still pertinent. In contrast, while the original insights of classical authors (Marshall 1890; Weber 1909) writing about agglomeration advantages still operate in many developing or transition countries, the rich country experience of globalization, multinationals and productivity is a very different story from the standard urban economics textbook description.

The globalization experience of developing countries nowadays depends critically on the modern technology transfer activities of multinationals, and these transfer activities depend on the two-way knowledge flows between MNE affiliated and the regions in which they are located, and in turn these flows depend on the two-way relationships between the MNE affiliates and their headquarter locations. In Chapter 8 we saw that the role of multinational affiliates has been increasing globally. In particular, the R&D and innovation generation roles of MNE affiliates are increasing in both the developed and developing economies, exactly as would be expected from the arguments we presented in Chapters 4 and 5. Moreover, and consistent with the investment development path thesis (IDP) discussed in Chapter 2 (Dunning 1981; Dunning and Narula 1996), global FDI inflows into developing and transition economies are far more important than the global outflows from such economies. While the outward flows of FDI from developing and transition countries are increasing rapidly, they still only account for a very small share of global FDI flows. In addition, these inflows to developing and transition economies are still dominated by greenfield FDI, and are therefore qualitatively very different to FDI flows between rich countries as discussed in Chapter 6. As such, the observations in Chapter 8 support the argument that the types of MNEs examined in the earlier chapters of this book, even when they belong simultaneously to all four broad categories described in Chapter 2, still differentiate their strategies across different levels of economic development and geography.

9.1 IMPLICATIONS OF OUR ANALYSIS FOR THE THEORY OF MULTINATIONALISM

In contrast to some of the various analytical positions taken by some observers, giant multinational corporations are nowadays often very far from being the inflexible dinosaurs some commentators have assumed, and instead appear now to be amongst the most flexible, adaptable and flourishing institutions on the planet. It is not a tautology to say that it is MNEs which benefit the most from globalization because they are already global players. In part, this is because many such firms have become global players since, or even well before, the advent of the modern era of globalization which has characterized by the most recent decades. Nor is it because MNEs are uniquely innovative that gives them their advantages. The basic technologies of modern globalization, such as semiconductors, computer operating systems, the Internet, the World Wide Web, and many new generations of aircraft technologies, to name a few, were more often than not developed by people working in laboratories subsidized and protected from competition by nation states. And, even if we see MNEs as diplomatic actors often capable of forcing down trade barriers (Strange 1992; Guy 2007), it would be going too far to give them credit for the fall of the Berlin Wall, the opening up of China, or the creation of the EU Single Market and the Euro. While these dramatic institutional developments were amongst the most important catalysts for the modern era of globalization, they were not precipitated primarily because of MNEs. It is simply that technology and these events have created a space in which the greatest rewards could be reaped by firms which either operated as, or developed into, multinationals. In terms of economic geography, the arguments in this book regarding the importance of the knowledge and technological capabilities of MNEs in shaping globalization therefore also imply that the geographical distribution of such assets across the global economy will increasingly depend on the location behaviour of multinational firms.

Such a rapidly changing global environment has implications for theory, empirical analysis and policy making. As discussed in the introductory chapter, a problem often indicated in the general literature on MNEs has been the lack of any holistic theoretical approach to multinational behaviour. We have shown in this book that such a universal framework is neither indispensable nor, in fact, at present, really conceivable. The MNE represents, ultimately, a certain stage in the evolution of an important economic agent, the business firm. Such a stage is not inevitable in a firm's growth path, nor is it predetermined: MNEs take different forms, and are highly sensitive to the changes in the external environment over time. Following Cantwell and Narula (2001), we have argued that the OLI eclectic framework continues to provide effective broad guidelines as to our economic understanding of MNEs. However, the constantly changing global economic context in which MNEs operate and thrive implies not only the ongoing adaptation of each of the OLI Ownership, Location and Internalization advantages to the new temporal, geographical and institutional settings, but also the reconsideration of their reciprocal influences. As we have seen in this book, the L has gained enormously in importance over recent years due to the transformations brought about by the current phase of globalization. Its status in the triadic set of explanations for multinationality has risen so much so that the typical treatment of L as being largely exogenous can no longer be justified. MNEs' location decisions are becoming increasingly complex and dependant on the *variety* and *quality* of highly localized assets. The heterogeneity of advantages across locations, both within and across national boundaries, is always intertwined with decisions regarding the internalization of particular L advantages and/or their externalization in networks via outsourcing or offshoring, and thus with the organization and structure of the MNE itself. As explained by Dunning, "With the gradual geographic dispersion of created assets, and as firms become more multinational by deepening or widening their cross-border value chains, then, . . . the structure and content of the location portfolio of firms becomes more critical to their global competitive positions" (Dunning 2003b, p. 63).

As we have endeavoured to demonstrate in this book the analysis of the specific multi-level geography of MNE activities, and therefore the search for a stronger L, can be substantially enhanced by bringing closer together already existing theories and explanations of the firm and its environment. We have learnt from location theories various ways of representing the fact that multinational firms are complex multiplant firms which operate in a range of different economic environments, depending on their location strategies. While these organizations are far more complex than the simple representative firm models employed in microeconomics, the arguments in this book suggest that a range of theoretical models of firm location behaviour could help gain insights into many aspects of such choices. Models of spatial profit and cost variations, spatial competition, geographical and product space, leader-follower and clustering behaviour, all allow us to gain particular understandings of the complex multidimensional nature of the relationship between geography and MNEs, and in particular, between geography and large MNEs.

As is clear, a firm's ability to generate profits depends on its competitiveness in its markets and this in turn depends on its ability to innovate and grow, and on the knowledge it can create, acquire and utilize. However, as we have also pointed out in this book, the presence of a multinational firm within a region, either within a 'home' region or a 'host' region or both, also often changes the industrial and competitive structure of that locality. MNEs are rarely passive participants in regions and clusters. They are more frequently active agents for local engagement and change as they provide new avenues for connecting the local context to outside sources of knowledge and resources. Yet, exactly how the firm's location strategies relate to these two-way interactions between the firm and the region also depends on the organizational aspects of the firm, the nature of the knowledge relations which operate in the local economy, and the systems, processes and structures which promote local innovation. Intra-firm, inter-firm and inter-organizational networks that link together different agents and economic systems are increasingly to be seen as dominant modes for the creation and diffusion of knowledge. Beugelsdijk et al. (2010) refer to those dimensions and relationships in terms of place, space and organization.

A productive cross-fertilization of rather disparate, and still mostly disconnected, theoretical strands may achieve, again in Dunning's words, an "evenly balanced three-legged stool" (Dunning 1998 and 2009b, p. 5), thus allowing a full understanding of the L and its interdependence with the O and I. We hope that the discussions in this book have provided some clues as to how these interdependencies might be better understood. The majority of innovation theories now suggest that corporate dynamic capabilities are associated with the openness of firms to their external knowledge environment (Fontana et al. 2006), as well as with their internal knowledge-generating capabilities. Moreover, as we have seen, the internal and external knowledge capabilities of the MNE are regarded as being increasingly interrelated, as manifested in new network forms of global organization. However, the simple host-home dichotomy still largely underpins many theoretical interpretations of firms and geography in terms of uni-directional flows, represented as outflows from location A which translates as inflows to location B. Further theoretical efforts towards two-way flows thinking are therefore necessary to integrate connectivity and bi-directional flows and spillovers into more mainstream international trade and international business thinking. As we have seen, the major locations of FDI outflows are also the same recipients of most FDI inflows. Theory also needs to better examine the different levels of geography in MNE operations. The different levels of geography often relate to the whole range of activities and functions in the value chain, and also intra-firm and inter-firm organization structures, and both theory and empirics need to find a better specification beyond simply that of the national unit, at both the supra-national level and, even more importantly, at the sub-national regional levels. The same multiple geographies should be considered in the light of the fact that multinational firms increasingly operate across technology and industry boundaries. Modern globalization is increasingly associated with ever greater differences in sectoral, functional and technological specialization across different firm types such as SMEs or MNEs, as well as within individual multiplant and multinational firms, as well as both across and within regions, cities and local systems.

A reassessment of MNEs and geography through an inductive learning process is, however, hampered by the extremely limited and fragmented empirical evidence currently available regarding these issues. In particular, there is a serious lack of comparable large censuses and surveys addressing MNEs and their sub-national location behaviour. The limited data which is available mostly compares sub-national units of analysis within one individual country. Even these limited data sources suggest, as shown in this book, that there are complex MNE patterns of differentiation, patterns which produce spatial hierarchies of activities which reflect and often reinforce existing territorial inequalities. Not only do MNEs appear to have a highly skewed spatial distribution of their operations within each of the main regional blocks of the European Union, North America and Asia-Pacific, but the relationship between these three regions and the rest of the world is also markedly uneven. As such, an increased focus on MNE regional strategies amongst this triad of macro-regions at a detailed spatial level may be more fruitful than focusing on global strategies (Rugman 2005).

This book, however, has offered ways forward for thinking about these issues by integrating various existing theoretical arguments with existing empirical evidence in a manner which has never been done before. From this perspective the taxonomy developed in Chapter 5 is very instructive. We know that the motives why multinational firms may locate facilities or establishments in different locations characterized by any of the typologies discussed in Chapter 5 can be for any of the reasons outlined in Chapter 2, including seeking assets, seeking resources and general assets (low cost land and labour), seeking efficiencies (global value chain efficiencies or scale and scope economy efficiencies), or seeking markets. However, the focus of our book has been very much on the modern importance of MNEs as being specific asset seekers or strategic asset seekers. If we recall, specific asset seekers are MNEs which are prompted to establish affiliates abroad specifically in response to the need to acquire particular technological capabilities, management and marketing expertise or organizational skills. Somewhat differently, strategic asset seekers are MNEs which are prompted to engage in foreign operations usually by acquiring the assets of foreign firms, thereby increasing their O advantages and reducing those of their competitors so as to advancing their long-term strategic objectives in terms of global capabilities and competitiveness. We obviously acknowledge that all of the multinational seeking motives listed above are associated with location decisions and considerations of economic geography. However, the argument in this book has been that in the modern era of globalization there are two motivations for multinational behaviour which are dominant, namely those of specific asset seeking and strategic asset seeking. We argue that these two motives are nowadays of paramount importance, because of the unprecedented global returns to knowledge which are available to MNEs.

Interpreting multinational location behaviour from this standpoint, we know from Chapters 3 and 5 that knowledge-related transactions costs are involved in MNE location decisions in two distinct senses: namely, in *searching* for investment locations and also in *maintaining* an ongoing presence in a location. Both of these dimensions are critical for MNE location decisions and the knowledge-related transactions costs involved may differ between the two dimensions. In particular, MNEs may locate in any of the different types of regions classified according to the typologies discussed in Chapter 5, subject to each of the organizational, knowledge spillover and contractual issues examined in Chapter 5. However, once an MNE is located in a region, the knowledge, information and ultimately the monetary transmission mechanisms linking an MNE establishment with other MNE establishments around the world, including those of other companies, are all mediated via formal corporate channels. In economic geography, it has recently become fashionable to discuss 'pipelines' (Bathelt et al. 2004) whereby informal knowledge flows between key individuals and actors, sometimes referred to as gatekeepers, are deemed to be critical for the knowledge-access of specific places. In contrast, the arguments in this book suggest that the various manifestations of global knowledge-transmission mechanisms, including the movements of money, investment capital, goods, and people, are overwhelmingly transmitted via multinational corporate and proprietary channels which operate within tight legal and formal contractual arrangements, irrespective of whether the multinational corporations are operating in the financial services, logistics, shipping, manufacturing, or airline sectors. These multinational intra-corporate as well as inter-corporate linkages are so powerful and rapid precisely because they are so tightly organized and controlled. The monetary aspects of the exploitation and communication of knowledge are characterized by formality rather than informality and are underpinned by global systems rather than tacit arrangements. As scholars of international business have long understood, structures and design are the hallmarks of ongoing multinational knowledge transmission mechanisms (McCann and Mudambi 2004, 2005), rather than the informal pipelines and tacit knowledge spillover arguments popular with economic geographers. Detailed case-studies offer the most fertile grounds for uncovering the specific geography of activities and functions within MNEs, and as we discuss below, a greater range of such studies would also help to provide a sounder knowledge base for the design of appropriate local economic development policies.

As repeatedly demonstrated in this book, evolutionary approaches to technological change and regional growth are very useful for considering the disruptive changes imposed by MNEs in terms of increasing the interdependence between the 'global' and the 'local'. Localized production systems are increasingly interdependent with globalized knowledge systems, and the potential dynamic advantages of clustering for MNEs to foster path dependent processes which themselves shape cluster features. When considering the geographical variations in local absorptive capacity and the potential for MNEs to tap into the local knowledge base in order to facilitate linkages and spillovers, it is necessary to look beyond firms' competence accumulation and to consider the likelihood of various potential endogenous or evolutionary 'feedback' mechanisms actually operating in local institutional settings (von Tunzelmann 2009b).

Evolutionary approaches to economic geography and MNEs therefore still face major challenges, particularly with respect to the formulation of policy recommendations. As we have seen in the taxonomical exercise on spatial configurations proposed in Chapter 5 of this book, the relationships between technology, knowledge, inter-firm and inter-organizational transactions, and the local innovation systems, are complex and extremely varied. Economic development and regional policies aimed at promoting regions by encouraging multinational investment must explicitly take into account such complexity and variety if they are to ensure a good fit between the features of the MNE and the features of the local economic context.

Following the logic of the taxonomic approach developed in this book, in situations where the focus of public policy is intended to be on cities, the *pure agglomeration* model is the most appropriate structural model of analysis. Where the policy attention is on more peripheral regions or on more mature oligopolistic industrial structures, the *industrial complex* would be the relevant model of reference. Where the policy effort is focussed towards new industries, the *competence-based social network* is the correct frame of reference, whereas when we are dealing with traditional local industries, the *old social network* model is likely to be the most suitable type of spatial configuration (Iammarino and McCann 2009). As we have seen from our analysis, the appropriate policies for growth and development in each case would be expected to be quite distinct according to the regional-industrial context. However, in reality different policies are often mostly conflated into a single stylized scheme which focuses indiscriminately on promoting particular agendas such as the emergence of technologically advanced clusters, linkages between firms and universities, R&D collaborations, or networks of 'trust' between small and medium sized enterprises. Conversely, in terms of MNEs our arguments suggest that more appropriately tailored regional policies in some cases would be better advised to mirror rather traditional development policies with their accent on infrastructural capital. In the case of the pure agglomeration model, for instance, flexible land use policies and infrastructure provision may be critical, along with a careful consideration of the city shadow effect on the hinterland (Swann 2006). In the case of the industrial complex, supply chain development, information provision and labour skills policies are all more likely to be the central to the policy-design. Trust-led social networks models of spatial clustering are in many ways the most complex cases, requiring integrated platforms for facilitating the convergence between old and new knowledge bases and competencies (Cooke 2007). Only in the case of competence-based social network may regional innovation policies specifically focussed on enhancing sophisticated intangible assets and spillovers be either appropriate or realistic (Iammarino and McCann 2009).

On the other hand, in general, local and regional economic development policies should go well beyond the traditional search for simply attracting

FDI. It is imperative to identify which types of regional structural conditions will be favourable to which types of MNE firms. We have seen that even within the Behrman-Dunning classification, further distinctions of MNE types are envisaged, and that contemporary MNEs may belong to more than one or even all types simultaneously. Today's MNEs are seeking markets, efficiency, as well as general, specific and strategic assets at once, and they utilize a range of modes of ownership, coordination and control that goes well beyond that of simply FDI. Techniques such as mergers and acquisitions (M&As); foreign investment in R&D; contract manufacturing; outsourcing and offshoring; franchising; licensing; and management contracts, among others (UNCTAD 2011) represent the arsenal of options available to modern MNEs.

The good 'fit' between firms and geography across as many of these options as possible, therefore, will help foster the development of innovative networks, and will help to promote the mutually beneficial local 'embeddedness' of MNEs (Iammarino et al. 2008; Kramer et al. 2011). In contrast, in the absence of any consideration of the knowledge base, technology and transactional aspects of both the firms and their spatial systems, it becomes extremely difficult for policy initiatives to help to engender any strong link between the MNEs and their local environment. In such situations, as we have also shown in this book, MNEs are far more likely to develop either as largely standalone operations with little or no local innovative linkages, or in the most extreme cases, as 'islands of innovation' (Simmie 1997).

Complementarity and relatedness between old and new knowledge and between local and extra-local capabilities and linkages between the alignment of institutional objectives and networks, and between openness and external integration, are all necessary conditions for ensuring a 'diversity for growth' (Jacobs 1961, p. 194) in regional systems, industrial clusters and city-regions. The balance between internal and external sources of innovation, and between the degree of closure versus openness of firm and regional innovation features, are a major explanation underlying the increasing intra- and interregional diversity and the formation of new regional hierarchies, leading to 'more similar but less equal' (Paci 1997) patterns of regional evolution. The acknowledgement and evaluation of diversity in firm and geographical space is most likely to improve the fairly modest achievement of traditional local economic development policies still firmly grounded on the maximization of FDI-inflows-no-matter-what. Employing the classification schemes proposed in this book as a blueprint for identifying the potential sources of innovation in each particular locality may be a step forward for designing policy responses which are able to better foster local MNE engagement and thereby spur the integration of the local context in the global economy. However, it is important to stress once more that, as for the firm, not only may different typologies partially overlap in the same geographical area, but also distinct models can be simultaneously at work within some regional boundaries. As different layers of geography can be employed for analytical and policy analysis, one of the major difficulties for policy makers lies precisely in identifying and applying the appropriate spatial scale of intervention. Recent evidence (OECD 2011a, b; Dijkstra et al. 2012) suggests that productivity growth in many countries is now starting to spread out from many of the core city-regions, regions which clearly dominated economic growth in the first decade of modern globalization, but which now are facing greater competition from outer-lying areas. As such, these emerging spatial trends may offer increased opportunities for MNEs to explore new locations away from the global city-regions over the coming decades. A dynamic view, however, is essential in order to understand and identify how these interregional changes impact on the knowledge characteristics of localities, and this is important both from the perspective of MNE decision-makers and also from the perspective of policy-makers. An evolutionary approach is very powerful where such an identification process is required, because as we have seen in this book these approaches focus on the ways in which the relations between the firm and spatial features are subject to change over time (Boschma 2009). Given that the limited stock of theoretical tools presently available to us for analysing these changing aspects of MNE-geography interactions, the arguments in this book suggest that the best way forward is to follow a rather heuristic evolutionary approach built around the knowledge taxonomy discussed in Chapter 5 and centring on the changes in the knowledge-technology-innovation relationships between the MNE affiliate and its surroundings. Therefore, in order to improve our understanding of these changes an increased stock of detailed and explicitly geographical case studies of MNEs would be very helpful indeed. This is because to a large extent, it is the regional historical evolution that determines both the available policy options and also the probable outcomes of regional development policies in today's integrated global economy.

9.2 IMPLICATIONS OF OUR ANALYSIS FOR THE POLITICAL ECONOMY OF MULTINATIONALISM

Following the arguments in this book the political economy of multinationalism over the coming decades would appear to depend much less on the relationships between MNEs and the parent nation state of the MNE, and far more on the relationships between the power of nation states and the power of the multinational transmission mechanisms driving globalization forces. Notwithstanding the recent turmoil on global markets, looking into the future it is possible on the basis of all of the material presented in this book to discern various patterns which suggest that the trends identified here are likely to continue for many decades into the future.

It is clear that the last few years have witnessed enormous turmoil on world markets since the global financial crisis of 2008, and this has had a significant and negative impact on global investment flows, and in particular on mergers and acquisitions, as did the Asian financial crisis of 1997. However, in spite of such short- and medium-term fluctuations, there are still four strong grounds for believing that the long term trends towards increasing international investment and global interdependence will continue over the coming decades.

Firstly, the pace of technological progress implies that international communication and transportation technologies are bound to be improved. These improvements are particularly rapid and marked in the case of information and communications technologies. Advances to long-distance transport technologies also continue, though it remains to be seen to what extent developments in fuel efficiency and also transport deregulation will be able to offset counteracting costs pressures determined not only by changes in the energy costs, but also by some form of carbon pricing from which, at present, shipping and aviation fuels remain exempt by international treaties. Secondly, over recent years there has been enormous institutional change in favour of international investment and cross-border exchanges, along with the emergence of new multi-regional and supra-national arrangements for economic integration. Thirdly, new forms of business organization which blend together hierarchical and network types of governance are spreading among firms other than large MNEs. These often involve a growing number of business units operating in a widening range of economic activities and functions, and frequently often spanning across industry boundaries. All these technological, institutional and organizational changes point to the conclusion that accessing greater geographical markets will continue to become easier and easier for an increasing number of firms, thereby fostering further rounds of globalization. Fourthly, the 1930s experience of the disastrous consequences of trade protectionism are not lost on policy–makers, and the imperative to maintain open markets internationally is widely shared by almost all governments and international institutions. However, that is not to say that the architecture of the global financial system should be left unchanged. Rather it should be redesigned to best avoid the worst impacts of unregulated global financial flows in a manner which is appropriate for twenty-first century realities of financial markets (Roubini and Mihm 2010), largely in a manner akin to what the Bretton-Woods system did for the mid twentieth century realities. Yet, the scale and power of modern multinationals in shaping globalization, and in particular the impacts that MNE location and investment behaviour play in determining the differential advantages enjoyed by certain places and regions over other regions, implies that any future changes in the global architecture of commerce are beyond the scope of individual countries to influence. Such changes in the architecture of the global trading system require new forms of international coordination, which may have common elements with Bretton-Woods, but will also necessarily be very different in many dimensions.

As we saw in Chapter 6, the four centuries spanning the sixteenth to the nineteenth century represent the period in which almost all parts of the world came to be dominated to some degree by the emergence of large European empire powers of France, Netherlands, Spain, Austria-Hungary, Portugal, Russia and most notably Great Britain (Findlay and O'Rourke 2007; Maddison 2007a). The principles governing the modern notion of sovereignty, which as we saw in Chapter 6 are typically attributed to the 1648 Treaty of Westphalia, were established at a time when there was little interaction between states. The vast majority of the commercial activities of states were dominated by the economic and trading relations which were primarily contained within their individual colonial systems (Findlay and O'Rourke 2007). This of course is in very marked contrast to the modern nation state, which as we saw in Chapter 6 is a relatively recent concept, and although the principles of statehood were established in the seventeenth century, de facto modern states are primarily a result of the nineteenth century structural and economic transformations. Apart from the specific political issues which were current at the time, at least in economic terms these larger states had a very concrete logic to them which is still very pertinent, in that larger countries imply larger economies, and the resulting 'home market effects' (Fujita et al. 1999) ensure that the economies of these larger states are more productive in comparison to smaller countries. These home market effects also foster greater internationalization advantages for MNEs deriving from the larger home market economies, exactly as we saw in Chapters 2 and 6.

However, when discussing the relationships between countries and the global economic system, it is important to recall here that in orthodox trade theory and international business theory the size of the country is treated as exogenous, being determined by history. While much international development research focuses on the role played by institutions and trade barriers in promoting growth (World Bank 2003b, 2007, 2008a), it is generally assumed that the size, and border geography of a country is given, at least in the short to medium term. Obviously through history we see that, over the long term, the number and size of countries is itself variable. Yet, while the formation and fragmentation of states is central to the work of many historians and political scientists, outside of economic history such issues have tended to play almost no role in modern economics, until very recently.

For the purposes of this book, the relevance of these issues relates to the modern-day challenges associated with fostering more appropriate multinational coordination over governance and regulatory matters. The power and speed of globalization forces are greater than ever and as we have seen in this book, modern MNEs operate according to their own corporate objectives in a largely stateless fashion independent of individual national political priorities. This is rather different to the nature of globalization in previous eras where the interests of the parent nation-state and that of the multinational organization were closely intertwined, irrespective of whether the context was the seventeenth century VOC Dutch East India Company or the mid-twentieth century American multinational manufacturing firms. In terms of monetary and investment flows, the global transmission mechanisms mediated via multinational banks, multinational mining companies, multinational service providers and multinational manufacturing conglomerates, are quite simply enormous, and can operate at devastating speeds. While the nature of global capitalism in some ways remains largely unchanged over four centuries of globalization (Ferguson 2008; Roubini and Mihm 2010), the technological and institutional changes of the 1990s means that in the modern era of globalization the scale of these stateless transmission mechanisms reflect entirely different orders of magnitude to earlier eras. One outcome of this is that under certain circumstances, these changes also require coordinated responses on the part of nation-states of a different order of magnitude to previous eras.

The increasing interest devoted to these issues has been spawned by the processes of modern globalization, because such processes also ask fundamental questions about the nature and role of the nation state. The recent global economic shocks in the aftermath of the 2008 global financial crisis, along with all of the many other but smaller crises (Roubini and Mihm 2010), mean that countries must cooperate and coordinate their macroeconomic and regulatory policies in new ways. Even the world's largest national economy, namely that of the USA, is unable to unilaterally influence or respond to these global shocks for three reasons. First, modern multinational transmission mechanisms are so powerful and rapid that governments in nation states are incapable of acting with sufficient speed to counter these transmission mechanisms; second, multinationals themselves are nowadays almost entirely independent on the nation-states from which they emerged and this is very different to the situation in the earlier decades of the twentieth century; third, even an enormous economy such as the US is very dependent on these multinational transmission mechanisms. Indeed, as we saw in Chapter 1, some 31 per cent of US economic growth since the 1990s is accounted for by multinational firms (McKinsey 2010a).

In terms of its power to influence its own economic destiny, these arguments suggest that even the US is in many ways too small as a nation to isolate itself from global shocks. However, shocks from the USA are rapidly transmitted to other generally much smaller countries, which are largely powerless to respond to these shocks. The arguments in this book all point to the conclusion that multinational and multi-lateral integration are therefore necessary today because unfettered globalization is simply too powerful and too unpredictable a force to be managed by individual nations. The 1930s experience demonstrated that a typical national political economy response to adverse global shocks is to strive for protection, which if widely repeated leads to wholesale protectionism and global depression. As such, international fragmentation and a lack of multinational coordination lead to a total inability on the part of all nations to influence or fashion the forces of globalization for the individual (national) or common (international) good.

It is this problem of the smallness of countries which requires new forms of multi-lateral multinational integration and coordination to be developed in order to correct for the international market failures associated with these political economy coordination problems. An irony of increased international openness and the stateless characteristics of MNEs in the modern era of globalization is therefore that it forces ever-increasing mutual interdependence between nation states, and ultimately leads to the secession of many aspects of national sovereignty. This ceding of national powers was obviously evident in the Bretton-Woods reforms as well as in the formation of other twentieth century bodies such as NATO and the United Nations, but it is most clearly manifested in the processes of European Union integration, in which selected powers are pooled for the common good. The pooling of national powers is required in order to both take advantage of the benefits of globalization, but also to provide a more powerful bulwark against its adverse implications. However, while these arguments follow naturally from analyses of the nature of the multinational transmission mechanisms, this observation appears to be rather at odds with some other currently-popular arguments positing the advantages associated with small nation-states.

The argument favouring small nation-states is most notably associated with Alesina and Spolaore (2005). According to Alesina and Spolaore (2005), the size of country depends on a trade-off between the benefits of size versus the costs of catering for cultural heterogeneity. On the one hand, in the Alesina and Spolaore (2005) schema the benefits of being a large country relate primarily to the efficiency gains in the provision of public goods, such as monetary and fiscal institutions, police, defence, health, embassies, national parks, transport infrastructure, the taxation system bureaucracy, all of which may be associated with economies of scale. Furthermore, national size also helps with interregional fiscal transfers, both of a temporary insurance nature or of a more systematic income distributional nature, since independent states cannot be partially stabilized by other countries. Indeed, empirical evidence suggests that the relative size of government spending is inversely related to the size of the country.

These public good scale benefits associated with large countries are in contrast with the benefits of being a small country, which arise primarily in an environment of preference heterogeneity. Such preference heterogeneity arises out of local variations in culture, language, ethnicity, and historical experience (Alesina and Spolaore 2005), and in a context of heterogeneous preferences the argument is that public goods provided by a distant capital city government centre will be unsuitable for local demands. As such, the provision of tailored and localized public goods which best respond to the needs and preferences of local people is argued to be much easier if it is determined and delivered locally. Given that smaller states are perceived to take decisions which are 'closer' to the people, the argument here therefore is that small states better cater for cultural preference variations.

From a welfare perspective, the optimum size of a country is therefore the particular size of the state which maximizes the average level of citizen welfare, by providing the optimal bundle of public goods which, subject to certain constraints, takes account of the trade-off between economies of scale and the need to cater for the heterogeneity of preferences. One of these constraints highlighted by Alesina and Spolaore (2005) is the fact that in welfare maximization terms, the optimal size of the state will generally differ for every good, service or policy. As such, this would lead to an overlapping maze of borders, which would be inefficient in the presence of economies of scope or scale. In general, therefore, because of these jurisdictional problems, higher economies of scale or scope tend to imply fewer centralized jurisdictions (Alesina and Spolaore 2005).

Allowing for these constraints and caveats, however, and also connecting the Alesina and Spolaore (2005) arguments with the broader economic geography and MNE arguments discussed in this book, the arguments of Alesina and Spolaore (2005) do lead to several direct and testable conclusions, all of which are important for our purposes here.

A most important conclusion arising from the Alesina and Spolaore (2005) framework is that the more open is the country, the smaller is its optimal size, since trade releases the commercial constraints associated with national size. Indeed, there is evidence to support this broad argument (World Bank 2007). The logic here is that under complete autarky, the size of a country and the size of its market are the same thing, while in a world of complete economic integration they are entirely independent of each other. In the more typical case where countries are neither perfectly closed nor perfectly open to one-another, more open economies will tend to be smaller. This relationship should also be accelerated as trade barriers fall and global economic integration increases, and the number of small countries should increase as these are the very countries which have relatively the most to gain from free trade. Given that the modern era of globalization has also been associated with increasing democratization and the decline of dictatorship empires, increasing democratization should therefore spur this fragmentation process, producing many smaller and heteregenous countries, each of which is more internally homogenous.

The evidence which Alesina and Spolaore (2005) point to in support of their arguments regarding preference heterogeneity favouring small countries comes from the fact that whereas in 1945 there were 74 independent countries, following the break-up of the Soviet Union, the dissolution of the European colonial systems, and the rapidly increasing process of globalization, by 2005 there were 193 countries (Alesina and Spolaore 2005). As of 1995, 87 countries had less than a million people, 58 had less than 2.5 million, and 35 countries had less than 500,000 inhabitants. By 2005, more than half of the world's countries had less than 6 million inhabitants. Similarly, the arguments of Alesina and Spolaore (2005) also predict that the greater regional variation in preferences, and also greater regional variations in income within a country, the greater will be the pressure for fragmentation and regional autonomy. Evidence in favour of fragmentation pressures emanating from the economic periphery comes from observations of Scottish, Catalan, Basque, Quebec, and Slovak nationalism, while national fragmentation pressures emanating from the economic centre comes from observations of Northern Italy, Belgium and the Czech Republic.

In marked contrast, however, there are powerful arguments which point to exactly the opposite conclusions of Alesina and Spolaore (2005). In particular, 58 of the countries which comprise the poorest billion of the global population are small countries (Collier 2007). These 58 countries have a combined income of less than that of India and most are actually declining in wealth in both absolute as well as relative terms (Collier 2007), while most of the fast-growing economies in the developing world have large populations with large cities. As such, the Alesina and Spolaore arguments in favour of smallness of countries therefore appear to be rather problematic, and it may be the case that their arguments relate primarily only to the rich countries. Yet, this raises the questions as to why the advantages of national smallness should be related to wealthy countries while the advantages of scale should be related to poorer countries. In a sense, these contrasting relationships between national scale and wealth mirror the contrasting relationships between urban scale and productivity discussed in Chapter 7.

On this point, much of the resolution of these apparently paradoxical findings arises naturally from the economic geography and multinational arguments in this book which suggest that it is the relationship between the size and openness of a country and its 'home market' size which is critical, and not just the relationship between size and openness (Venables 2010). This is because the arguments in this book actually imply that the importance of the effect of country size depends not on the size of the country, but rather on the home market effect attainable to its domestic firms. This in turn is seen to depend not just on the degree of trade openness of the country to the global economy – defined in terms of having low trade barriers – but on two additional features. These are first, the levels of economic integration (including all regulatory and legal enforcement issues) that a country shares with all of its adjacent and neighbouring countries, and second, also on having sufficiently large agglomerations above a certain threshold in which global firms can invest.

As Venables (2010) and Collier (2007) point out, having good institutions, open markets and investing in infrastructure are not sufficient for growth, because the outcomes of good institutional and infrastructure policies also depend on what your neighbouring countries do. As such, many small and poor countries are highly vulnerable in the current phase of globalization for two reasons. First, they lack the economies of scale, agglomeration and home market effects which are necessary for growth in today's global environment (Collier 2007; Venables 2010). Almost all of the fast growing developing countries are large and contain large cities. Second, many of these countries suffer from a chronic lack of geographical accessibility which they are largely incapable of overcoming by their own efforts (Collier 2007).

For small and geographically isolated island nations, many of which voted for independence in the post-war era and are included in the evidence cited by Alesina and Spolaore (2005) in support of their arguments, access to seaport facilities is of little use without domestic scale. As such, economic geography means that many of these countries will remain trapped in poverty unless these island nations find ways to be deeply integrated into a much larger economic system. Many isolated islands and archipelagos suffer these problems, and they are particularly acute for many islands in the Pacific, Indian and South Atlantic Oceans. Only those whose political economy and institutional frameworks allow them to be deeply integrated into a much larger home market, such as Hawaii or Bermuda, are able to break this vicious economic geography circle.

Meanwhile, for landlocked countries, there is little point in trying to improve infrastructure such as road or rail systems in an effort to make the country more attractive for inward FDI as a means of fostering domestic growth, unless the neighbouring countries also undertake complementary and coordinated investments. Without absolute guarantees that these multinational and cross-border investments will be undertaken and maintained, no individual country has any incentive to develop such infrastructures. These types of complementary investments require not only institutional and governance coordination between adjacent countries in order to develop infrastructure, but also long term multinational agreements regarding the pricing arrangements and the guaranteeing of enforceable cross-border contracts regarding the distribution of the associated costs and rents. The more fragmented is the cartographical geography in terms of the number of countries which border each other and the complexity of the geographical borders, the more difficult it is to achieve these multinational agreements which would be beneficial to all. These coordination problems are particularly acute in the case of countries with potential sectors for producing manufactured goods or extracting land-based raw materials, because these are commodities which require road and rail transportation to port facilities, and cannot easily be profitably bypassed by air-freight. In addition, small countries are particularly prone to fostering narrowly-defined governing elites with powerful monopoly and monopsony interests. Many members of these elites will have little appetite for multinational agreements which potentially may undermine their dominant social, political and economic positions, and those that do support institutional innovations will find themselves under intense peer-pressure.

These fragmentation-coordination problems have many elements which are familiar in the 'tragedy of the commons' problem but they are institutional problems which have explicitly economic geography underpinnings. Yet, these problems are strangely absent in the analysis of Alesina and Spolaore (2005). Indeed, the weakness of Alesina and Spolaore's arguments is that they are asking questions about institutions, and in particular, institutions concerning governance and trade, without taking into account the role of economic geography. The reason Alesina and Spolaore (2005) are able to ignore these economic geography and fragmentation-coordination arguments, is that the types of cases they discuss are implicitly countries or regions which are already located in rich, open and integrated super-regional and multinational markets. In the modern global context, in many ways these types of places are actually the exception, not the rule. It is where such markets operate (such as EU, NAFTA and so on) that the advantages of globalization are most evident, rather than in highly cartographically fragmented environments such as the continent of Africa (Venables 2010). Indeed, an enormous achievement of the European Union has been to integrate so many different countries into something akin to a unified system in which small countries can still thrive. Comparing the EU with Africa demonstrates that unless we explicitly consider the underlying economic geography logic of the super-regional context in which an individual country is located, any arguments concerning the institutional advantages of large or small countries are of little value.

Most recent work increasingly suggests that there are no simple ideal institutional typologies for promoting national economic development (Rodrik 2007), and as we see here, the development possibilities logic resulting from national and multinational institutional arrangements will themselves also depend heavily on both the existing economic and political geography (Collier 2007). In today's global economy, multinational coordination of institutional and regulatory regimes between neighbouring countries is critical in order to foster home market and agglomeration effects. Free trade relations are no longer simply seen as a matter of comparative advantage and specialization, but rather in terms of the need to foster market and production scale effects, and especially those built around the development of knowledge, technology and innovation. As we have seen throughout this book, multinational firms are critical drivers of these international technology transfer processes, and the economic geography of globalization depends crucially on how and where MNEs decide to invest. Indeed, the need to foster the cross-border investment flows essential for promoting home market scale effects underpins the logic of the creation of the EU Single Market. Similar considerations also apply to different degrees in the case of all of the various super-regional arrangements which have emerged in the modern era of globalization, including EU, NAFTA, CER, APECD, ASEAN and MERCOSUR. Although quite different in nature, each of these multinational groupings is designed precisely to change the institutional relationships between the member countries in a manner which reduces their separateness and increases their mutual openness and interdependence.

As we have already seen, in the modern era of globalization the cross-border transmission mechanisms mediated overwhelmingly via multinational enterprises are more rapid and powerful than ever. This also implies that coordinated multinational institutional reforms are required both to maximize the advantages of globalization and also to limit its adverse consequences. Indeed, the 2008 global financial crisis and its aftermath have clearly demonstrated the need for multinational coordination of policy between countries (Roubini and Mihm 2010). Smaller numbers of larger countries facilitate multi-lateral policy coordination because the number of governments which are required to agree to a particular policy response is reduced. In contrast, greater numbers of smaller countries imply a fragmentation of financial firepower and greater policy-coordination possibilities (Roubini and Mihm 2010). As we have seen in this book, in today's global marketplace, multinational firms are increasingly independent of any individual countries, and the need to ensure that the benefits of globalization are dispersed to as many parts of the world as possible therefore requires both a greater and more nuanced pooling of multinational sovereignty. Ironically, therefore, from a global system-wide perspective, increased openness in many ways reduces the benefits of greater numbers of small countries, unless national smallness is also associated with the pooling of sovereignty, exactly as is the case with the European Union.

These same global-local tensions faced by nation states are also inherent in all of the investment and location decisions made by modern (and largely-stateless) MNEs. In many ways the global-local challenges faced today by both MNEs and nation-states somewhat mirror one another, as they did in earlier eras. However, the difference today is that their interests are largely decoupled from one another in a manner which previously was not the case. As Chapters 6 and 7 have demonstrated, the nature and role of the nation state has evolved continuously over the four centuries of globalization, as have the relationships between nation states, regions, and the multinational enterprises which have emerged from them or locate in them. In the past their interests tended to be much more closely aligned, whereas today their interests may often largely diverge. Nations and regions benefit greatly from the presence of MNEs as they provide the connectivity which is essential in today's global economy, but the correct geographical and institutional set-up is essential in order to ensure that these firms remain. As we have seen in this book, the relationships between institutions and geography not only remain complex but they also differ between places, and as localities evolve these relationships may need to be continually monitored and whenever necessary revised as required. The efficacy of local decisions depends on global trends, and not the other way around, and this is particularly marked in the case of small countries and regions. To the extent that the world is culturally becoming 'flatter' (Friedman 2005), in the sense that democratic expectations and demands for greater governance accountability are increasing, the increasing statelessness of MNEs points in quite another direction. As such, while the drive for better definitions of FDI (OECD 2008) along with better corporate responsibility goals (OECD 2010) will help to improve nation-MNE governance relations, the competition between regions and nations to attract MNEs in order to foster innovation will remain fierce. As we have seen in this book, economic geography tells us that this competition favours particular types of places over others, and therefore carefully developed and integrated strategies are required (OECD 2011) in order to allow less-advantageous regions and nations to gain a foothold in the global MNE-mediated innovation networks. Evidently, the relations between regions, nations and MNEs in today's global marketplace feature both competition and coordination aspects, and balancing these somewhat conflicting demands will remain a difficult task.

The global-local tensions regarding the coordination challenges faced by nation states and regions in the modern era of globalization have been magnified by the global financial crisis of 2008, and the resulting Euro crisis of 2011. Yet, as the 2011 Euro crisis has demonstrated, even the European Union is too small to respond to some aspects of globalization, responses which require coordination on an even larger scale. Movements from the G7 to the G20 are small positive steps in this direction, but the coming decades are likely to witness much greater changes in the architecture of global trade (Reyes et al. 2010) and finance from the structure we have inherited from the second half of the twentieth century (Roubini and Mihm 2010). The arguments in this book suggest that a proper understanding of these changes cannot be achieved without explicitly considering the interrelationships between economic geography and the emerging spatial distributions of knowledge, technology and innovation which govern multinational investment decisions.

Bibliography

Abramovitz, M. (1956), 'Resource and output trends in the United States since 1870', *American Economic Review*, 46(2), 5–23, May.

Abramovitz, M. (1986), 'Catching up, forging ahead, and falling behind', *Journal of Economic History*, 46, 385–406.

Abramovitz, M. (1999), 'What economists don't know about growth', *Challenge*, 42(1), 81–91, January/February.

Acemoglu, D. and Johnson, S.H. (2006a), 'De facto political power and institutional persistence', *American Economic Review*, 96(2), 325–330.

Acemoglu, D. and Johnson, S.H. (2006b), 'Unbundling institutions', *Journal of Political Economy*, 113(5), 949–995.

Acemoglu, D. and Robinson, J.A. (2000), 'Political losers as a barrier to economic development', *American Economic Review*, 90(2), 126–130.

Acs, Z.J. (2002), *Innovation and the Growth of Cities*, Cheltenham, UK and Northampton, MA, USA: Edward Elgar Publishing.

Acs, Z.J. and Armington, C. (2004a), 'The impact of geographic differences in human capital on service firm formation rates', *Journal of Urban Economics*, Elsevier, 56(2), 244–278, September.

Acs, Z.J. and Armington, C. (2004b), 'New firm survival and human capital', Papers on Entrepreneurship, Growth and Public Policy 2004-14, Max Planck Institute of Economics, Entrepreneurship, Growth and Public Policy Group.

Acs, Z.J., Bosma, N. and Sterberg, R. (2008), 'The entrepreneurial advantage of world cities: evidence from global entrepreneurship monitor data', Working Paper SCALES, University of Utrecht and Netherlands Ministry of Economic Affairs.

Aguilera, A. (2008), 'Business travel and mobile workers', *Transportation Research A*, 42, 1109–1116.

Aharoni, Y. (1966), *The Foreign Investment Decision Process*, Boston, MA: Division of Research, Graduate School of Business Administration, Harvard University.

Aitken, B., Hanson, G. and Harrison, A.E. (1997), 'Spillovers, foreign investment, and export behaviour', *Journal of International Economics*, 43, 103–132.

Akerlof, G.A. (1970), 'The market for "lemons": quality uncertainty and the market mechanism', *The Quarterly Journal of Economics*, 84(3), 488–500, August.

Alcácer, J. (2006), 'Location choices across the value chain: how activity and capability influence colocation', *Management Science*, 52(10), 1457–1471.

Alchian, A.A. (1950), 'Uncertainty, evolution, and economic theory', *Journal of Political Economy*, 58(3), 211–221, June.

Alesina, A. and Spolaore, E. (2005), *The Size of Nations*, Cambridge, MA: The MIT Press.

Alessi, L., Barigozzi, M. and Capasso, M. (2011), 'Nonfundamentalness in structural econometric models: a review', *International Statistical Review*, 79(1), 16–47.

Alfaro, L. and Chen, M. (2009), 'The global agglomeration of multinational firms', Harvard Business School Working Papers 10-043, Harvard Business School, revised April 2010.

Aliber, R.Z. (1970), 'A theory of direct foreign investment', in Kindleberger, C.P. (ed.), *The International Corporation*, Cambridge, MA: The MIT Press, pp. 17–34.

Ali-Yrkkö, J. (2001), 'Nokia's network – gaining competitiveness from co-operation', The Research Institute of the Finnish Economy (ETLA), Series B174, Taloustieto Ltd, Helsinki.

Almeida, P. (1996), 'Knowledge sourcing by foreign multinationals: patent citation analysis in the US semiconductor industry', *Strategic Management Journal*, 17, winter special issue, 155–165.

Almeida, P. and Kogut, B. (1997), 'The exploration of technological diversity and the geographic localization of innovation', *Small Business Economics*, 9(1), 21–31.

Almeida, P. and Kogut, B. (1999), 'Localization of knowledge and the mobility of engineers in regional networks', *Management Science*, 45(7), 905–915.

Amighini, A., Rabellotti, R. and Sanfilippo, M. (2009), 'The rise of multinationals from emerging countries. A review of the literature, emerging economic regional powers and local systems of production: new threats or new opportunities?', Working Paper Series, 4/09.

Amin, A. and Thrift, N. (1992), 'Neo-Marshallian nodes in global networks', *International Journal of Urban and Regional Research*, 16, 571–587.

Amin, S. (1997), *Capitalism in the Age of Globalization*, London: Zed Press.

Amiti, M. and Cameron, L. (2007), 'Economic geography and wages', *Review of Economics and Statistics*, 89(1), 15–29.

Amiti, M. and Konings, J. (2007), 'Trade liberalization, intermediate inputs, and productivity: evidence from Indonesia', *American Economic Review*, American Economic Association, 97(5), 1611–1638, December.

Andersson, M. (2009), 'External trade and internal geography', CESIS Working Paper, Jonkoping University, Jonkoping and Royal Institute of Technology, Stockholm, Sweden.

Andersson, U., Forsgren, M. and Holm, U. (2002), 'The strategic impact of external networks: subsidiary performance and competence development in the multinational corporation', *Strategic Management Journal*, 23(11), 979–996.

Angang, H., Jakobson, L. and Mingming, S. (2005), China's transforming society and foreign policy. Accessed in March 2007 from January 12, 2006, from http://www.upi-fiia.fi/doc/UPI-raportti_11.pdf.

Antonelli, C. (2000), 'Collective knowledge communication and innovation: the evidence of technological districts', *Regional Studies*, 34(6), 535–547.

Antonelli, C. (2007), 'Technological knowledge as an essential facility', *Journal of Evolutionary Economics*, 17(4), 451–471.

Antonelli, C. (2008), *Localized Technological Change. Towards the Economics of Complexity*, London: Routledge.

Archibugi, D. (2001), 'Pavitt's taxonomy sixteen years on: a review article', *Economics of Innovation and New Technology*, 10, 415–425.

Archibugi, D. and Iammarino, S. (1998), 'The policy implications of the globalisation of innovation', in Archibugi, D., Howells, J. and Michie, J. (eds) (1999) *Innovation Policy in a Global Economy*, Cambridge, UK: Cambridge University Press.

Archibugi, D. and Iammarino, S. (1999), 'The policy implications of the globalisation of technology', *Research Policy*, 28(2–3), 317–336.

Archibugi, D. and Iammarino, S. (2002), 'The globalization of technological innovation: definition and evidence', *Review of International Political Economy*, 9(1), 98–122. ISSN 0969-2290.

Archibugi, D. and Michie, J. (1995), 'The globalisation of technology: a new taxonomy', *Cambridge Journal of Economics*, 19(1), 121–140.

Archibugi, D., Cesaratto, S. and Sirilli, G. (1991), 'Sources of innovative activities and industrial organisation in Italy', *Research Policy*, 20, 299–313.

Arita, T., Iguchi, C. and McCann, P. (2011), 'Foreign direct investment, knowledge assets and the economic geography of growth in the Asian BRIICS countries', in Stimson, R., Stough, R.R. and Nijkamp, P. (eds) (2011), *Endogenous Regional Development*, Cheltenham, UK and Northampton, MA, USA: Edward Elgar Publishing.

Arita, T. and McCann, P. (2000), 'Industrial alliances and firm location behaviour: some evidence from the US semiconductor industry', *Applied Economics*, 32, 1391–1403.

Arita, T. and McCann, P. (2002), 'The relationship between the spatial and hierarchical organization of multiplant firms; observations from the global semiconductor industry', in McCann, P. (ed.), *Industrial Location Economics*, Cheltenham, UK and Northampton, MA, USA: Edward Elgar Publishing.

Arita, T. and McCann, P. (2004), 'A comparison of industrial location behaviour within the US and European semiconductor industries', ERSA Conference Papers ersa04p450, European Regional Science Association.

Arita, T. and McCann, P. (2006), 'Clusters and regional development: some cautionary observations from the semiconductor industry', *Information Economics and Policy*, 18(2), 157–180.

Arora, A. and Athreye, S. (2002), 'The software industry and India's economic development', *Information Economics and Policy*, Elsevier, 14(2), 253–273, June.

Arora, A. and Forman, C. (2007), 'How local are IT services markets: proximity and IT outsourcing', *Journal of Management Information Systems*, 24(2), 73–102.

Arora, A. and Gambardella, A. (1995), 'Division of innovative labour in biotechnology', in Rosenberg, N. and Geijins, A. (eds), *University-Industry Interface and Medical Innovation*, Washington, DC: National Academy of Sciences.

Arthur, W.B. (1989), 'Competing technologies, increasing returns, and lock-in by historical events', *Economic Journal*, 97, 642–665.

Asakawa, K. (2001a), 'Evolving headquarters-subsidiary dynamics in international R&D: the case of Japanese multinationals', *R&D Management*, 31(1), 1–14.

Asakawa, K. (2001b), 'Organizational tension in international R&D management: the case of Japanese firms', *Research Policy*, Elsevier Science, 30(5), 735–757.

Asheim, B.T. (1995), 'Industrial districts as learning regions: a condition for prosperity?', Step Report, no. 3, Oslo: Step Group.

Asheim, B.T. and Gertler, M.S. (2003), 'Regional innovation systems and the geographical foundations of innovation', Teari Working Paper, no. 11, October.

Asheim, B.T. and Isaksen, A. (2002), 'Regional innovation systems: the integration of local "sticky" and global "ubiquitous" knowledge', *Journal of Technology Transfer*, 27, 77–86.

Athreye, S.S. (2003), 'Agglomeration and growth: a study of the Cambridge hi-tech cluster', Urban/Regional 0308001, EconWPA.

Athreye, S.S. (2004), 'Role of transnational corporations in the evolution of a high-tech industry: the case of India's software industry – a comment', *World Development*, Elsevier, 32(3), 555–560, March.

Athreye, S.S. (2005), 'The Indian software industry and its evolving service capability', *Industrial and Corporate Change*, Oxford University Press, 14(3), 393–418, June.

Athreye, S. and Cantwell, J. (2007), 'Creating competition?: Globalisation and the emergence of new technology producers', *Research Policy*, Elsevier, 36(2), 209–226, March.

Athreye, S. and Kapur, S. (2009), 'Introduction: the internationalization of Chinese and Indian firms – trends, motivations and strategy', *Industrial and Corporate Change*, 18(2), 209–221.

Athreye, S.S. and Srinivasan, V. (2005), 'Managerial synergies and related diversification: software services and the business process outsourcing sector in India', in Thatchenkery, T. and Stough, R. (eds), *Information Communication Technology and Economic Development: Learning from the Indian Experience*, Cheltenham, UK and Northampton, MA, USA: Edward Elgar Publishing.

Audretsch, D.B. (1997), 'Technological regimes, industrial demography and the evolution of industrial structures', *Industrial and Corporate Change*, 6(1), 49–82.

Audretsch, D.B. (1998), 'Agglomeration and the location of innovative activity', *Oxford Review of Economic Policy*, 14(2), 18–29.

Audretsch, D.B. and Dohse, D. (2007), 'Location: a neglected determinant of firm growth', *Review of World Economics* (Weltwirtschaftliches Archiv), Springer, 143(1), 79–107, April.

Audretsch, D.B. and Feldman, M.B. (1996), 'R&D spillovers and the geography of innovation and production', *American Economic Review*, 86, 630–640.

Aw, B.Y. and Lee, Y. (2008), 'Firm heterogeneity and location choice of Taiwanese multinationals', *Journal of International Economics*, 76, 403–415.

Aydalot, P. (ed.) (1986), *Milieux Innovateurs in Europe*, Paris: Gremi.

Bacci, L., Labory, S. and Lombardi, M. (2010), 'The evolution of external linkages and relational density in the Tuscan leather industry', in Belussi F. and Sammarra, A. (eds), *Business Networks in Clusters and Industrial Districts. The Governance of the Global Value Chain*, London: Routledge, 146–171.

Badinger, H. and Tondl, G. (2003), 'Trade, human capital and innovation', in Fingleton, B. (ed.), *European Regional Growth*, Heidelberg: Springer.

Bagnasco, A. and Le Gales, P. (2000), *Cities in Contemporary Europe*, Cambridge, UK: Cambridge University Press.

Bailey, D. and Driffield, N. (2002), 'Hymer and uneven development revisited: foreign direct investment and regional inequalities', *Contributions to Political Economy*, Oxford University Press, 21(1), 55–68, December.
Bain, J.S. (1956), *Barriers to New Competition*, Cambridge, MA: Harvard University Press.
Baldwin, R. (2006), 'Multilateralising regionalism: spaghetti bowls as building blocs on the path to global free trade', Cambridge, MA: NBER Working Paper 12545.
Baldwin, R. and Okubo, T. (2006), 'Heterogeneous firms, agglomeration and economic geography: spatial selection and sorting', *Journal of Economic Geography*, 6, 323–346.
Ball, M., Lizieri, C. and MacGregor, B. (1998), *The Economics of Commercial Property Markets*, London: Routledge.
Baptista, R. and Swann, P. (1998), 'Do firms in clusters innovate more?', *Research Policy*, 27(5), 525–540.
Barba Navaretti, G. and Venables, A.J. (2004), *Multinational Firms in the World Economy*, Princeton, NJ: Princeton University Press.
Barca, F. (2009), 'An agenda for a reformed cohesion policy: a place-based approach to meeting European Union challenges and expectations', Independent Report Prepared at the Request of the European Commissioner for Regional Policy, Danuta Hübner, European Commission, Brussels.
Barca, F. and McCann, P. (2010), 'The place-based approach: a response to Mr Gill', accessed at http://www.voxeu.org/index.php?q5 node/5644.
Barnet, R.J. and Cavanagh, J. (1994), *Global Dreams: Imperial Corporations and the New World Order*, New York: Simon and Schuster.
Barrell, R. and Pain, N. (1999), 'Domestic institution, agglomerations and foreign direct investment in Europe', *European Economic Review*, 43, 925–934.
Barthel, F., Busse, M. and Neumayer, E. (2010), 'The impact of double taxation treaties on foreign direct investment: evidence from large dyadic panel data', *Contemporary Economic Policy*, 28(3), 366–377.
Bartlett, A. and Ghoshal, S. (1989), *Managing Across Borders. The Transnational Solution*, Boston, MA: Harvard Business School Press.
Basile, R., Castellani, D. and Zanfei, A. (2008), 'Location choices of multinational firms in Europe: the role of EU cohesion policy', *Journal of International Economics*, 74(2), 328–340.
Basile, R., de Nardis, S. and Girardi, A. (2009), 'Pricing to market when quality matters', ISAE Working Papers 123, ISAE, Institute for Studies and Economic Analyses, Rome, Italy.
Bathelt, H., Malmberg, A. and Maskell, P. (2004), 'Clusters and knowl-

edge: local buzz, global pipelines and the process of knowledge creation', *Progress in Human Geography*, 28(1), 31–56.
Baumol, W.J. (1962), 'On the theory of expansion of the firm', *The American Economic Review*, 52(5), 1078–1087, December.
Beata, S.J. and Spatareanu, M. (2008), 'Liquidity constraints and linkages with multinationals', Working Papers Rutgers University, Newark 2008-005, Department of Economics, Rutgers University, Newark.
Beaverstock, J.V., Smith, R.G. and Taylor, P.J. (1999), 'A roster of world cities', *Cities*, 16(6), 445–458.
Beaverstock, J.V., Smith, R. and Taylor, P.J. (2000), 'World city network: a new metageography?', *Annals of the Association of American Geographers*, 90(1), 123–134.
Becattini, G. (ed.) (1987), *Mercato e forze locali. Il distretto industriale*, Bologna: Il Mulino.
Beck, U. (2000), *What is Globalization?*, Cambridge, UK: Polity Press.
Becker, G.S. (1965), 'A theory of the allocation of time', *The Economic Journal*, 75(299), 493–517, September.
Beckman, M. and Thisse, J. (1986), 'The location of production activities', in Nijkamp, P. (ed.), *Handbook of Regional and Urban Economics*, Amsterdam: North-Holland.
Behrman, J.N. (1972), *The Role of International Companies in Latin America: Autos and Petrochemicals*, Lexington, MA: Lexington Books.
Bel, G. and Fageda, X. (2008), 'Getting there fast: globalization, intercontinental flights and location of headquarters', *Journal of Economic Geography*, 8, 471–495.
Bell, M. and Albu, M. (1999), 'Knowledge systems and technological dynamism in industrial clusters in developing countries', *World Development*, 27(9), 1715–1734.
Belussi, F. (2010), 'The evolution of a technologically dynamic district: the case of Montebelluna', in Belussi, F. and Sammarra, A. (eds), *Business Networks in Clusters and Industrial Districts*, London: Routledge.
Belussi, F. and Sammarra, A. (eds) (2010), *Business Networks in Clusters and Industrial Districts*, London: Routledge.
Belussi, F., Sammarra, A. and Sedita, S.S. (2008), 'Managing long distance and localized learning in the Emilia Romagna life science cluster', *European Planning Studies*, 16(5), 665–692.
Belussi, F., Sammarra, A. and Sedita, S.R. (2010), 'Learning at the boundaries in an "open regional innovation system": a focus on firms' innovation strategies in the Emilia Romagna life science industry', *Research Policy*, 39(6), 710–721.
Benini, R. and Czyzewski, A. (2007), 'Regional disparities and economic

growth in Russia: net growth patterns and catching up', *Economic Change and Restructuring*, 40, 91–135.

Bernard, A., Eaton, J., Bradford Jensen, J. and Kortum, S. (2003), 'Plants and productivity in international trade', *American Economic Review*, 93, 1268–1290.

Berry, C.R. and Glaeser, E.L. (2005), 'The divergence of human capital levels across cities', Papers in Regional Science, 84(3), 407–444.

Best, M.H. (1990), *The New Competition: Institutions of Industrial Restructuring*, Cambridge, UK: Polity Press.

Best, M.H. (1998), 'Production principles, organisational capabilities and technology management', in Michie, J. and Grieve-Smith, J. (eds), *Globalisation, Growth, and Governance*, Oxford: Oxford University Press.

Beugelsdijk, S. (2007), 'The regional environment and a firm's innovative performance: a plea for a multilevel interactionist approach', *Economic Geography*, 83, 181–199.

Beugelsdijk, S. and Frijns, B. (2010), 'A cultural explanation of the foreign bias in international asset allocation', *Journal of Banking and Finance*, doi:10.1016/j.jbankfin.2010.01.020.

Beugelsdijk, S., McCann, P. and Mudambi, R. (2010), 'Place, space and organisation: economic geography and the multinational enterprise', *Journal of Economic Geography*, 10, 485–493.

Birkinshaw, J. (1996), 'How multinational subsidiaries mandates are gained and lost', *Journal of International Business Studies*, 27(3), 467–496.

Birkinshaw, J. (2000a), 'Determinants and consequences of subsidiary initiative in multinational corporations', *Entrepreneurship Theory and Practice*, 24(1), 9–35.

Birkinshaw, J. (2000b), *Entrepreneurship in the Global Firm*, London: Sage.

Birkinshaw, J. (2001), 'Strategies for managing internal competition', *California Management Review*, 44(1), 24–38.

Birkinshaw, J. and Hagstrom, P. (eds) (2000c), *Flexible Firm: Capability Management in Network Organizations*, Oxford: Oxford University Press.

Birkinshaw, J. and Hood, N. (1998), 'Multinational subsidiary evolution: capability and charter change in foreign-owned subsidiary companies', *Academy of Management Review*, 23(4), 773–795.

Birkinshaw, J. and Hood, N. (2000), 'Characteristics of foreign subsidiaries in industrial clusters', *Journal of International Business Studies*, 31(1), 141–154.

Birkinshaw, J., Batenburg, R. and Murray, G. (2002), 'Venturing to succeed', *Business Strategy Review*, 13(4), 10–17.

Birkinshaw, J., Hood, N. and Jonsson, S. (1998), 'Building firm-specific advantages in multinational corporations: the role of subsidiary initiative', *Strategic Management Journal*, 19(3), 221–242, March.
Birkinshaw, J., Hood, N. and Young, S. (2005), 'Subsidiary entrepreneurship, internal and external competitive forces, and subsidiary performance', *International Business Review*, 14(2), 227–248.
Blanc, H. and Sierra, C. (1999), 'The internationalisation of R&D by multinationals: a trade-off between external and internal proximity', *Cambridge Journal of Economics*, 23(2), 187–203.
Blien, U. and Suedekum, J. (2005), 'Local economic structure and industry development in Germany, 1993–2001', IAB Discussion Paper 200501, Institut für Arbeitsmarkt- und Berufsforschung (IAB), Nürnberg [Institute for Employment Research, Nuremberg, Germany].
Blomström, M. (1989a), *Labor Productivity Differences between Foreign and Domestic Firms in Mexico*, Stockholm: Department of International Economics [and] Geography, Stockholm School of Economics.
Blomström, M. (1989b), *Foreign Investment and Spillovers*, London: Routledge.
Blomström, M. (1990), *Transnational Corporations and Manufacturing Exports from Developing Countries*, Stockholm.
Blomström, M. and Kokko, A. (1996), 'Multinational corporations and spillovers', CEPR Discussion Papers 1365, CEPR Discussion Papers.
Blomström, M. and Kokko, A. (1997), 'Regional integration and foreign direct investment', CEPR Discussion Papers 1659, CEPR Discussion Papers.
Blomström, M. and Kokko, A. (1998), 'Multinational corporations and spillovers', *Journal of Economic Surveys*, 12(3), 247–277.
Blomström, M. and Persson, H. (1983), 'Foreign investment and spillover efficiency in an underdeveloped economy: evidence from the Mexican manufacturing industry', *World Development*, Elsevier, 11(6), 493–501, June.
Blum, B.B. and Goldfarb, A. (2006), 'Does the Internet defy the law of gravity', *Journal of International Economics*, 70(2), 384–405.
Bobonis, G.J. and Shatz, H.J. (2007), 'Agglomeration, adjustment, and state policies in the location of foreign direct investment in the United States', *Review of Economics and Statistics*, 89(1), 30–43.
Boldrin, M. and Canova, F. (2001), 'Inequality and convergence in Europe's regions: reconsidering European regional policies', *Economic Policy*, 16, 207–253.
Bolton, R. (1992), 'Place prosperity vs people prosperity revisited: an old issue with a new angle', *Urban Studies*, 29(2), 185–203.
Bordo, M., Taylor, A.M. and Williamson, J. (eds) (2003), 'Globalization

in historical perspective', National Bureau of Economic Research Conference Report, Chicago: University of Chicago Press.
Borja, J., Castells, M., Belil, M. and Benner, C. (1997), *Local and Global, Management of Cities in the Information Age*, London: Earthscan.
Boschma, R.A. (2004), 'Competitiveness of regions from an evolutionary perspective', *Regional Studies*, 38(9), 1001–1014.
Boschma, R.A. (2005), 'Proximity and innovation: a critical assessment', *Regional Studies*, 39(1), 61–74.
Boschma, R.A. (2009), *Evolutionary Economic Geography and its Implications for Regional Innovation Policy*, Paris: OECD.
Boschma, R.A. and Iammarino, S. (2009), 'Related variety, trade linkages and regional growth', *Economic Geography*, 85(3), 289–311.
Boschma, R.A. and Lambooy, J.G. (1999), 'Evolutionary economics and economic geography', *Journal of Evolutionary Economics*, 9, 411–429.
Boschma, R.A. and Ter Wal, A. (2011), 'Coevolution of firms, industries and networks in space', *Regional Studies*, 47(7), 919–933.
Boschma, R.A. and Wenting, R. (2007), 'The spatial evolution of the British automobile industry in Great Britain, 1895–1968', *Industrial and Corporate Change*, 16(2), 213–238.
Boschma, R.A., Eriksson, R. and Lindgren, U. (2009), 'How does labour mobility affect the performance of plants? The importance of relatedness and geographical proximity', *Journal of Economic Geography*, 9(2), 169–190.
Bottazzi, G., Dosi, G., Fagiolo, G. and Secchi, A. (2008), 'Sectoral and geographical specificities in the spatial structure of economic activities', *Structural Change and Economic Dynamics*, 19, 189–202.
Bottazzi, L., Da Rin, M. and Hellman, T. (2005), 'What role of legal systems in financial intermediation? Theory and evidence', European Corporate Governance Institute.
Bouba-Olga, O. and Grossetti, M. (2005), 'Une (ré)definition des notions de proximié', CRIEF Working Paper T2007-02.
Boulhol, H. and de Serres, A. (2010), 'Have developed countries escaped the curse of distance?', *Journal of Economic Geography*, Oxford University Press, 10(1), 113–139, January.
Braczyk, H., Cooke, P. and Heinderinch, M. (eds) (1998), *Regional Innovation Systems*, London: University College Press.
Brainard, S.L. (1993), 'A simple theory of multinational corporations and trade with a trade-off between proximity and concentration', NBER Working Papers 4269, National Bureau of Economic Research.
Brainard, S.L. (1997), 'An empirical assessment of the proximity-concentration trade-off between multinational sales and trade', *American Economic Review*, 87, 520–544.

Brakman, S. and van Marrewijk, C. (2008), 'It's a big world after all: on the economic impact of location and distance', *Cambridge Journals of Regions, Economy and Society*, 1(3), 411–437.
Brecher, J. and Costello, T. (1995), *Global Village or Global Pillage*, Boston, MA: South End Press.
Breschi, S. (2000), 'The geography of innovations: a cross-sector analysis', *Regional Studies*, 34(3), 213–229.
Breschi, S. and Lissoni, F. (2001a), 'Knowledge spillovers and local innovation systems: a critical survey', *Industrial and Corporate Change*, 10(4), 975–1005.
Breschi, S. and Lissoni, F. (2001b), 'Localised knowledge spillovers versus innovative milieux: knowledge "tacitness" reconsidered', *Papers in Regional Science*, 80, 255–273.
Breschi, S. and Lissoni, F. (2003), 'Mobility and social networks: localised knowledge spillovers revisited', KITeS Working Papers 142, KITeS, Centre for Knowledge, Internationalization and Technology Studies, Universita' Bocconi, Milano, Italy, revised March 2003.
Breschi, S. and Lissoni, F. (2004), 'Knowledge networks from patent data: methodological issues and research targets', KITeS Working Papers 150, KITeS, Centre for Knowledge, Internationalization and Technology Studies, Universita' Bocconi, Milano, Italy.
Breschi, S. and Malerba, F. (1997), 'Sectoral systems of innovation: technological regimes, Schumpeterian dynamics and spatial boundaries', in Edquist, C. (ed.), *Systems of Innovation*, London: F. Pinter, pp.130–156.
Breschi, S. and Malerba, F. (2005), *Clusters, Networks and Innovation*, Oxford: Oxford University Press.
Breschi, S., Lissoni, F. and Montobbio, F. (2005a), 'From publishing to patenting: do productive scientists turn into academic inventors?', *Revue d'Economie Industrielle*, 110(2), 75–102.
Breschi, S., Lissoni, F. and Montobbio, F. (2007), 'The scientific productivity of academic inventors: new evidence from Italian data', *Economics of Innovation and New Technology*, 16(2), 101–118.
Breschi, S., Malerba, F. and Orsenigo, L. (2000), 'Technological regimes and Schumpeterian patterns of innovation', *Economic Journal*, Royal Economic Society, 110(463), 388–410, April.
Brooks, N., Donaghy, K. and Knaap, G. (eds) (2012), *The Oxford Handbook of Urban Economics and Planning*, Oxford: Oxford University Press.
Brouwer, E. and den Hertog, P. (2000), 'Key data on the Dutch information and communication cluster', Paper for the OECD Cluster Focus Group Workshop 'Do Clusters Matter in Innovation Policy?' in Utrecht, Holland, 8–9 May 2000.

BTRE (2004), 'Focus on Regions 3: Taxable Income', Canberra, Australia: Bureau of Transport and Regional Economics.

Buckley, P.J. and Casson, M.C. (1976), *The Future of Multinational Enterprise*, London: Macmillan.

Buckley, P.J. and Ghauri, P.N. (2004), 'Globalisation, economic geography and the strategy of multinational enterprises', *Journal of International Business Studies*, 35, 81–98, March.

Buckley, P.J., Devinney, T.M. and Louviere, J.J. (2007), 'Do managers behave the way theory suggests? A choice-theoretic examination of foreign direct investment location decision making', *Journal of International Business Studies*, 38, 1069–1094.

Burghouwt, G. (2005), *Airline Network Development in Europe and its Implications for Airport Planning*, Netherlands: Utrecht University Press.

Button, K., Lall, S., Stough, R. and Trice, M. (1999), 'High-technology employment and hub airports', *Journal of Air Transport Management*, 5, 53–59.

Button, K., Stough, R., Bragg, M. and Taylor, S. (2006), *Telecommunications, Transportation and Location*, Cheltenham, UK and Northampton, MA, USA: Edward Elgar Publishing.

Cairncross, F. (1997), *The Death of Distance: How Communications revolution will Change our Lives*, London: Orion Business Books.

Cameron, R.E. (1993), *Concise Economic History of the World: From Paleolithic Times to the Present*, Oxford: Oxford University Press.

Caniëls, M.C.J. (2000), *Knowledge Spillovers and Economic Growth: Regional Growth Differences across Europe*, Cheltenham, UK and Northampton, MA, USA: Edward Elgar Publishing.

Caniëls, M. and Verspagen, B. (2003), 'Spatial distance in a technology gap model', in Fingleton, B. (ed.), *European Regional Growth*, Heidelberg: Springer.

Cantwell, J. (1987a), 'The reorganization of European industries after integration: selected evidence on the role of multinational enterprise activities', *Journal of Common Market Studies*, Wiley Blackwell, 26(2), 127–151, December.

Cantwell, J. (1987b), 'Standing guard: protecting foreign capital in the nineteenth and twentieth centuries', *Journal of Development Economics*, Elsevier, 26(1), 190–193, June.

Cantwell, J. (1989), *Technological Innovation and Multinational Corporations*, Oxford: Basil Blackwell.

Cantwell, J. (1991), 'The theory of technological competence and its application to international production', in McFetridge, D. (ed.), *Foreign Investment, Technology and Economic Growth*, Calgary: University of Calgary Press.

Cantwell, J. (1992), 'The internationalisation of technological activity and its implications for competitiveness', in Granstrand, O., Håkanson, L. and Sjölander, S. (eds), *Technology Management and International Business: Internationalisation of R&D and Technology*, Chichester: John Wiley.

Cantwell, J. (ed.) (1994), *Transnational Corporations and Innovatory Activities*, London: Routledge.

Cantwell, J. (1995), 'The globalisation of technology: what remains of the product cycle model?', *Cambridge Journal of Economics*, Oxford University Press, 19(1), 155–174, February.

Cantwell, J. (2001), 'Innovation and information technology in MNE', in Rugman, A.M. and Brewer, T.L. (eds), *The Oxford Handbook of International Business*, Oxford: Oxford University Press.

Cantwell, J. (2009), 'Location and the multinational enterprise', *Journal of International Business Studies*, Palgrave Macmillan, 40(1), 35–41, January.

Cantwell, J. and Fai, F. (1999), 'Firms as the source of innovation and growth: the evolution of technological competence', *Journal of Evolutionary Economics*, Springer, 9(3), 331–366.

Cantwell, J. and Iammarino, S. (1998), 'MNCs, technological innovation and regional systems in the EU: some evidence in the Italian case', *International Journal of the Economics of Business*, Special Issue, 5 (3), 383–408.

Cantwell, J. and Iammarino, S. (2000), 'Multinational corporations and the location of technological innovation in the UK Regions', *Regional Studies*, 34(4), 317–332.

Cantwell, J. and Iammarino, S. (2001), 'EU regions and multinational corporations: change, stability and strengthening of technological comparative advantages', *Industrial and Corporate Change*, Special Issue on Geography of Innovation and Economic Clustering, 10(4), 1007–1037.

Cantwell, J. and Iammarino, S. (2002), 'The technological relationships between indigenous firms and foreign-owned multinational corporations in the European regions', in McCann, P. (ed.), *Industrial Location Economics*, Cheltenham, UK and Northampton, MA, USA: Edward Elgar Publishing.

Cantwell, J. and Iammarino, S. (2003), *Multinational Corporations and European Regional Systems of Innovation*, London and New York: Routledge.

Cantwell, J. and Iammarino, S. (2005), 'The technological innovation of multinational corporations in the French regions', *Revue d'Économie Industrielle*, 109(1), 9–28.

Cantwell, J. and Janne, O.E.M. (1999), 'Technological globalisation and

the innovative centres: the role of corporate technological leadership and locational hierarchy', *Research Policy*, 28(2–3), 119–144.
Cantwell, J. and Kosmopoulou, E. (2002), 'What determines the internationalisation of corporate technology?', in Havila, V., Forsgren, M. and Håkanson, H. (eds), *Critical Perspectives on Internationalisation*, Oxford: Pergamon, pp. 305–334.
Cantwell, J. and Mudambi, R. (2005), 'MNE competence-creating subsidiary mandates', *Strategic Management Journal*, 26(12), 1109–1128.
Cantwell, J. and Narula, R. (2001), 'The eclectic paradigm in the global economy', *International Journal of the Economics of Business*, Taylor and Francis Journals, 8(2), 155–172.
Cantwell, J. and Noonan, C.A. (2001), 'The regional distribution of technological development – evidence from foreign-owned firms in Germany', in Feldman, M.P. and Massard, N. (eds), *Knowledge Spillovers and the Geography of Innovation*, Dordrecht: Kluwer Academic Publishing.
Cantwell, J. and Noonan, C.A. (2002), 'How localized is knowledge sourcing? – Evidence from foreign-owned firms in Germany', European International Business Association (EIBA) annual conference, Athens, 8–10 December.
Cantwell, J. and Piscitello, L. (1999), 'The emergence of corporate international networks for the accumulation of dispersed technological capabilities', *Management International Review*, 39 (Special Issue 1), 123–147.
Cantwell, J. and Piscitello, L. (2000), 'Accumulating technological competence – its changing impact on corporate diversification and internationalisation', *Industrial and Corporate Change*, 9, 21–51.
Cantwell, J. and Piscitello, L. (2002), 'The location of technological activities of MNCs in European regions: the role of spillovers and local competencies', *Journal of International Management*, 8, 69–96.
Cantwell, J. and Piscitello, L. (2005), 'Recent location of foreign-owned research and development activities by large multinational corporations in the European regions: the role of spillovers and externalities', *Regional Studies*, 39(1), 1–16.
Cantwell, J. and Piscitello, L. (2007), 'Attraction and deterrence in the location of foreign-owned R&D activities: the role of positive and negative spillovers', *International Journal of Technological Learning, Innovation and Development*, 1(1), 83–111, January.
Cantwell, J, and Piscitello, L. (2010), 'Historical changes in knowledge sourcing by MNCs: the balance between localized vs. non-localized knowledge', Paper presented at the annual meeting of the European International Business Academy Annual Conference, Porto, Portugal.
Cantwell, J. and Santangelo, G.D. (1999), 'The frontier of international

technology networks: sourcing abroad the most highly tacit capabilities', *Information Economics and Policy*, Elsevier, 11(1), 101–123, March.

Cantwell, J. and Santangelo, G.D. (2006), 'The boundaries of firms in the new economy: M&As as a strategic tool toward corporate technological diversification', *Structural Change and Economic Dynamics*, Elsevier, 17(2), 174–199, June.

Cantwell, J., Iammarino, S. and Noonan, C.A. (2001), 'Sticky places in slippery space – the location of innovation by MNCs in the European regions', in Pain, N. (ed.), *Inward Investment, Technological Change and Growth. The Impact of Multinational Corporations on the UK Economy*, Basingstoke: Palgrave.

Capasso, M., Cefis, E. and Frenken, K. (2011a), 'Spatial differentiation in industrial dynamics. A core-periphery analysis based on the Pavitt-Miozzo-Soete taxonomy', LEM Working Paper, 2011/02, Sant'Anna School of Advanced Studies.

Capasso, M., Cusmano, L. and Morrison, A. (2011b), 'The determinants of outsourcing and offshoring strategies in industrial districts: evidence from Italy', *Regional Studies*, 1–15.

Cappelen, A., Castellacci, F., Fagerberg, J. and Verspagen, B. (2003), 'Regional disparities in income and unemployment in Europe', in Fingleton, B. (ed.), *European Regional Growth*, Heidelberg: Springer.

Carlaw, K.I. and Lipsey, R.G. (2006), 'GPT-driven, endogenous growth', *Economic Journal*, 116, 155–174.

Carlino, G.A., Chatterjee, S. and Hunt, R.M. (2007), 'Urban density and the rate of invention', *Journal of Urban Economics*, 61(3), 389–419.

Carlos, A.M. and Nicholas, S. (1988), 'Giants of an earlier capitalism: the chartered trading companies as modern multinationals', *Business History Review*, 62, 450–471.

Carlsson, B. (2003), 'Innovation systems: a survey of the literature from a Schumpeterian perspective', Case Western Reserve University, mimeo.

Carlsson, B. and Stankiewicz, R. (1991), 'On the nature, function, and composition of technological systems', *Journal of Evolutionary Economics*, 1(2), 93–118.

Carlton, D. and Perloff, J. (2005), *Modern Industrial Organization*, Boston, MA: Pearson/Addison Wesley.

Carr, D.L., Markusen, J.R. and Maskus, K.E. (2001), 'Estimating the knowledge-capital model of the multinational enterprise', *American Economic Review*, 91, 693–708.

Casson, M.C. (1987), *The Firm and the Market: Studies in Multinational Enterprises and the Scope of The Firm*, Cambridge, MA: The MIT Press and Oxford: Blackwell.

Casson, M.C. and McCann, P. (1999), 'Globalisation, competition, and the corporation: the UK experience', in Whitman M. (ed.), *The Evolving Corporation: Global Imperatives and National Responses*, Washington, DC: Group of Thirty.

Castellacci, F. (2008a), 'Technological paradigms, regimes and trajectories: manufacturing and service industries in a new taxonomy of sectoral patterns of innovation', *Research Policy*, Elsevier, 37(6–7), 978–994, July.

Castellacci, F. (2008b), 'Technology clubs, technology gaps and growth trajectories', *Structural Change and Economic Dynamics*, Elsevier, 19(4), 301–314, December.

Castellani, D. and Zanfei, A. (2002), 'Multinational experience and the creation of linkages with local firms: evidence from the electronic industry', *Cambridge Journal of Economics*, 26(1), 1–25.

Castellani, D. and Zanfei, A. (2004), 'Choosing international linkage strategies in the electronics industry: the role of multinational experience', *Journal of Economic Behavior & Organization*, Elsevier, 53(4), 447–475, April.

Castellani, D. and Zanfei, A. (2006), 'Multinational firms, innovation and productivity', *Economia Politica*, Le edizioni del Mulino, 0(2), 351–354, August.

Castells, M. and Hall, P. (1994), *Technopoles of the World*, London: Routledge.

Castells, M. and Hall, P. (1995), *Technopoles: Mines and Foundries of the Information Economy*, London: Routledge.

Caves, R.E. (1971), 'International corporations: the industrial economics of foreign direct investment', *Economica*, 38, 1–27.

Caves, R.E. (1974), 'Multinational firms, competition, and productivity in host-country markets', *Economica*, 41, 176–193.

Caves, R.E. (1982a), *Multinational Enterprise And Economic Analysis*, Cambridge, UK: Cambridge University Press.

Caves, R.E. (1982b), 'Multinational enterprises and technology transfer', in Rugman, A. (ed.), *New Theories of The Multinational Enterprise*, London: Croom Helm, 254–279.

Caves, R.E. (1996), *Multinational Enterprise and Economic Analysis*, Cambridge, UK: Cambridge University Press.

Cefis, E. (1996), 'Is there any persistence in innovative activities?', Discussion Paper no. 6, Department of Economics, University of Trento.

Centre for International Competitiveness (2008), *World Knowledge Competitiveness Index 2008*, Cardiff: Cardiff School of Management, University of Wales Institute.

Chaminade, C. and Edquist, C. (2006), 'From theory to practice. The use of the systems of innovation approach in innovation policy' in Hage, J. and De Meeus, M. (eds), *Innovation, Learning and Institutions*, Oxford: Oxford University Press.
Chaminade, C. and Vang, J. (2008), 'Globalisation of knowledge production and regional innovation policy: supporting specialized hubs in the Bangalore software industry', *Research Policy*, 37(10), 1684–1697.
Chandler, A.D., Jr (1959), 'The beginnings of "big business" in American industry', in McCraw, T.K. (ed.) (1988), *Essential Alfred Chandler*, Boston, MA: Harvard Business School Press.
Chandler, A.D. (1962), *Strategy and Structure: Chapters in the History of Industrial Enterprise*, Cambridge, MA: The MIT Press.
Chandler, A.D., Jr (1977), *The Visible Hand*, Cambridge, MA and London, UK: The Belknap Press of Harvard University Press.
Chandler, A.D. and Mazlish, B. (2005), *Leviathans: Multinational Corporations and the New Global History*, Cambridge, UK Cambridge University Press.
Chandler, A.D., Hagström, P. and Sölvell, O. (eds) (1998), *The Dynamic Firm. The Role of Technology, Strategy, Organisation and Regions*, Oxford: Oxford University Press.
Chandler, T. (1987), *Four Thousand Years of Urban Growth: An Historical Census*, Lampeter, UK: The Edwin Mellen Press.
Chatterji, L. and Tsai, C-M. (2006), 'Transport logistics in the global economy: spatial implications', in Kobayashi, K., Lakshmanan, T.R. and Anderson, W.P. (eds), *Structural Change in Transportation and Communications in the Knowledge Society*, Cheltenham, UK and Northampton, MA, USA: Edward Elgar Publishing.
Chen, M. and Moore, M. (2009), 'Location decision of heterogeneous multinational firms', *Journal of International Economics*, doi:10.1016/j.jinteco.2009.08.007.
Chen, Y-C., Vang, J. and Chaminade, C. (2012), 'Globalisation of innovation in knowledge intensive industries: lessons from the new China', *International Journal Technology and Globalization*, 6(4), 264–284.
Chesbrough, H. (2006), *Open Business Models: How to Thrive in the New Innovation Landscape*, Boston, MA: Harvard Business School Press.
Cheshire, P.C. and Gordon, I.R. (eds) (1995), *Territorial Competition in an Integrating Europe*, Aldershot: Avebury.
Chesnais, F. (1988), 'Technical co-operation agreements between firms', *Science Technology Industry Review*, 4, 57–119.
Chesnais, F. (1994), *La Mondialisation du Capital*, Paris: Syros.
Chesnais J.C. (1992), *The Demographic Transition Stages, Patterns and*

Economic Implications. A Longitudinal Study of Sixty-Seven Countries Covering the Period 1720–1984, Oxford: Clarendon Press.
Chien, S.-S. (2008), 'The isomorphism of local development policy: a case study of the formation and transformation of national development zones in post-Mao Jiangsu, China', *Urban Studies*, 45(2), 273–294.
Chisholm, M. (1990), *Regions in Recession and Resurgence*, London: Unwin Hyman.
Christopherson, S., Garretsen, H. and Martin, R. (2008), 'The world is not flat: putting globalization in its place', *Cambridge Journal of Regions, Economy and Society*, 1(3), 343–349.
Ciapetti, L. (2010), Politiche pubbliche per lo sviluppo locale e pianificazione strategica. A che punto siamo?, Foedus no. 26.
Ciapetti, L. (2011), 'Technological change, knowledge integration, and adaptive processes under globalization. The mechatronic evolution of the Reggio Emilia district', in Jacobson, D., Robertson, P. (eds), *Knowledge Transfer and Technological Diffusion*, Cheltenham, UK and Northampton, MA, USA: Edward Elgar Publishing.
Ciccone, A. and Hall, R.E. (1996), 'Productivity and the density of economic activity', *American Economic Review*, 86, 54–70.
Coase, R.H.(1937), 'The nature of the firm', *Economica*, 4(16), 386–405.
Coe, N.M., Dicken, P. and Hess, M. (2008), 'Global production networks: realizing the potential', *Journal of Economic Geography*, 8(3), 271–295.
Coe, N.M., Hess, M., Yeung, H.W-C., Dicken, P. and Henderson, J. (2004), 'Globalizing regional development: a global production networks perspective', *Transactions of the Institute of British Geographers*, 29, 468–484.
Coenen, L., Moodysson, J. and Asheim, B. (2004), 'Nodes, networks and proximities: on the knowledge dynamics of the Medicon Valley biotech cluster', *European Planning Studies*, 12(7), 1003–1018.
Cohen, B.J. (1998), *The Geography of Money*, Ithaca, NY: Cornell University Press.
Cohen, W. and Levinthal, D. (1989), 'Innovation and learning: the two faces of R&D', *The Economic Journal*, 99, 569–596, September.
Cohen, W. and Levinthal, D. (1990), 'Absorptive capacity: a new perspective on learning and innovation', *Administrative Science Quarterly*, 35, 128–152.
COL (2002), *Aviation Services for London*, London: Corporation of London.
COL (2005a), *The Competitive Position of London as a Global Financial Centre*, London: Corporation of London.

COL (2005b), *Off-shoring and the City of London*, London: Corporation of London.
COL (2006a), *London's Place in the UK Economy 2006–07*, London: Corporation of London.
COL (2006b), *The Importance of Wholesale Financial Services to the EU Economy 2006*, London: Corporation of London.
COL (2007a), *The Global Financial Centres Index 1*, London: Corporation of London.
COL (2007b), *The Global Financial Centres Index 2*, London: Corporation of London.
COL (2007c), *The Impact of Recent Immigration on the London Economy*, London: Corporation of London.
COL (2008), *The Global Financial Centres Index 3*, London: Corporation of London.
COL (2009), *The Global Financial Centres Index 6*, London: Corporation of London.
Coleman, W.D. (1996), *Financial Services, Globalization and Domestic Policy Change*, London: Macmillan.
Collier, P. (2007), *The Bottom Billion: Why the Poorest Countries are Failing and What Can be Done About It*, Oxford: Oxford University Press.
Combes, P-P. (2000), 'Economic structure and local growth: France, 1984–1993', *Journal of Urban Economics*, 47.
Combes, P-P. and Lafourcade, M. (2005), 'Transport costs: measures, determinants, and regional policy implications for France', *Journal of Economic Geography*, 319–349.
Combes, P-P., Duranton, G., Gobillon, L., Puga, D. and Roux, S. (2009), 'The productivity advantages of large cities: distinguishing agglomeration from selection', Center for Economic Policy Research, Discussion Paper 7191.
Cooke, P. (2001), 'Regional innovation systems, clusters, and the knowledge economy', *Industrial and Corporate Change*, 10(4), 945–974.
Cooke, P. (2007), 'To construct regional advantage from innovation systems first build policy platforms', *European Planning Studies*, 15, 179–194.
Cooke, P., Boekholt, P. and Tödtling, F. (2000), *The Governance of Innovation in Europe. Regional Perspective on Global Competitiveness*, London: Pinter.
Cooke, P., Gomez Uraga, M. and Etxebarria, G. (1997), 'Regional innovation systems: institutional and organisational dimensions', *Research Policy*, 26, 475–491.
Cooke, P., Gomez Uraga, M. and Etxebarria, G. (1998), 'Regional

innovation systems: an evolutionary perspective', *Environment and Planning A*, 30, 1563–1584.
Coriat, B. and Weinstein, O. (2002), 'Organizations, firms and institutions in the generation of innovation', *Research Policy*, 31, 273–290.
Corporación Andina de Fomento (2010), Desarrollo Local: Hacia un Nuevo Protagonismo de las Ciudades y Regiones. Caracas: Corporación Andina de Fomento.
Cowan, R. and Foray, D. (1997), 'The economics of codification and the diffusion of knowledge', *Industrial and Corporate Change*, 6(3), 595–622.
Cowan, R.K., Lwiza, K.M.M., Sponaugle, S., Paris, B. and Olson, D.B. (2000), 'Connectivity of marine populations: open or closed?', *Science*, 287, 857–859.
Cowling, K. and Sugden, R. (1987a), *Transnational Monopoly Capitalism*, Brighton: Wheatsheaf.
Cowling, K. and Sugden, R. (1987b), 'Market exchange and the concept of a transnational corporation: analysing the nature of the firm', *British Review of Economic Issues*, 9(20), 57–68.
Coyle, D. (1997), *The Weightless World*, London: Capstone, and Cambridge, MA: The MIT Press.
Crafts, N. (2004), 'Globalization and economic growth: a historical perspective', *World Economy*, 27(1), 45–58.
Crafts, N. and Venables, A. (2003), 'Globalization in history. A geographical perspective', in *Globalization in Historical Perspective*, National Bureau of Economic Research, pp. 323–370.
Crescenzi, R. and Rodríguez-Pose, A. (2009), 'Systems of innovation and regional growth in the EU: endogenous vs. external innovative efforts and socio-economic conditions', in Fratesi, U. and Senn, L. (eds), *Growth and Innovation of Competitive Regions: The Role of Internal and External Connections. Advances in Spatial Science*, Berlin: Springer-Verlag, pp. 167–192.
Crescenzi, R. and Rodríguez-Pose, A. (2011a), *Innovation and Regional Growth in the European Union*, Berlin: Springer.
Crescenzi, R. and Rodríguez-Pose, A. (2011b), 'Reconciling top-down and bottom-up development policies', *Environment and Planning A*, 43(4), 773–780.
Crescenzi, R., Rodríguez-Pose, A. and Storper, M. (2007), 'The territorial dynamics of innovation: a Europe–United States comparative analysis', *Journal of Economic Geography*, 7(6), 673–709.
Crescenzi, R. and Percoco, M. (eds) (2012), *Geography, Institutions and Regional Economic Performance*, Springer, forthcoming.
Criscuolo, C., Haskel, J. and Slaughter, M. (2005), 'Global engagement

and the innovation activities of firms', National Bureau of Economic Research Working Paper No. 11479.

Crozet, M., Mayer, T. and Mucchielli, J. (2004), 'How do firms agglomerate? A study of FDI in France', *Regional Science and Urban Economics*, 34, 27–54.

Culem, C. (1988), 'Direct investment among industrialized countries', *European Economic Review*, 32, 885–904.

Cusmano, L., Mancusi, M.L. and Morrison, A. (2010), 'Globalization of production and innovation: how outsourcing is reshaping an advanced manufacturing area', *Regional Studies*, 44(3), 235–252.

Cyert, R.M. and March, J.G. (1963/1992), *A Behavioral Theory of the Firm*, 2nd edn, Englewood Cliffs, NJ: Prentice Hall.

D'Aspremont, C., Cremer, J. and Gerard-Varet, L-A. (1998), 'Unique implementation in auctions and in public goods problems', CORE Discussion Papers 1998010, Université catholique de Louvain, Center for Operations Research and Econometrics (CORE).

D'Aspremont, C., Gabszewicz, J.J. and Thisse, J-F. (1979), 'On Hotelling's "Stability in Competition"', *Econometrica*, Econometric Society, 47(5), 1145–1150, September.

D'Este, P., Guy, F. and Iammarino, S. (2012), 'Shaping the formation of university-industry research collaborations: what type of proximity does really matter?', *Journal of Economic Geography*, May, doi:10.1093/jeg/lbs010.

D'Este, P., Iammarino, S., Savona, M. and von Tunzelmann, N. (2012), 'What hampers innovation? Revealed versus deterring barriers', *Research Policy*, 41, 482–488.

De Vor, F. and de Groot, H.L.F. (2008), 'Agglomeration Externalities and Localized Employment Growth', Tinbergen Institute Discussion Paper, TI 2008–033/3.

Dahmén, E. (1988), 'Development blocks in industrial economics', *Scandinavian Economic History Review*, 1.

Dall'Erba, S. and Le Gallo, J. (2008), 'Regional convergence and the impact of structural funds over 1989–1999: a spatial econometric analysis', *Papers in Regional Science*, 87, 219–244.

Daudin, G., Morys, M. and O'Rourke, K. (2008), 'Globalization, 1870–1914', University of Oxford Department of Economics Discussion Paper Series No. 395.

Davis, M. (2006). 'Planet of slums', *New Perspectives Quarterly*, 23, 6–11.

Day, G. and Reibstein, D. (1997), 'Introduction: the dynamic challenges for theory and practice', in Day, G.S. and Reibstein, D.J. (eds), *Wharton on Dynamic Competitive Strategy*, New York: John Wiley and Sons, 1–18.

Deardorff, A. (2003), 'Time and trade: the role of time in determining the structure and effects of international trade with an application to Japan', in Stern, R.M. (ed.), *Analytical Studies in US–Japan International Economic Relations*, Cheltenham, UK and Northampton, MA, USA: Edward Elgar Publishing.

Defever, F. (2006), 'Functional fragmentation and the location of multinational firms in the enlarged Europe', *Regional Science and Urban Economics*, 36(5), 658–677.

Deichmann, U., Kaiser, K., Lall, S.V. and Shalizi, Z. (2005), 'Agglomeration, transport, and regional development in Indonesia', Policy Research Working Paper 3477, World Bank.

Dei Ottati, G. (2003), 'Exit, voice and the evolution of industrial districts: the case of the post-World War II economic development of Prato', *Cambridge Journal of Economics*, 27, 501–522.

Delamaide, D. (1994), *The New Superregions of Europe*, New York: Dutton/Penguin.

Delbridge, R. and Lowe, J. (eds) (1998), *Manufacturing in Transition*, London: Routledge.

Derudder, B., Taylor, P.J., Witlox, F. and Catalano, G. (2003), 'Hierarchical tendencies and regional patterns in the world city network', *Regional Studies*, 37(9), 875–886.

Desai, M.A. (2009), 'The decentering of the global firm', *The World Economy*, 32, 1271–1290.

Devereux, M.P., Griffith, R. and Simpson, H. (2000), 'The geographic distribution of production activity in the UK', Econometric Society World Congress 2000 Contributed Papers 1397, Econometric Society.

Dewhurst, J.H.L.L. and McCann, P. (2002), 'A comparison of measures of industrial specialisation for travel-to-work areas in Great Britain 1981–97', *Regional Studies*, 36(5), 541–551.

Dewhurst, J.H.L.L. and McCann, P. (2007), 'Specialisation and regional size', in Fingleton, B. (ed.), *New Directions in Economic Geography*, Cheltenham, UK and Northampton, MA, USA: Edward Elgar Publishing.

Dicken, P. (1977), 'A note on location theory and the large business enterprise', *Area*, 9(2), 138–143.

Dicken, P. (1988), 'The changing geography of Japanese foreign direct investment in manufacturing industry: a global perspective', *Environment and Planning A*, 20, 633–653.

Dicken, P. (1994), 'Global-local tensions: firms and states in the global space-economy', *Economic Geography*, 70, 101–128.

Dicken, P. (1998), *Global Shift: Transforming the World Economy*, 3rd edn, London: Paul Chapman.

Dicken, P. (2003a), 'Global production networks in Europe and East Asia: the automobile components industries', GPN Working Paper 7.
Dicken, P. (2003b), *Global Shift: Reshaping the Global Economic Map in the 21st Century*, 4th edn, London: Sage and New York: Guildford.
Dicken, P. (2004), 'Geographers and "globalization": (yet) another missed boat?', *Transactions of the Institute of British Geographers*, 29, 5–26.
Dicken, P. (2005), 'Tangled webs: transnational production networks and regional integration', *Spaces*, 4, Marburg: Philipps-University of Marburg.
Dicken, P. (2007), *Global Shift: Mapping the Changing Contours of the World Economy*, 5th edn, London: Sage and New York: Guilford.
Dicken, P. and Henderson, J. (2003), 'Making the connections: global production networks in Britain, East Asia and Eastern Europe', Final Report on ESRC Project R000238535.
Dicken, P. and Malmberg, A. (2001), 'Firms in territories: a relational perspective', *Economic Geography*, 77, 345–363.
Dicken, P., Peck, J.A. and Tickell, A. (1997), 'Unpacking the global', in Lee, R. and Wills, J. (eds), *Geographies of Economies*, London: Arnold, 158–66.
Dijkstra, L., Garcilazo, E. and McCann, P. (2012), 'The economic performance of European cities and city-regions: myths and realitities', *European Planning Studies*, August, doi: 10.1080/09654313.2012.716245.
Disdier, A-C. and Head, K. (2008), 'The puzzling persistence of the distance effect on bilateral trade', *Review of Economics and Statistics*, 90(1), 37–48.
Disdier, A.-C. and Head, K. (2009), 'The puzzling persistence of the distance effect on bilateral trade', *Review of Economics and Statistics*, 90, 37–48.
Doloreux, D. and Parto, S. (2004), 'Regional innovation systems: a critical synthesis', United Nations University, INTECH, Discussion Paper Series.
Doloreux, D. and Parto, S. (2005), 'Regional innovation systems: current discourse and unresolved issues', *Technology in Society*, 27, 133–153.
Dopfer, K., Foster, J. and Potts, J. (2004), 'Micro-meso-macro', *Journal of Evolutionary Economics*, 14, 263–279.
Dosi, G. (1984), 'Technological paradigms and technological trajectories. The determinants and directions of technical change and the transformation of the economy' in Freeman, C. (ed.), *Long Waves in the World Economy*, 2nd edn, London: Francis Pinter.
Dosi, G. (1988), 'Sources, procedures and microeconomic effects of innovation', *Journal of Economic Literature*, 26, 1120–1171.

Dosi, G. (1996), 'The contribution of economic theory to the understanding of a knowledge-based economy', in Foray, D. and Lundvall, B-Å, *Employment and Growth in the Knowledge-Based Economy*, Paris: OECD, 81–92.

Dosi, G., Nelson, R.R. and Winter, S. (1999), *The Nature and Dynamics of Organisational Capabilities*, Oxford: Oxford University Press.

Dosi, G., Silverberg, G. and Orsenigo, L. (1988), 'Innovation, diversity and diffusion: a self-organising model', *The Economic Journal*, 98(393), 1032–1054.

Dosi, G., Malerba, F., Marsili, O., Orsenigo, L. (1997), 'Industrial structures and dynamics: evidence, interpretations and puzzles', *Industrial and Corporate Change*, 6(1), 3–24.

Dosi G., Marsili, O., Orsenigo, L. and Salvatore, R. (1995), 'Learning, market selection and the evolution of industrial structures', *Small Business Economics*, 7, 411–436.

Douglas, S.P. and Wind, Y. (1987), 'The myths of globalization', *Columbia Journal of World Business*, 22(4), 19–29.

Dumais, G., Ellison, G. and Glaeser, E.L. (2002), 'Geographic concentration as a dynamic process', *Review of Economic and Statistics*, 84(2), 193–204.

Dunford, M. and Smith, A. (2000), 'Catching up or falling behind? Economic performance and regional trajectories in the "new" Europe', *Economic Geography*, 76, 169–195.

Dunning, J.H. (1970), *Studies in International Investment*, London: Allen & Unwin.

Dunning, J.H. (1977), 'Trade, location of economic activity', Proceedings of a Nobel Symposium Held at Stockholm, London: Macmillan, 395–418.

Dunning, J.H. (1979), 'Explaining changing patterns of international production: in defence of the eclectic theory', *Oxford Bulletin of Economics and Statistics*, 41(4), 269–295.

Dunning, J.H. (1981), *International Production and the Multinational Enterprise*, London: Allen & Unwin.

Dunning, J.H. (1988a), 'The eclectic paradigm of international production: a restatement and some possible extensions', *Journal of International Business Studies*, 19, 1–31.

Dunning, J.H. (1988b), 'The theory of international production', *The International Trade Journal*, 3(1), 21–66.

Dunning, J.H. (1988c), *American Investment in British Manufacturing Industry*, London and New York: Routledge.

Dunning, J.H. (1993), *Multinational Enterprises and the Global Economy*, Workingham: Addison-Wesley.

Dunning, J.H. (1994), 'Re-evaluating the benefits of foreign direct investment', *Transnational Corporations*, 3(1), 27–51.
Dunning, J.H. (1995), 'Reappraising the eclectic paradigm in an age of alliance capitalism', *Journal of International Business Studies*, 26(3), 461–491.
Dunning, J.H. (1997), *Alliance Capitalism and Global Business*, London: Routledge.
Dunning, J.H. (1998), 'Location and the multinational enterprise: a neglected factor?', *Journal of International Business Studies*, 29(1), 45–66.
Dunning, J.H. (ed.) (2000), *Regions, Globalisation, and the Knowledge-based Economy*, Oxford: Oxford University Press.
Dunning, J.H. (2001), 'The eclectic (OLI) paradigm of international production: past, present and future', *International Journal of the Economics of Business*, 8(2), 173–190.
Dunning, J.H. (2003a), 'The moral imperatives of global capitalism: an overview', in Dunning, J.H. (ed.), *Making Globalization Good*, Oxford: Oxford University Press, 11–40.
Dunning, J.H. (2003b), *Making Globalization Good*, Oxford: Oxford University Press.
Dunning, J.H. (2009a), *Seasons of A Scholar: Some Personal Reflections of an International Business Economist*, Cheltenham, UK and Northampton, MA, USA: Edward Elgar Publishing.
Dunning, J.H. (2009b), 'Location and the multinational enterprise: John Dunning's thoughts on receiving the Journal of International Business Studies 2008 Decade Award, *Journal of International Business Studies*, 40(1), 20–34.
Dunning, J.H. and Lundan, S.M. (2007), *Multinational Enterprises and the Global Economy*, Cheltenham, UK and Northampton, MA, USA: Edward Elgar Publishing.
Dunning, J.H. and Lundan, S.M. (2008), *Multinational Enterprises and the Global Economy*, 2nd edn, Cheltenham, UK and Northampton, MA, USA: Edward Elgar Publishing.
Dunning, J.H. and Narula, R. (eds) (1996), *Foreign Direct Investment and Governments: Catalysts for Economic Restructuring*, London and New York: Routledge.
Dunning, J.H. and Pitelis, C.N. (2008), 'Stephen Hymer's contribution to international business scholarship: an assessment and extension', *Journal of International Business Studies*, 39(1), 167–176.
Dunning, J.H. and Wymbs, C. (1999), 'The geographical sourcing of technology based assets by multinational enterprises', in Archibugi, D.,

Howells, J. and Michie, J. (eds), *Innovation Policy in a Global Economy*, Cambridge, UK: Cambridge University Press.

Dunning, J.H., Fujita, M. and Yakova, N. (2007), 'Some macrodata on the regionalization/globalization debate: a comment on the Rugman/Verbeke analysis', *Journal of International Business Studies*, 38(1), 177–199.

Duranton, G. and Puga, D. (2001), 'Nursery cities: urban diversity, process innovation, and the life cycle of products', *American Economic Review*, American Economic Association, 91(5), 1454–1477, December.

Duranton, G. and Puga, D. (2003), 'Microfoundations of urban agglomeration economies', CEPR Discussion Papers 4062, CEPR Discussion Papers.

Duranton, G. and Puga, D. (2004), 'Micro-foundations of urban agglomeration economies', in Henderson, J.V. and Thisse, J.F. (eds), *Handbook of Regional and Urban Economics*, 1st edn, vol. 4, chapter 48, 2063–2117, Elsevier.

Duranton, G. and Puga, D. (2005), 'From sectoral to functional urban specialisation', *Journal of Urban Economics*, 57(2), 343–370, March.

Duranton, G. and Storper, M. (2008), 'Rising trade costs? Agglomeration and trade with endogenous transaction costs', *Canadian Journal of Economics*, 41(1), 292–319.

Easton, B. (2007), *Globalisation and the Wealth of Nations*, Auckland, New Zealand: Auckland University Press.

Edquist, C. (1997), *Systems of Innovation: Technologies, Institutions and Organisations*, London: Pinter.

Edquist, C. (2005), 'Systems of innovation: perspectives and challenges', in Fagerberg, J., Mowery, D.C. and Nelson, R.R. (eds), *The Oxford Handbook of Innovation*, Oxford: Oxford University Press, 181–208.

Eisenhardt, K. and Martin, J. (2000), 'Dynamic capabilities: what are they?', *Strategic Management Journal*, 21, 1105–1121.

Ellison, G. and Glaeser, E.L. (1997), 'Geographic concentration in U.S. manufacturing industries: a dartboard approach', *Journal of Political Economy*, 105, 889–927.

Enright, M.J. (1998), 'Regional clusters and firm strategy', in Chandler, A.D., Hagström, P. and Sölvell, O. (eds), *The Dynamic Firm*, Oxford: Oxford University Press.

Enright, M.J. (2000a), 'The globalization of competition and the localization of competition: policies toward regional clustering', in Hood, N. and Young, S. (eds), *The Globalization of Multinational Enterprise Activity and Economic Development*, London: Macmillan.

Enright, M.J. (2000b), 'Regional clusters and multinational enterprises:

independence, dependence, or interdependence?', *International Studies of Management and Organization*, 30(2), 114–138.

Enright, M.J. (2003), 'Regional clusters: what we know and what we should know', in Bröcker, J., Dohs, D. and Soltwedel, R. (eds), *Innovation Clusters and Interregional Competition*, Berlin: Springer, 99–129.

Ernst, D. (1997), 'High-tech competition puzzles. how globalization affects firm behavior and market structure in the electronics industry', DRUID Working Papers 97-9, DRUID, Copenhagen Business School, Department of Industrial Economics and Strategy/Aalborg University, Department of Business Studies.

Ernst, D. (1998), 'What permits small firms to compete in high-tech industries? Inter-organizational knowledge creation in the Taiwanese computer industry', DRUID Working Papers 98–3.

Ernst, D. (2002a), 'Global production networks in East Asia's electronics industry and upgrading perspectives in Malaysia', Economics Study Area Working Papers wp44, East-West Center, Economics Study Area, revised October 2003.

Ernst, D. (2002b), 'Digital information systems and global flagship networks: how mobile is knowledge in the global network economy', Economics Study Area Working Papers 48, East-West Center, Economics Study Area.

Ernst, D. (2005), 'Limits to modularity: reflections on recent developments in chip design', *Industry and Innovation*, 12(3), 303–335, September.

Ernst, D. (2010), 'Upgrading through innovation in a small network economy: insights from Taiwan's IT industry', *Economics of Innovation and New Technology*, 19(4), 295–324.

Ernst, D. and Guerrieri, P. (1998), 'International production networks and changing trade patterns in East Asia: the case of the electronics industry', *Oxford Development Studies*, 26 (2).

Ernst, D. and Kim, L. (2002), 'Global production networks, knowledge diffusion, and local capability formation', *Research Policy*, Elsevier, 31(8–9), 1417–1429, December.

Ernst, D., Martin, S. and East-West Center (2010), *The Common Criteria for Information Technology Security Evaluation: Implications for China's Policy on Information Security Standards*, Honolulu, HI: East-West Center.

Ernst, D., Fagerberg, J., Hildrum, J. and East-West Center (2002), *Do Global Production Networks and Digital Information Systems make Knowledge Spatially Fluid?*, Honolulu, HI: East-West Center.

Ernst, D., Guerrieri P., Iammarino, S. and Pietrobelli, C. (2001), 'SME Clusters facing global restructuring: what can be learnt comparing

Italy and Taiwan?', in Guerrieri, P., Iammarino, S. and Pietrobelli, C. (eds), *The Global Challenge to Industrial Districts. The case of Italy and Taiwan*, Cheltenham, UK and Northampton, MA, USA: Edward Elgar Publishing.

Estevadeordal, A., Frantz, B. and Taylor, A.M. (2003), 'The rise and fall of world trade: 1870–1939', *Quarterly Journal of Economics*, 118, 359–407.

Eswaran, M., Kanemoto, Y. and Ryan, D. (1981), 'A dual approach to the location decision of the firm', *Journal of Regional Science*, 21, 469–490.

Ethier, W.J. (1986), 'The multinational firm', *The Quarterly Journal of Economics*, The MIT Press, 101(4), 805–833, November.

Ethier, W.J. and Markusen, J.R. (1996), 'Multinational firms, technology diffusion and trade', *Journal of International Economics*, Elsevier, 41(1–2), 1–28, August.

European Commission (2010), 'Investing in Europe: Fifth Cohesion Report on Economic, Social and Territorial Cohesion', Brussels: European Commission.

Evangelista, R. (1999), *Knowledge and Investment. The Sources of Innovation in Industry*, Cheltenham, UK and Northampton, MA, USA: Edward Elgar Publishing.

Evangelista, R. (2000), 'Sectoral patterns of technological change in services', *Economics of Innovation and New Technology*, 9(3).

Evangelista, R., Iammarino, S., Mastrostefano, V. and Silviani, A. (2002), 'Looking for regional systems of innovation. Evidence from the Italian innovation survey', *Regional Studies*, 36(2), 173–186.

Fagerberg, J. (1987), 'A technology gap approach to why growth rates differ', *Research Policy*, 16(2–4), 87–99.

Fąggerberg, J. (1994), 'Technology and international differences in growth rates', *Journal of Economic Literature*, 32, 1147–1175.

Fagerberg, J. and Godinho, M.M. (2005), 'Innovation and catching-up', in Fagerberg, J., Mowery, D. and Nelson, R. (eds), *The Oxford Handbook of Innovation*, Oxford: Oxford University Press.

Fagerberg, J. and Srholec, M. (2008), 'National innovation systems, capabilities and economic development', *Research Policy*, Elsevier, 37(9), 1417–1435, October.

Fagerberg, J. and Verspagen, B. (1996), 'Heading for divergence? Regional growth in Europe reconsidered', *Journal of Common Market Studies*, Wiley Blackwell, 34(3), 431–448.

Fagerberg, J. and Verspagen, B. (2002), 'Technology-gaps, innovation-diffusion and transformation: an evolutionary interpretation', *Research Policy*, 31, 1291.

Fagerberg, J. and Verspagen, B. (2007), 'Innovation, growth and economic development: have the conditions for catch-up changed?', *International Journal of Technological Learning, Innovation and Development*, 1(1), 13–33.

Fagerberg, J., Srholec, M. and Knell, M. (2007), 'The competitiveness of nations: why some countries prosper while others fall behind', *World Development*, Elsevier, 35(10), 1595–1620, October.

Fagerberg, J., Verspagen, B. and Caniëls, M. (1997), 'Technology, growth and unemployment across European regions', *Regional Studies*, Taylor and Francis Journals, 31(5), 457–466.

Fagerberg, J., Verspagen, B. and von Tunzelmann, N. (eds) (1994), *The Dynamics of Technology, Trade and Growth*, Aldershot, UK and Brookfield, VT, USA: Edward Elgar Publishing.

Faggian, A. and McCann, P. (2006), 'Human capital flows and regional knowledge assets: a simultaneous equation approach', *Oxford Economic Papers*, 58(3), 475–500.

Farole, T., Rodríguez-Pose, A. and Storper, M. (2011), 'Cohesion policy in the European Union: growth, geography, institutions', *Journal of Common Market Studies*, 49(5), 1089–1111, doi: 10.1111/j.1468-5965.2010.02161.x.

Farshchi, M.A., Janne, O.E.M. and McCann, P. (2009), *Technological Change and Mature Industrial Regions: Firms, Knowledge and Policy*, New Horizons in Regional Science Series, Cheltenham, UK and Northampton, MA, USA: Edward Elgar Publishing.

Feldman, M.P. and Audretsch, D.B. (1999), 'Innovation in cities: science-based diversity, specialization and localized competition', *European Economic Review*, Elsevier, 43(2), 409–429, February.

Ferguson, N. (2003), *Empire: How Britain made the Modern World*, London: Penguin Press/Allen Lane.

Ferguson, N. (2004), *Colossus: The Price of America's Empire*, London: Penguin Press/Allen Lane.

Ferguson, N. (2006), *The War of the World: History's Age of Hatred*, London: Allen Lane/Penguin Press.

Ferguson, N. (2008), *The Ascent of Money: A Financial History of the World*, London: Allen Lane.

Ferguson, N., Mair, C.S., Manela, E. and Sargent, D.J. (eds) (2010), *The Shock of the Global: The 1970s in Perspective*, Cambridge, MA: Harvard University Press.

Fifarek, B. and Veloso, F.M. (2010), 'Offshoring and the geography of innovation', *Journal of Economic Geography*, 10(4), 559–578, doi:10.1093/jeg/lbq013.

Findlay, R. and O'Rourke, K. (2007), *Power and Plenty: Trade, War, and*

the World Economy in the Second Millennium, Princeton, NJ: Princeton University Press.
Fine, B. (2003), 'Neither the Washington nor the post-Washington consensus: an introduction', in Fine, B., Lapavitsas, C. and Pincus, J. (eds), *Development Policy in the Twenty-first Century: Beyond the Post-Washington Consensus*, London: Routledge, 1–27.
Fingleton, B. (ed.) (2003a), *European Regional Growth*, Heidelberg: Springer.
Fingleton, B. (2003b), 'Increasing returns: evidence from local wage rates in Great Britain', *Oxford Economic Papers*, 55, 716–739.
Fingleton, B. (ed.) (2007), *New Directions in Economic Geography*, Cheltenham, UK and Northampton, MA, USA: Edward Elgar Publishing.
Fingleton, B. and McCann, P. (2004), 'Sinking the iceberg? On the treatment of transport costs in new economic geography', in Fingleton, B. (ed.), *New Directions in Economic Geography*, Cheltenham, UK and Northampton, MA, USA: Edward Elgar Publishing.
Fisher, S. (2003), 'Globalization and its challenges', *American Economic Review: Papers and Proceedings*, 93(2), 1–30.
Flaherty, D, (1995), 'Regional inequality in South Africa: issues, measurements and policy implications', Occasional Paper no. 70, Development Bank of Southern Africa, Halfway House, February.
Florida, R (1995), 'Toward the learning region', *Futures*, 27(5), 527–536.
Florida, R. (2005a), *Cities and the Creative Class*, London: Routledge.
Florida, R. (2005b), *The Flight of the Creative Class: The New Global Competition for Talent*, New York: Harper Business Books.
Florida. R. and Kenney, M. (1994), 'The globalization of Japanese R&D: the economic geography of Japanese R&D in the US', *Economic Geography*, 70(4), 344–369. 16.
Fontana, R., Geuna, A. and Matt, (2006), 'Factors affecting university – industry R&D projects: the importance of searching, screening and signaling', *Research Policy*, 35, 309–323.
Foray, D. and Freeman, C. (eds) (1993), *Technology and the Wealth of Nations*, London: Pinter.
Forni, M. and Paba, S. (2001), 'Knowledge spillovers and the growth of local industries', CEPR Discussion Papers 2934, CEPR Discussion Papers.
Fors, G. (1998), 'Locating R&D abroad: the role of adaptation and knowledge-seeking', in Braunerhjelm, P. and Ekholm, K. (eds), *The Geography of Multinational Firms*, Boston, Dordrecht and London: Kluwer Academic Publishers.
Foss, N.J. (1997a), 'Equilibrium vs evolution in the resource-based per-

spective the conflicting legacies of Demsetz and Penrose', DRUID Working Papers 97-10.
Foss, N.J. (1997b), 'Incomplete contracts and economic organization Brian Loasby and the theory of the firm', DRUID Working Papers 97-11.
Foss, N.J. (1998), 'The resource-based perspective an assessment and diagnosis of problems', *Scandinavian Journal of Management*, 14(3), 133–149.
Fothergill, S., Kitson, M. and Monk, S. (1985), *Urban Industrial Decline: The Causes of Urban/Rural Contrasts in Manufacturing Employment Change*, London: HMSO.
Frankel, J.A. and Romer, D. (1999), 'Does trade cause growth', *American Economic Review*, 89(3), 379–399.
Franko, L.G. (1974), 'The origins of multinational manufacturing by continental European firms', *The Business History Review*, 48(3), 227–302.
Franko, L.G. (1976), *The European Multinationals*, London: Harper and Row and Stamford, CT: Greylock Press.
Fransman, M. (1990), *The Market and Beyond: Cooperation and Competition in Information Technology Development in the Japanese System*, Cambridge, UK Cambridge University Press.
Freeman, C. (1982a), *The Economics of Industrial Innovation*, 2nd edn, Cambridge, MA: The MIT Press.
Freeman, C. (1982b), 'Technological infrastructure and international competitiveness', *Industrial and Corporate Change*, 13(3), 541–569.
Freeman, C. (1986), 'The role of technical change in national economic development', in Amin, A. and Goddard, J. (eds), *Technological Change, Industrial Restructuring and Regional Development*, London: George Allen & Unwin, 100–114.
Freeman, C. (1987), *Technology Policy and Economic Performance*, London: Pinter.
Freeman, C. (1988), *Technology Policy and Economic Performance: Lessons From Japan*, London: Pinter.
Freeman, C. (2002), 'Continental, national and sub-national innovation systems – complementarity and economic growth', *Research Policy*, 31(2), 191–211.
Freeman, C. and Perez, C. (1988), 'Structural crises of adjustment, business cycles and investment behaviour', in Dosi, G. et al. (eds), *Technical Change and Economic Theory*, London: Pinter, 38–66.
Freeman, C. and Perez, C. (1998), 'Structural crises of adjustment, business cycles and investment behaviour', reprinted in Hanusch, H. (ed.), *The Economic Legacy of Joseph Schumpeter*, Cheltenham, UK and Lyme, NH, USA: Edward Elgar Publishing.

Freeman, C. and Soete, L. (1997), *The Economics of Industrial Innovation*, 3rd edn, London: Pinter.

Freeman, C., Clark, J. and Soete, L. (1982), *Unemployment and Technical Innovation: A Study of Long Waves in Economic Development*, London: Pinter.

Frenken, K. and Boschma, R.A. (2007), 'A theoretical framework for economic geography: industrial dynamics and urban growth as a branching process', *Journal of Economic Geography*, 7(5), 635–649.

Frenken, K., van Oort, F.G. and Verburg, T. (2007), 'Related variety, unrelated variety and regional economic growth', *Regional Studies*, 41(5), 685–697.

Frenz, M. and Ietto-Gillies, G. (2007), 'Does multinationality affect the propensity to innovate? An analysis of the third UK Community Innovation Survey', *International Review of Applied Economics*, 21(1), 99–117.

Friedman, T.L. (2005), *The World is Flat: A Brief History of the Twenty-First Century*, New York: Farrar, Straus, and Giroux.

Friedman, T.L. (2006), *The World Is Flat: The Globalized World In The Twenty-First Century*, London: Penguin.

Friedman, T.L. (2007), *The World is Flat: A Brief History of the Twenty-First Century*, 3rd edn, New York: Picador.

Friedmann, J. (1986), 'The world city hypothesis', *Development and Change*, 17, 69–83.

Friedmann, J. (1995), 'The world city hypothesis', in Knox, P.L. and Taylor, P.J. (eds), *World Cities in a World System*, Cambridge, UK: Cambridge University Press, 317–331.

Friedmann, J. and Wolff, G. (1982), 'World city formation: an agenda for research and action', *International Journal of Urban and Regional Research*, 6, 309–344.

Fritsch, M. (1997), 'New firms and regional employment change', *Small Business Economics*, 9(5), 437.

Fritsch, M. (2001), 'Cooperation in regional innovation systems', *Regional Studies*, 35, 297–307.

Frost T.S. (1996), 'From exploitation to exploration: the geographic sources of subsidiary innovations and the evolutionary theory of the multinational enterprise', mimeo, University of Western Ontario.

Frost, T.S. (2001), 'The geographic sources of foreign subsidiaries' innovation', *Strategic Management Journal*, 22, 101–123.

Frost, T.S., Birkinshaw, J.M and Ensign, P.C. (2002), 'Centers of excellence in multinational corporations', *Strategic Management Journal*, 23, 997–1018.

Fuchs, G. and Koc, A. (2005), 'Biotechnology and multimedia: cluster development in new industries', in Karlsson, C., Johansson, B. and

Stough, R.R. (eds), *Industrial Clusters and Inter-Firm Networks*, Cheltenham, UK and Northampton, MA, USA: Edward Elgar Publishing.

Fujimoto, T. (1999), *The Evolution of a Manufacturing System at Toyota*, Oxford: Oxford University Press.

Fujimoto, T. (2001), 'The Japanese automobile parts supplier system: the triplet of effective inter-firm routines', *International Journal of Automotive Technology and Management*, 1, 1–34.

Fujita, M. (ed.) (2007a), *Regional Integration in South East Asia from the Viewpoint of Spatial Economics*, Palgrave, London.

Fujita, M. (2007b), 'Towards the new economic geography in the brain power society', *Regional Science and Urban Economics*, Elsevier, 37(4), 482–490, July.

Fujita, M., Krugman, P.R. and Venables, A.J. (1999), *The Spatial Economy; Cities, Regions, and International Trade*, The MIT Press.

Galbraith, J.K (1987), 'The 1929 Parallel', *The Atlantic Monthly*, 62–66, January.

Garcilazo, E., Oliveira Martins, J. and Tompson, W. (2010), 'Why policies may need to be place-based in order to be people-centred', Paris: OECD Regional Development Policy Division. Accessed at http://www.voxeu.org/.

Garofoli, G. (1983), *Industrializzazione Diffusa in Lombardia*, Milan: FrancoAngeli.

Garofoli, G. (2003), 'Introduzione' in Garofoli G. (ed.), *Impresa e Territorio*, Bologna: Il Mulino, 9–39.

Gaspar, J. and Glaeser, E.L. (1998), 'Information technology and the future of cities', *Journal of Urban Economics*, 43, 136–156.

Gereffi, G. (2005), 'The global economy: organization, governance, and development', in Smelser, N.J. and Swedberg, R. (eds), *The Handbook of Economic Sociology*, 2nd edn, Princeton, NJ: Princeton University Press and Russell Sage Foundation.

Gereffi, G. and Kaplinsky, R. (eds) (2001), 'The value of value chains', *IDS Bulletin*, 32(3), special issue.

Gereffi, G. and Korzeniewicz, M. (eds) (1994), *Commodity Chains and Global Capitalism*, Westport, CT, Praeger.

Gereffi, G., Humphrey, J. and Sturgeon, T. (2005), 'The governance of global value chains', *Review of International Political Economy*, 12(1), 78–104.

Gertler, M. (2003), 'Tacit knowledge and the economic geography of context, or the undefinable tacitness of being (there)', *Journal of Economic Geography*, 3, 75–99.

Ghemawat, P. (2001), 'Distance still matters', *Harvard Business Review*, 79(8), 137–147, September.

Ghemawat, P., and Altman, S.A. (2011), *DHL Global Connectedness Index 2011*, IESE Business School, Barcelona, Spain.

Ghemawat, P., and Altman, S.A. (2012), *DHL Global Connectedness Index 2012*, IESE Business School, Barcelona, Spain.

Ghironi, F. and Melitz, M. (2005), 'International trade and macroeconomic dynamics with heterogeneous firms', *Quarterly Journal of Economics*, 120, 865–915.

Ghoshal, S. and Bartlett, C.A. (1990), 'The multinational corporation as an interorganizational network', *Academy of Management Review*, 15(4), 603–625.

Giddens, A. (1990), *The Consequences of Modernity*, Stanford, CA: Stanford University Press.

Gill, I. (2010), 'Regional development policies: place-based or people-centred?', Paris: OECD Regional Development Policy Division. Accessed at http://www.voxeu.org/index.php?q=node/5644.

Giuliani, E. (2005), 'The structure of cluster knowledge networks: uneven and selective, not pervasive and collective', DRUID Working Paper, 2005–11.

Giuliani, E. (2007), 'The selective nature of knowledge networks in clusters: evidence from the wine industry', *Journal of Economic Geography*, 7(2) 139–168.

Giuliani, E. (2008),'Multinational corporations and patterns of local knowledge transfer in Costa Rican high-tech industries', *Development and Change*, 39(3), 385–407.

Glaeser, E.L. (2005), 'Urban colossus: why is New York America's largest city?', *Economic Policy Review*, Urban Dynamics in New York, Federal Reserve Bank of New York, New York.

Glaeser, E.L. and Kohhase, J. (2004), 'Cities, regions and the decline of transport costs', *Papers in Regional Science*, 83, 197–228.

Glaeser, E.L. and Shapiro, J.M. (2003), 'Urban growth in the 1990s: is city living back?', *Journal of Regional Science*, 43(1), 139–165.

Glaeser, E.L., Scheinkman, J.A. and Shleifer, A. (1995), 'Economic growth in a cross section of cities', *Journal of Monetary Economics*, 36, 117–143.

Glaeser, E.L., Kallal, H.D., Scheinkman, J.A. and Shleifer, A. (1992), 'Growth in cities', *Journal of Political Economy*, University of Chicago Press, 100(6), 1126–1152, December.

Godfrey, B. and Zhou, Y. (1999), 'Ranking world cities: multinational corporations and the global city hierarchy', *Urban Geography*, 20(3), 268–281.

Golley, J. (2007), 'China's Western development strategy and nature versus nurture', *Journal of Chinese Economic and Business Studies*, Taylor and Francis Journals, 5(2), 115–129.
Gordon, I. (2002), 'Global cities, internationalisation and urban systems', in McCann, P. (ed.), *Industrial Location Economics*, Cheltenham, UK and Northampton, MA, USA: Edward Elgar Publishing, pp. 187–206.
Gordon, I.R. and McCann, P. (2000), 'Industrial clusters, complexes, agglomeration and/or social networks?', *Urban Studies*, 37, 513–532.
Gordon, I.R. and McCann, P. (2005), 'Clusters, innovation and regional development: an analysis of current theories and evidence', in Johansson, B., Karlsson, C. and Stough, R. (eds), *Entrepreneurship, Spatial Industrial Clusters and Inter-Firm Networks*, Cheltenham, UK and Northampton, MA, USA: Edward Elgar Publishing.
Graham, E.M. (1978), 'Transatlantic investment by multinational firms: a rivalistic phenomenon?', *Journal of Post Keynesian Economics*, 1, 82–99.
Graham, E.M. (1982), 'The terms of transfer of technology to the developing nations: a survey of the major issues', OECD, North/South Technology Transfer, OECD, Paris.
Grandadam, D., Cohendet, P. and Simon, L. (2012), 'Places, spaces and the dynamics of creativity: the video game industry in Montreal', *Regional Studies*, doi: 10.1080/00343404.2012.699191.
Granovetter, M.S. (1973), 'The strength of weak ties', *American Journal of Sociology*, 78(6), 1360–1380.
Granstrand, O. and Sjölander, S. (1992), 'Internationalisation and diversification of multitechnology corporations', in Grandstrand, O., Håkanson, L. and Sjölander, S. (eds), *Technology Management and International Business: Internationalisation of R&D and Technology*, London: Wiley.
Greunz, L. (2003), 'The technology gap and European regional growth dynamics', in Fingleton, B. (ed.), *European Regional Growth*, Heidelberg: Springer.
Griffith, D.A. and Harvey, M.G. (2001), 'A resource perspective of global dynamic capabilities, *Journal of International Business Studies*, 32(3), 597–606.
Griliches, Z. (1986), 'Economic data issues', in Griliches, Z. and Intriligator, M.D. (eds), *Handbook of Econometrics*, vol. 3, Amsterdam: Elsevier Science, pp. 1465–1514.
Grindley, P.C. and Teece, D.J. (1997), 'Managing intellectual capital: licensing and cross-licensing in semiconductors and electronics', *California Management Review*, 39(2), 8–41.
Grossman, G. and Helpman, E. (1991), *Innovation and Growth in the Global Economy*, Cambridge, MA: The MIT Press.

Grossman, G., Helpman, E. and Szeidl, A. (2003), 'Optimal integration strategies for the multinational firm', NBER Working Paper 10189.

Grossman, G., Helpman, E. and Szeidl, A. (2006), 'Optimal strategies for the multinational firm', *Journal of International Economics*, 70, 216–238.

Grubesic, T.H., Matisziw, T.C. and Zook, M.A. (2008), 'Global airline networks and nodal regions', *GeoJournal*, 71(1), 53–66.

Grubesic. T.H., Matisziw, T.C. and Zook, M.A. (2009), 'Spatio-temporal fluctuations in the global airport hierarchies', *Journal of Transport Geography*, 17(4), 264–275.

Guerrieri, P., Iammarino, S. and Pietrobelli, C. (eds) (2001), *The Global Challenge to Industrial Districts: SMEs in Italy and Taiwan*, Cheltenham, UK and Northampton, MA, USA: Edward Elgar Publishing.

Guiso L., Sapienza, P. and Zingales, L. (2010), 'Civic capital as the missing link', NBER Working Paper No 15845, Cambridge, MA.

Guy, F. (2007), 'Strategic bundling: information products, market power, and the future of globalization', *Review of International Political Economy*, 14(1), 26–48, February.

Guy, F. (2009), *The Global Environment of Business*, Oxford: Oxford University Press.

Guy, F. (2010), 'Tortoise and hare: regional institutions and the retreat of global liberalization', mimeo, Department of Management, Birkbeck, University of London.

Hall, P. (1966), *The World Cities*, London: Weidenfeld & Nicolson.

Hall, P. (1998), *Cities and Civilization*, New York: Pantheon Books.

Hanson, P. and Bradshaw, M.J. (eds) (2000), *Regional Economic Change in Russia*, Cheltenham, UK and Northampton, MA, USA: Edward Elgar Publishing.

Harrigan, J. and Venables, A.J. (2006), 'Timeliness and agglomeration', *Journal of Urban Economics*, Elsevier, 59(2), 300–316, March.

Hart, G. (2010), 'Redrawing the map of the world? Reflections on the World Development Report 2009', *Economic Geography*, 86, 341–350.

Harvey, D. (2006), 'Editorial: the geographies of critical geography', *Transactions* IBG 31, 409–412.

Hatton, T.J. and Williamson, J.G. (2005), *Global Migration and the World Economy*, Cambridge, MA: The MIT Press.

Haynes, K., Lall, S., Stough, R. and Yilmaz, S. (2006), 'Network usage patterns and the substitution and complementarity effects between telecommunications and transportation: a demand-side approach', in Kobayashi, K., Lakshmanan, T.R. and Anderson, W.P.(eds), *Structural Change in Transportation and Communications in the Knowledge*

Society, Cheltenham, UK and Northampton, MA, USA: Edward Elgar Publishing.

Head, K. and Ries, J. (1996), 'Inter-city competition for foreign investment: static and dynamic effects of China's incentive areas', *Journal of Urban Economics*, 40(1), 38–60.

Healey, M. and Watts, H.D. (1987), 'The multiplant enterprise', in Lever, W.F. (ed.), *Industrial Change in the United Kingdom*, Harlow: Longman, pp. 149–166.

Hedlund, G. (1986), 'The hypermodern MNC – a heterarchy?', *Human Resource Management*, 25, 9–35, doi: 10.1002/hrm.3930250103.

Heidenreich, M. (2005), 'The renewal of regional capabilities: experimental regionalism in Germany', *Research Policy*, 34(5), 739–757.

Helleiner, E. (1994), 'The world of money: the political economy of international capital mobility', *Policy Sciences*, 27(4), 295–298.

Heitor, M.V. (2004), 'Rethinking technology policy and innovation', *Technological Forecasting and Social Change*, 71(6), 549–551.

Helpman, E. (1984), 'A simple theory of international trade and multinational corporations', *Journal of Political Economy*, 93(3), 451–471.

Helpman, E. (1985), 'Multinational corporations and trade structure', *Review of Economic Studies*, 52(3), 443–457.

Helpman, E. and Krugman, P. (1985), *Market Structure and Foreign Trade: Increasing Returns, Imperfect Competition and the World Economy*, Cambridge MA: The MIT Press.

Helpman, E., Melitz, M. and Yeaple, S. (2004), 'Exports versus FDI with heterogeneous firms', *The American Economic Review*, 94, 300–316.

Henderson, J., Dicken, P., Hess, M., Coe, N.M. and Yeung, H.W-C. (2002), 'Global production networks and the analysis of economic development', *Review of International Political Economy*, 9, 436–464.

Henderson, J.V., Kuncoro, A. and Turner, M. (1995), 'Industrial development in cities', *Journal of Political Economy*, 103, 1067–1085.

Henderson, R. and Clark, K.B. (1990), 'Architectural innovation: the reconfiguration of existing product technologies and the failure of established firms', *Administrative Science Quarterly*, 35, 9–30.

Hennart, J-F. (1977), 'A theory of foreign direct investment', University of Maryland, PhD dissertation.

Hennart, J-F. (1982), *A Theory of Multinational Enterprise*, Ann Arbor, MI: University of Michigan Press.

Hennart, J-F. (2000), 'The transaction cost theory of the multinational enterprise', in Pitelis, C. and Sugden, R. (eds), *The Nature of the Transnational Firm*, 2nd edn, London: Routledge, 81–116.

Hess, M. and Yeung, H. (2006), 'Whither global production networks in economic geography? Past, present, and future', *Environment and Planning A*, 38(7), 1193–1204.

Higgins, M.J., Levy, D. and Young, A.T. (2006), 'Growth and convergence across the United States: evidence from county-level data', *Review of Economics and Statistics*, 88(4), 671–681.

Hill, H. (1991), 'The Emperor's clothes can now be made in Indonesia', *Bulletin of Indonesian Economic Studies*, 27(3).

Hirsch, S. (1967), *Location of Industry and International Competitiveness*, Oxford: Clarendon Press.

Hirsch, S. (1976), 'An international trade and investment theory of the firm', *Oxford Economic Papers*, 28, 258–270. Hirsch, W.Z. (1965), 'Cost functions of an urban government service: refuse collection', *Review of Economics & Statistics*, 47(1), 87.

Hirschmann, A.O. (1958), *The Strategy of Economic Development*, New Haven, CT: Yale University Press.

Hirst, P. and Thompson, G. (1996), *Globalization in Question: the International Economy and the* Possibilities *of Governance*, Cambridge, UK: Polity.

Hirst, P. and Zeitlin, J. (1992), 'Flexible specialization versus post-fordism', in Storper, M. and Scott, A. (eds), *Pathways to Industrialization and Regional Development*, London: Routledge, 70–115.

HMT-DTI (2001), *Productivity in the UK: The Regional Dimension*, London: HM Treasury and Department of Trade and Industry.

Hobday, M. (1995), *Innovation in East Asia: The Challenge to Japan*, Aldershot, UK and Brookfield, VT, USA: Edward Elgar Publishing.

Hobday, M. (1996), 'Innovation in South-East Asia: lessons for Europe?', *Management Decision*, 34 (9), 71–81.

Hobday, M., Davies, A. and Prencipe, A. (2005), 'Systems integration: a core capability of the modern corporation', *Industrial and Corporate Change*, 14(6), 1109–1143.

Hoover, E.M. (1937), *Location Theory and the Shoe and Leather Industries*, Cambridge, MA: Harvard University Press.

Hoover, E.M. (1948), *The Location of Economic Activity*, New York: McGraw Hill.

Hoover, E. and Vernon, R. (1959), *Anatomy of a Metropolis*, Cambridge, MA: Harvard University Press.

Hoover, E.M. and Vernon, R. (1962), *Anatomy of a Metropolis*, New York: Doubleday Anchor Books.

Horsley, W., Buckley, R. (1990), *Nippon: New Superpower: Japan Since 1945*, London: BBC Books.

Horst, T. (1972), 'The industrial composition of U.S. exports and subsidiary sales to the Canadian market', *American Economic Review*, 62, 37–45.

Horstmann, I.J. and Markusen, J.R. (1987a), 'Strategic investments and the development of multinationals', *International Economic Review*, 28(1), 109–121, February.

Hotelling, H. (1921), 'A mathematical theory of migration', MA thesis presented at the University of Washington; re-published in 1978 in *Environment and Planning*, 10, 1223–1239.

Hounshell, D.A. (1984), *From the American System to Mass Production 1800–1932*, Baltimore, MD: John Hopkins University Press.

Howells, J. (1996), 'Tacit knowledge, innovation and technology transfer', *Technology Analysis & Strategic Management*, 8, 91–106.

Howells, J. (1999), 'Regional systems of innovation?', in Archibugi, D., Howells, J. and Michie, J. (eds), *Innovation Policy in a Global Economy*, Cambridge: Cambridge University Press, 67–93.

Howells, J. and Wood, M. (1993), *The Globalisation of Production and Technology*, London: Wiley.

Howells, J.R.L. (2002), 'Tacit knowledge, innovation and economic geography', *Urban Studies*, 39, 871–884.

Hufbauer, G.C. (1966), *Synthetic Materials and the Theory of International Trade*, Cambridge, MA: Harvard University Press.

Hufbauer, G. and Brunel, C. (2008), 'Economic Integration in North America', in *One Issue: Two Voices*, The Canada Institute, Woodrow Wilson International Center for Scholars, February, 2–7.

Hufbauer, G.C. and Findlay, C. (1996), 'Flying high: liberalizing civil aviation in the Asia Pacific', Peterson Institute Press, Peterson Institute for International Economics, 27, March.

Hufbauer, G.C. and Schott, J.J. (2005), *NAFTA Revisited: Achievements and Challenges*, Washington, DC: Institute for International Economics.

Hummels, D. (1999), 'Toward a geography of trade costs', mimeo, University of Chicago, Chicago.

Hummels, D. (2001), 'Time as a trade barrier', mimeo, Purdue University, Indiana.

Hurter, A.P. and Martinich, J. (1989), *Facility Location and the Theory of Production*, Boston, MA: Kluwer Academic Press.

Hymer, S. (1968), *The Political Economy of the Gold Coast and Ghana*, New Haven, CT: Economic Growth Center, Yale University.

Hymer, S. (1970a), 'The efficiency contradictions of multinational corporations', *American Economic Review*, 60, 441–448.

Hymer, S. (1970b), *A Model of an Agrarian Economy with Nonagricultural Activities*, New Haven, CT: Yale University, Economic Growth Center.

Hymer, S. (1972), 'The multinational corporation and the law of uneven development', in Bhagwati, J.N. (ed.), *Economics and World Order: From the 1970s to the 1990s*, New York: The Free Press, 113–140.

Hymer, S. (1960/1976), *The International Operations of National Firms: A study of direct Foreign Investment*, Cambridge, MA: The MIT Press.

Iammarino, S. (2005), 'An evolutionary integrated view of regional systems of innovation. Concepts, measures and historical perspectives', *European Planning Studies*, 13(4), 495–517.

Iammarino, S. (2011), 'Regional innovation and diversity', in Cooke, P., Asheim, B.T. and Boschma, R. (eds), *Handbook of Regional Innovation And Growth*, Cheltenham, UK and Northampton, MA, USA: Edward Elgar Publishing.

Iammarino, S. and McCann, P. (2006), 'The structure and evolution of industrial clusters: transactions, technology and knowledge spillovers', *Research Policy*, 35(7), 1018–1036.

Iammarino, S. and McCann, P. (2010), 'The relationship between multinational firms and innovative clusters', in Boschma, R. and Martin, R.L. (eds), *The Handbook of Evolutionary Economic Geography*, Cheltenham, UK and Northampton, MA, USA: Edward Elgar Publishing.

Iammarino, S., Jona Lasinio, C. and Mantegazza, S. (2001), 'The regional dimension of the ICT industry in Italy: a preliminary description', *Economia e Politica Industriale*, 110, October.

Iammarino, S., Padilla, R. and von Tunzelmann, N. (2008), 'Technological capabilities and global-local interactions: the electronics industry in two Mexican regions', *World Development*, 36, 1980–2003.

Iammarino, S., Sanna-Randaccio, F. and Savona, M. (2009), 'The perception of obstacles to innovation: foreign multinationals and domestic firms in Italy', *Revue d'Économie Industrielle*, 125.

Iammarino, S., Piva, M., Vivarelli, M. and von Tunzelmann, N. (2012), 'Patterns of cooperation and technological capabilities of UK firms: a regional investigation', *Regional Studies*, May, doi:10.1080/00343404.2012.679259.

Iansiti, M. (1995a), 'Technology integration: managing technological evolution in a complex environment', *Research Policy*, 24(2), 521–542, March.

Iansiti, M. (1995b), 'Technology development and integration: an empirical study of the interaction between applied science and product development', *IEEE Transactions on Engineering Management*, 42, 259–269.

Iansiti, M. and Clark, K.B. (1994), 'Integration and dynamic capability: evidence from development in automobiles and mainframe computers', *Industrial and Corporate Change*, 3(3), 557–605.

Ietto-Gillies, G. (2002). *Transnational Corporations: Fragmentation Amidst Integration*, London: Routledge.

Ietto-Gillies, G. (2005), *Transnational Corporations and International Production. Concepts, Theories and Effects*, Cheltenham, UK and Northampton, MA, USA: Edward Elgar Publishing.

Ietto-Gillies, G. (2012, 2nd edition), *Transnational Corporations and International Production. Concepts, Theories and Effects*, Cheltenham, UK and Northampton, MA, USA: Edward Elgar Publishing.

Ignatius H. and Markusen, J.R. (1987), 'Licensing versus direct investment: a model of internalization by the multinational enterprise', *Canadian Journal of Economics*, Canadian Economics Association, 20(3), 464–481, August.

Ioannides, Y.M., Overman, H.G., Rossi-Hansberg, E. and Schmidheiny, K. (2007), 'The effect of information and communication technologies on urban structure', Paper presented at 54th North American Meeting of the Regional Science Association International, Savannah, Georgia USA, 7–10 November.

Isard, W. and Kuenne, R.E. (1953), 'The impact of steel upon the Greater New York–Philadelphia region', *Review of Economics and Statistics*, 35, 289–301.

Jacobs, J. (1961), *Death and Life of Great American Cities*, New York: Random House.

Jacobs, J. (1969), *Economy of Cities*, New York: Vintage.

Jacobs, J. (1984), *Cities and the Wealth of Nations: Principles of Economic Life*, New York: Vintage.

Jaffe, A.B., Trajetenberg, M. and Henderson, R. (1993), 'Geographic localization of knowledge spillovers as evidenced by patent citations', *Quarterly Journal of Economics*, 63(3), 577–598.

James, W.E. (2006), 'Structural change and competitiveness in Indonesia's garment industry in the post-quota era', Jakarta: World Bank, July 10.

Janne, O.E.M. (2002a), 'The emergence of corporate integrated innovation systems across regions. The case of the chemical and pharmaceutical industry in Germany, the UK and Belgium', *Journal of International Management*, 8, 1–23.

Janne, O.E.M. (2002b), 'Agglomeration and comparative advantage in innovative activities: the impact of multinationals in Europe', Conference Paper presented at the 28th European International Business Academy (EIBA) annual conference, Athens, 8–10 December.

Javorcik, B.S. (2004), 'Does foreign direct investment increase the productivity of domestic firms? In search of spillovers through backward linkages', *The American Economic Review*, 94(3), 605–627, June.

Jensen, M.C. and Meckling, W.H. (1976), 'Theory of the firm: managerial behavior, agency costs and ownership structure', *Journal of Financial Economics*, 3(4), 305–360.

Johansson, B. and Loof, H. (2009), 'How are non-MNEs affected by globalization?', Centre of Excellence for Science and Innovation.

Johansson, B. and Quigley, J.M. (2004), 'Agglomeration and networks in spatial economics', *Papers in Regional Science*, 83, 165–176, doi: 10.1007/s10110-003-0181-z.

Johansson, J. and Vahlne, E. (1977), 'The internationalization process of the firm: a model of knowledge development and increasing foreign market commitments', *Journal of International Business Studies*, 8, 23–32.

Johnson, H.G. (1970), 'The efficiency and welfare implications of the multinational corporation', in Kindleberger, C.P. (ed.), *The International Corporation: A Symposium*, Cambridge, MA: The MIT Press, Chapter 2.

Johnston, S. and Menguc, B. (2007), 'Subsidiary size and the level of subsidiary autonomy in multinational corporations: a quadratic model investigation of Australian subsidiaries', *Journal of International Business Studies*, 38(5) , 787–801.

Jones, A. (2007a), 'More than managing across borders? The complex role of face-to-face interaction in globalising law firms', *Journal of Economic Geography*, 7, 223–246.

Jones, A. (2007b), 'The rise of global work', *Transactions of the Institute of British Geographers*, 33, 12–26.

Jones, G. (1996), *The Evolution of International Business: An Introduction*, London: Routledge.

Jones, G. (2005), *Multinationals and Global Capitalism: From the Nineteenth to the Twenty-First Century*, Oxford: Oxford University Press.

Jones, G. and Khanna, T. (2006), 'Bringing history (back) into international business', *Journal of International Business Studies*, 37(4), 453–488.

Jones, G. and Schroeter, H.G. (eds) (1993), *The Rise of Multinationals in Continental Europe*, Aldershot: Edward Elgar Publishing.

Kali, R. and Reyes, J. (2006), 'The architecture of globalization: a network approach to international economic integration', Working Paper: Department of Economics, University of Arkansas.

Kambhampati, U. and McCann, P. (2007), 'Regional performance and characteristics of Indian manufacturing industry', *Regional Studies*, Taylor and Francis Journals, 41(3), 281–294.

Kamien, M.I. and Schwartz, N.L. (1982), *Market Structure and Innovation*, Cambridge, UK: Cambridge University Press.

Kamnungwut, W. and Guy, F. (2012), 'Knowledge in the air and cooperation among firms: traditions of secrecy and the reluctant emergence of specialization in the ceramic manufacturing district of Lampang, Thailand', *Environment and Planning A*, 44, doi: 10.1068/a44522.

Kaplan, R.S. and Johnson, H.T. (1987), *Relevance Lost: The Rise and Fall of Management Accounting*, Boston, MA: Harvard Business School.

Keegan, J. (1993), *A History of Warfare*, New York: Knopf Doubleday Publishing Group.

Khalili, A., Vijay, K. and Bodenborn, D. (1974), 'Location and the theory of production: a generalization', *Journal of Economic Theory*, 9, 467–475.

Kirat, T. and Lung, Y. (1999), 'Innovation and proximities: territories as a loci of collective learning', *European Urban and Regional Studies*, 6(1), 27–38.

Klein, N. (2000), *No Logo*, New York: Picador.

Klein, N. (2002), *No Logo: No Space, No Choice, No Jobs*, New York: Picador.

Klepper, S. (1996), 'Entry, exit, growth, and innovation over the product life cycle', *American Economic Review*, 86(3), 562–583.

Klier, T.H. and Rubenstein, J.M. (2008), 'Who really made your car?', Chicago Fed Letter, Federal Reserve Bank of Chicago, issue October.

Kline, S.J. and Rosenberg, N. (1986), 'An overview of innovation', in Landau, R. and Rosenberg, N. (eds), *The Positive Sum Strategy: Harnessing Technology for Economic Growth*, Washington, DC: National Academy Press, pp. 275–306.

Knickerbocker, F.T. (1973), *Oligopolistic Reaction and Multinational Enterprise*, Cambridge, MA: Harvard University Press.

Knox, P.L. and Taylor, P.J. (1995), *World Cities in a World-System*, Cambridge, UK: Cambridge University Press.

Kodama, F. (1995), *Emerging Patterns of Innovation: Sources of Japan's Technological Edge*, Boston, MA: Harvard Business School Press.

Kogut, B. and Chang, S.J. (1991), 'Technological capabilities and Japanese foreign direct investment in the United States', *Review of Economics and Statistics*, 73(3), 401–113.

Kogut, B. and Zander, U. (1993), 'Knowledge of the firm and the evolutionary-theory of the multinational corporation', *Journal of International Business Studies*, 24(4), 625–645.

Kojima, K. (1973), *Economic Cooperation in the Western Pacific*, Tokyo: Japan Economic Research Center.

Kojima, K. (1982a), *Development Oriented Direct Foreign Investment and the Role of Asian Development Bank*, Report No. 4, Manila: Asian Development Bank.

Kojima, K. (1982b), 'Macroeconomic versus international business approach to direct foreign investment', *Hitotsubashi Journal of Economics*, 231,1–19.

Kojima, K. (1990), 'Monograph series 1', Social Science Research Institute, International Christian University, Tokyo.

Kokko, A. (1992a), *Foreign Direct Investment, Host Country Characteristics, and Spillovers*, Stockholm: Stockholm School of Economics, Economic Research Institute.

Kokko, A. (1992b), *Technology, Market Characteristics, and Spillovers*, Stockholm: EFI.

Kokko, A. (1994), 'Sweden: effects of EU membership on investment and growth', *The World Economy*, 17(5), 667–677.

Kokko, A. (1996), *Productivity Spillovers from Competition between Local Firms and Foreign Affiliates*, Stockholm: Department of Economics, Stockholm School of Economics.

Kotabe, M. (2001), 'Contemporary research trends in international marketing: the 1990s', in Rugman, A.M. and Brewer, T.L. (eds), *The Oxford Handbook of International Business*, Oxford: Oxford University Press.

Kottaridi, C. (2005a), 'The "core-periphery" pattern of FDI-led growth and production structure in the EU', *Applied Economics*, Taylor and Francis Journals, 37(1), 99–113.

Kottaridi, C. (2005b), 'FDI, growth and the role of governance: changing the rules of the game', *Contributions to Political Economy*, Oxford University Press, 24(1), 79–98, August.

Koutsoyiannis, A. (1982), *Non-Price Decisions: The Firm in a Modern Context*, London: Macmillan.

Kramer, J.-P., Marinelli, E., Iammarino, S. and Revilla Diez, J. (2011), 'Intangible assets as drivers of innovation: empirical evidence on multinational enterprises in German and UK regional systems of innovation', *Technovation*, 31(9), 447–458.

Krugman, P. (1991a), *Geography and Trade*, Cambridge, MA: The MIT Press.

Krugman, P. (1991b), 'Increasing returns and economic geography', *Journal of Political Economy*, 99, 483–499.

Krugman, P. (1991c), 'Cities in space: three simple models', NBER Working Papers 3607, National Bureau of Economic Research.

Krugman, P.R. (1991d), 'First nature, second nature, and metropolitan location', NBER Working Papers 3740, National Bureau of Economic Research.

Krugman, P. (1994), *Rethinking International Trade*, Cambridge, MA: The MIT Press.

Krugman, P. (1995), *Development, Geography and Economic Theory*, Cambridge, MA : The MIT Press.

Krugman, P. (2007), 'The "new" economic geography: where are we?', in Fujita, M., *Regional Integration in East Asia from the Viewpoint of Spatial Economics*, Basingstoke: Palgrave.

Krugman, P. (2008), 'The great illusion', *The New York Times*, 14 August.

Krugman, P. and Venables, A.J. (1995), 'Globalization and the inequality of nations', *Quarterly Journal of Economics*, 110(4), 857–880.

Kuemmerle, W. (1999), 'The drivers of foreign direct investment into research and development: an empirical investigation', *Journal of International Business Studies*, 30(1), 1–24.

Kumar, N. (1996), 'Multinational enterprises, new technologies and export-oriented industrialization in developing countries: trends and prospects', The United Nations University, INTECH, Discussion Paper Series no 9602, Maastricht.

Kumar, N. (1999), 'Competition policy', *The Economic Times*, 1 September.

Kumar, N. (2001), 'Determinants of location of overseas R&D activity of multinational enterprises: the case of US and Japanese corporations', *Research Policy*, Elsevier, 30(1), 159–174, January.

Kutznets, S. (1953), *Economic Change*, New York: W.W. Norton.

Lagendijk, A. and Cornford, J. (2000), 'Regional institutions and knowledge – tracking new forms of regional development policy', *Geoforum*, 31(2), 209–218.

Lai, H. and Zhu, S.C. (2006), 'US exports and multinational production', *Review of Economics and Statistics*, 88(3), 531–548.

Lall, S. (1993), 'Promoting technology development: the role of technology transfer and indigenous effort', *Third World Quarterly*, 141, 95–108.

Lall, S. (1993a), 'The interrelationship between investment flows and technology transfer: an overview of the main issues', Geneva: UNCTAD Secretariat.

Lall, S. (1993b), 'Understanding technology development', *Development and Change*, 24, 719–753.

Lall, S. (2004), 'Reinventing industrial strategy: the role of government policy in building industrial competitiveness', G-24 Discussion Papers 28, United Nations Conference on Trade and Development.

Lall, S., Albaladejo, M. and Zhang, J. (2004), 'Mapping fragmentation: electronics and automobiles in East Asia and Latin America', *Oxford Development Studies*, 32(3), 407–432.

Lall, S. and Deichmann, U. (2004), 'Agglomeration economies and productivity in Indian industry', *Journal of Development Economics*, 73(3), 643–673.

Lall, S., Funderburg, R. and Yepes, T. (2004), 'Location, concentration, and performance of economic activity in Brazil', Policy Research Working Paper 3268, World Bank.

Lall, S., Koo, J. and Chakravorty, S. (2003), 'Diversity matters: the economic geography of industry location in India', Policy Research Working Paper 3072, World Bank.

Lall. S. and Shalizi, Z. (2003), 'Location and growth in the Brazilian northeast', *Journal of Regional Science*, 43(4), 1–19.

Lambooy, J.G. and Boschma, R.A. (2001), 'Evolutionary economics and regional policy', *The Annals of Regional Science*, 35, 113–131.

Lancaster, K. (1966), 'A new approach to consumer theory', *Journal of Political Economy*, 74(1), 132–157.

Lancaster, K. (1971), *Consumer Demand: A New Approach*, New York: Columbia University Press.

Lancaster, K.J. (1979), *Variety, Equity and Efficiency*, Oxford: Basil Blackwell.

Landes, D.S. (1998), *The Wealth and Poverty of Nations: Why Some Are So Rich and Some So Poor*, New York: W.W. Norton.

Lankhuizen, M. (2009), 'Determinants of foreign direct investment: an empirical investigation into sources of attraction', dissertation, Radboud Universiteit Nijmegen, accessed on 05th November 2012 at: http://dare.ubn.kun.nl/bitstream/2066/82567/1/82567.pdf.

Lardy, N.R. (2005), 'China: the great new economic challenge?', in Bergsten, C.F. (ed.), *The United States and the World Economy: Foreign Economic Policy for the Next Decade*, Institute for International Economics, 121–141.

Laursen, K., Masciarelli, F. and Prencipe, A. (2007), 'Regions matter: how regional characteristics affect external knowledge acquisition and innovation', DRUID Working Paper No. 07-20.

Laursen, K., Masciarelli, F. and Prencipe, A. (2012), 'Regions matter: how localized social capital affects innovation and external knowledge acquisition', *Organization Science*, 23(1), 177–193.

Lazonick, W. (2008), 'Entrepreneurial ventures and the developmental state: lessons from the advanced economies', World Institute of Development Economics Research Discussion Paper dp2008-01.

Le Gales, P. (2002), *European Cities*, Oxford: Oxford University Press.

Leamer, E. (2007), 'A flat world, a level playing field, a small world after all, or none of the above? A review of Thomas L. Friedman's *The World is Flat*', *Journal of Economic Literature*, 83–126.

Leinbach, T.R. and Capineri, C. (2007), *Globalized Freight Transport: Intermodality, E-Commerce, Logistics and Sustainability*, Cheltenham, UK and Northampton, MA, USA: Edward Elgar Publishing.

Leonard-Barton, D. (1995), *Wellsprings of Knowledge: Building and Sustaining the Sources of Innovation*, Boston, MA: Harvard Business School Press.

Leoncini, R., Montresor, S., Vertova, G. (2006), 'Dynamic capabilities between firm organization and local development: a critical survey', *Economia Politica*, 3, 475–502.

Lever, W.F. (ed.) (1987), *Industrial Change in the United Kingdom*, Harlow: Longman.

Levitt, T. (1983), 'The globalization of markets', *Harvard Business Review*, May–June.

Leyshon, A. and Thrift, N. (1997), *Money/Space: Geographies of Monetary Transformation*, London and New York: Routledge.
Limtanakool, N., Schwanen, T. and Dijst, M. (2007), 'Ranking functional urban regions: a comparison of interaction and node attribution data', *Cities*, 24(5), 26–42.
Lin, P. and Saggi, K. (2004), 'Ownership structure and technological upgrading in international joint ventures', *Review of Development Economics*, Wiley Blackwell, 8(2), 279–294.
Lipsey, R.E. (2001a), 'Foreign direct investment and the operations of multinational firms: concepts, history, and data', NBER Working Papers 8665, National Bureau of Economic Research.
Lipsey, R.E. (2001b), 'Foreign direct investors in three financial crises', NBER Working Papers 8084, National Bureau of Economic Research.
Lipsey, R.E. (2002), 'Home and host country effects of FDI', NBER Working Papers 9293, National Bureau of Economic Research.
Long Finance (2011), The Global Financial Centres Index 9, London.
Longhi, C. (2005), 'A French revolution: technology management in the aerospace industry. The case of Toulouse', *International Journal of Technology Management*, 29(3), 194–215.
Lu, J., Liu, X. and Wang, H. (2011), 'Motives for Chinese outward FDI: firm resources, industry dynamics, and government policies', *Management and Organization Review*, 7(2), 223–248.
Lucas, R.E. (1988), 'On mechanisms of economic development', *Journal of Monetary Economics*, 22(1), 3–42.
Lucas, R.E. (2000), 'Some macroeconomics for the twenty-first century', *Journal of Economic Perspectives*, 14(1), 159–178.
Lundvall, B-Å. (1992), *National Systems of Innovation: Towards a Theory of Innovation and Interactive Learning*, London: Pinter.
Lundvall, B-A. and Johnson, B. (1994), 'The learning economy', *Journal of Industry Studies*, 1(2), 23–42, December.
Lundvall, B-Å., Johnson, B., Andersen, E.S. and Dalum, B. (2002), 'National systems of production, innovation and competence building', *Research Policy*, 31, 213–231.
Lundvall, B-Å., Joseph, K.J., Chaminade, C. and Vang, J. (2009), 'Introduction' in *Handbook of Innovation Systems and Developing Countries*, Cheltenham, UK and Northampton, MA, USA: Edward Elgar Publishing, pp. 1–30.
MacGillivray, A. (2006), *A Brief History of Globalization*, London: Robinson.
Mack, R.S. and Schaeffer, P.V. (1993), 'Nonmetropolitan manufacturing in the United States and product cycle theory: a review of the literature', *Journal of Planning Literature*, 8, 124–139.

Maddison, A. (2006), *The World Economy. Volume 1: A Millennial Perspective Volume 2: Historical Statistics*, Paris: OECD Organisation for Economic Cooperation and Development.

Maddison, A. (2007a), *Contours of the World Economy 1000-2030: Essays in Macro-Economic History*, Oxford: Oxford University Press.

Maddison, A. (2007b), *Chinese Economic Performance in the Long Run Second Edition Revised and Updated 960-2030AD*, Paris: OECD Organisation for Economic Cooperation and Development.

Magee, S.P. (1977), 'Information and the multinational corporation: an appropriability theory of direct foreign investment', in Bhagwati, J.N. (ed.), *The New International Economic Order*, Cambridge, MA: The MIT Press.

Malerba, F. (1992), 'Learning by firms and incremental technical change', *Economic Journal*, 102, 845–859.

Malerba, F. (2002), 'Sectoral systems of innovation and production', *Research Policy*, Elsevier, 31(2), 247–264, February.

Malerba, F. and Orsenigo, L. (1990), 'Technological regimes and patterns of innovation: a theoretical and empirical investigation of the Italian case', in Heertje, A. and Perlman, M. (eds), *Evolving Technology and Market Structure*, Ann Arbor, MI: Michigan University Press, pp. 283–306.

Malerba, F. and Orsenigo, L. (1992), 'Schumpeterian patterns of innovation', mimeo, Paper presented to the Schumpeter Society Conference, Kyoto.

Malerba, F. and Orsenigo, L. (1993), 'Technological regimes and firm behavior', *Industrial and Corporate Change*, 2, 45–74.

Malerba, F. and Orsenigo, L. (1995), 'Schumpeterian patterns of innovation', *Cambridge Journal of Economics*, 19, 47–65.

Malerba, F. and Orsenigo, L. (1996), 'The dynamics and evolution of firms', *Industrial and Corporate Change*, 5(1), 51–88.

Malerba, F. and Orsenigo, L. (1997), 'Technological regimes and sectoral patterns of innovative activities', *Industrial and Corporate Change*, Oxford University Press, 6(1), 83–117.

Malerba, F., Nelson, R., Orsenigo, L. and Winter, S. (2006), 'Vertical integration and dis-integration of computer firms: a history friendly model of the co-evolution of the computer and semiconductor industries', Papers on Economics and Evolution 2006-19, Max Planck Institute of Economics, Evolutionary Economics Group.

Malmberg, A. and Maskell, P. (2002), 'The elusive concept of localisation economies: towards a knowledge-based theory of spatial clustering', *Environment and Planning A*, 34, 429–449.

Malmberg, A., Sölvell, O. and Zander, I. (1996), 'Spatial clustering, local

accumulation of knowledge and firm competitiveness', *Geografiska Annaler*, 78B(2), 85–97.
Mameli, F. (2007), 'Agglomeration economies and local growth in Italy', unpublished PhD dissertation, University of Reading.
Mameli, F., Faggian, A. and McCann, P. (2008), 'Employment growth in Italian local labour systems: issues of model specification and sectoral aggregation', *Spatial Economic Analysis*, 3, 343–360.
Manning, A. (2003), 'The real thin theory: monopsony in modern labour markets', *Labour Economics*, Elsevier, 10(2), 105–131, April.
Mansfield, E.D. and Milner, H.V. (1999), 'The New Wave of Regionalism', *International Organization*, 53(3), 589–627.
Marin, A. and Giuliani, E. (2007), 'Global and local knowledge linkages: the case of MNE subsidiaries in Argentina', SPRU Electronic Working Paper Series 155, University of Sussex, SPRU – Science and Technology Policy Research.
Marin, A. and Bell, M. (2010), 'The local/global integration of MNC subsidiaries and their technological behaviour: Argentina in the late 1990s', *Research Policy*, 39(7), 919–931.
Mariotti, S., Piscitello, L. and Elia, S. (2010), 'Spatial agglomeration of multinational enterprises: the role of information externalities and knowledge spillovers', *Journal of Economic Geography*, 10(4), 519–538.
Markusen, A. (1985), *Profit Cycles, Oligopoly and Regional Development*, Cambridge, MA: The MIT Press.
Markusen, A. (1996), 'Sticky places in slippery spaces: a typology of industrial districts', *Economic Geography*, 72, 293–313.
Markusen, J.R. (1984), 'Multinationals, multi-plant economies and the gains from trade', *Journal of International Economics*, 16, 205–226.
Markusen, J.R. (1995), 'The boundaries of multinational enterprises and the theory of international trade', *Journal of Economic Perspectives*, American Economic Association, 9(2), 169–189, Spring.
Markusen, J.R. (2002), *Multinational Firms and the Theory of International Trade*, Cambridge, MA: The MIT Press.
Markusen, J.R. and Venables, A.J. (1996), 'Multinational production, skilled labor and real wages', NBER Working Papers 5483, National Bureau of Economic Research.
Markusen, J.R. and Venables, A.J. (1998), 'Multinational firms and the new trade theory', *Journal of International Economics*, Elsevier, 46(2), 183–203, December.
Markusen, J.R. and Venables, A.J. (2000), 'The theory of endowment, intra-industry and multi-national trade', *Journal of International Economics*, Elsevier, 52(2), 209–234, December.

Martin, R, and Sunley, P. (2003), 'Deconstructing clusters: chaotic concept or policy panacea?', *Journal of Economic Geography*, 3(1), 5–35.

Marshall, A. (1890), *Principles of Economics*, London: Macmillan.

Marshall, A. (1920), *Principles of Economics*, 8th edn. London: Macmillan.

Marsili, O. (2001), *The Anatomy and Evolution of Industries: Technological Change and Industrial Dynamics*, Cheltenham UK and Northampton, MA, USA: Edward Elgar Publishing.

Marsili, O. and Verspagen, B. (2002), 'Technology and the dynamics of industrial structures: an empirical mapping of Dutch manufacturing', *Industrial and Corporate Change*, Oxford University Press, 11(4), 791–815, August.

Maskell, P. (2001), 'Towards a knowledge-based theory of the geographical cluster', *Industrial and Corporate Change*, 10(4), 921–943.

Mastercard (2007), Worldwide Centers of Commerce Index, accessed at www.mastercardworldwide.com.

Mastercard (2008), Worldwide Centers of Commerce Index, accessed at: http://www.mastercardworldwide.com.

McCann, P. (1993), 'The logistics-costs location-production problem', *Journal of Regional Science*, 33, 503–516.

McCann, P. (1995), 'Rethinking the economics of location and agglomeration', *Urban Studies*, 32, 563–577.

McCann, P. (1996), 'Rethinking the theory of industrial location: the logistics-costs argument', presented at the 5th World Congress of the Regional Science Association International, Tokyo.

McCann, P. (1997), 'How deeply embedded is Silicon Glen? A cautionary note', *Regional Studies*, Taylor and Francis Journals, 31(7), 695–703.

McCann, P. (1998), *The Economics of Industrial Location: A Logistics Costs Approach*, Heidelberg: Springer.

McCann, P. (2001), *Urban and Regional Economics*, Oxford: Oxford University Press.

McCann, P. (2005), 'Transport costs and new economic geography', *Journal of Economic Geography*, 5(3), 305–318.

McCann, P. (2007), 'Sketching out a model of innovation, face-to-face interaction and economic geography', *Spatial Economic Analysis*, 2(2), 117–134.

McCann, P. (2008), 'Globalization and economic geography: the world is curved, not flat', *Cambridge Journal of Regions, Economy and Society*, 1(3), 351–370.

McCann, P. (2009a), 'Globalisation, multinationals and the BRIICS countries', in Lattimore, R. and Safadi, R. (eds), *Globalisation and Emerging Economies*, Paris: OECD, pp. 78–124.

McCann, P. (2009b), 'Economic geography, globalisation and New

Zealand's productivity paradox', *New Zealand Economic Papers*, 43(3), 279–314.

McCann, P. (2010), 'The role of industrial clustering and increasing returns to scale in economic development and urban growth', in Brooks, N., Donaghy, K. and Knaap, G. (eds), *The Oxford Handbook of Urban Economics and Planning*, Oxford: Oxford University Press.

McCann, P. and Acs, Z.J. (2011), 'Globalization: countries, cities and multinationals', *Regional Studies*, 45(1), 17–32.

McCann, P. and Arita, T. (2006), 'Clusters and regional development: some cautionary observations from the semiconductor industry', *Information Economics and Policy*, Elsevier, 18(2), 157–180, June.

McCann, P. and Fingleton, B. (1996), 'The regional agglomeration impact of just-in-time input linkages: evidence from the Scottish electronics industry', *Scottish Journal of Political Economy*, 43(5), 493–518.

McCann, P. and Mudambi, R. (2004), 'The location behaviour of the multinational enterprise: some analytical issues', *Growth and Change*, 35(4), 491–524.

McCann, P. and Mudambi, R. (2005), 'Analytical differences in the economics of geography: the case of the multinational firm', *Environment and Planning A*, 37(10), 1857–1876.

McCann, P. and Sheppard, S. (2003), 'The rise, fall and rise again of industrial location theory', *Regional Studies*, 37, 6&7, 649–663.

McCann, P., Arita, T. and Gordon, I.R. (2002), 'Industrial clusters, transactions costs and the institutional determinants of MNE behaviour', *International Business Review*, 11, 647–663.

McCann, P., Poot, J. and Sanderson, L. (2010), 'Migration, relationship capital and international travel: theory and evidence', *Journal of Economic Geography*, 10(3), 361–387.

Melitz, M. (2003), 'The impact of trade on intra-industry reallocation and aggregate industry productivity', *Econometrica*, 71, 1695–1725.

Melitz, M. and Ottaviano, G. (2008), 'Market size, trade, and productivity', *Review of Economic Studies*, 75(1), 295–316.

Menghinello, S., De Propis, L. and Driffield, N. (2010), 'Industrial districts, inward foreign investment and regional development', *Journal of Economic Geography*, 10(4), 539–558.

Metcalfe, S. (1995), 'Technology systems and technology policy in an evolutionary framework', *Cambridge Journal of Economics*, 19(1), 25–46.

MGI (2009), 'Preparing for China's urban billion', McKinsey Global Institute.

MGI (2010a), 'Growth and competiveness in the United States: the role of its multinational companies', McKinsey Global Institute.

MGI (2010b), 'India's urban awakening: building inclusive cities, sustaining economic growth', McKinsey Global Institute.
MGI (2011a), 'Urban world: mapping the economic power of cities', McKinsey Global Institute.
MGI (2011b), 'Building globally competitive cities: the key to Latin American growth', McKinsey Global Institute.
MGI (2012), 'Urban America: US cities in the global economy', McKinsey Global Institute.
Miles, D. (1992), 'Housing markets, consumption and financial liberalization in the major economies', *European Economic Review*, 36, 1093–1135.
Miller, R. (1994), 'Global R&D networks and large-scale innovations: the case of the automobile industry', *Research Policy*, 23, 27–46.
Miller, S.M. and Jensen, D.M. (1978), 'Location and the theory of production: a review, summary and critique of recent contributions', *Regional Science and Urban Economics*, 8, 117–128.
Mills, D.Q. (1972a), 'Construction wage stabilization: a historic perspective', *Industrial Relations*, 11(3), 350–365.
Mills, D.Q. (1972b), *Industrial Relations and Manpower in Construction*, Cambridge, MA: The MIT Press.
Mills, E.S. and Hamilton, B. (1994), *Urban Economics*, 5th edn, New York: Harper-Collins.
Miozzo, M. and Soete, L. (2001), 'Internationalization of services: a technological perspective', Open Access publications from Maastricht University urn:nbn:nl:ui:27-6255, Maastricht University.
Mitchener, K.J. and Weidenmier, M. (2008), 'Trade and empire', *Economic Journal*, 118, 1805–1834.
Michie, J. and Grieve-Smith, J. (eds) (1999), *Globalisation, Growth, and Governance*, Oxford: Oxford University Press.
Molina-Morales, F.X. and Martínez-Fernández, M.T. (2008), 'Does homogeneity exist within industrial districts? A social capital-based approach', *Papers in Regional Science*, 88(1), 209–229.
Monitor Data. Working Paper SCALES, University of Utrecht, Utrecht, and Netherlands Ministry of Economic Affairs, Amsterdam.
Monteiro Monasterio, L. (2008), 'Clusters and the spatial structure of wages in Rio Grande do Sul (Brazil): a multilevel approach', Chapter 7 in Blien, U. and Maier, G. (eds), *The Economics of Regional Clusters: Networks, Technology and Policy*, Cheltenham, UK and Northampton, MA, USA: Edward Elgar Publishing.
Moodysson, J. and Jonsson, O. (2007), 'Knowledge collaboration and proximity', *European Urban and Regional Studies*, 14(2), 115–131.
Morck, R. and Yeung, B. (2007), 'History in perspective: comment on

Jones and Khanna "Bringing history (back) into international business"', *Journal of International Business Studies*, 38, 357–360.
Morgan, K. (1997), 'The learning region: institutions, innovation and regional renewal', *Regional Studies*, 31, 491–504.
Morgan, K. (2004), 'The exaggerated death of geography: learning, proximity and territorial innovation systems', *Journal of Economic Geography*, 4, 3–21.
Morrison, A. (2008), 'Gatekeepers of knowledge within industrial districts: who they are, how do they interact?', *Regional Studies*, 42(6), 817–835.
Morroni, M. (1992), *Production Process and Technical Change*, Cambridge: Cambridge University Press, reprinted 2009.
Moses, L. (1958), 'Location and the theory of production', *Quarterly Journal of Economics*, 72, 259–272.
Moulart, F. and Sekia, F. (2003), 'Territorial innovation models: a critical survey', *Regional Studies*, (37), 289–302.
Mowery, D.C. (1983), 'Economic theory and government technology policy', *Policy Sciences*, 16, 27–43.
Mowery, D.C (1989), 'Collaborative ventures between US and foreign manufacturing firms', *Research Policy*, 18(1), 19–32.
Mudambi, R. (2002), 'Knowledge management in multinational firms', *Journal of International Management*, Elsevier, 8(1), 1–9.
Mudambi, R. (2008), 'Location, control and innovation in knowledge-intensive industries', *Journal of Economic Geography*, 8, 699–725.
Mudambi, R., Mudambi, S. and Navarra, P. (2007), 'Global innovation in MNCs: the effects of subsidiary self-determination and teamwork', *Journal of Product Innovation Management*, 24, 442–455.
Myrdal, G. (1957), *Economic Theory and Underdeveloped Regions*, London: Hutchinson.
Nachum, L. and Zaheer, S. (2005), 'The persistence of distance? The impact of technology on MNE motivations for foreign investment', *Strategic Management Journal*, 26, 747–767.
Nadin, V. and Stead, D. (2008), 'European spatial planning systems: social models and learning', disP 172(1), 25–47.
Narula, R. (1996), *Multinational Investment and Economic Structure*, London: Routledge.
Narula, R. (2004), 'R&D collaboration by SMEs: new opportunities and limitations in the face of globalisation', *Technovation*, 24(2), 153–161.
Narula, R. and Dunning, J.H. (2000), 'Industrial development, globalization and multinational enterprises: new realities for developing countries', *Oxford Development Studies*, 28(2), 141–167.
Narula, R. and Santangelo, G.D., (2010), 'Location, collocation and

R&D alliances in the European ICT industry', *Research Policy*, 38(2), 393–403.

Narula, R. and Zanfei, A. (2005), 'Globalization of innovation: the role of multinational enterprises', in Fagerberg, J., Mowery, D. and Nelson, R.R. (eds), *The Oxford Handbook of Innovation*, Oxford: Oxford University Press.

Naude, W.A. and Krugell, W.F. (2003), 'An inquiry into cities and their role in subnational economic growth in South Africa', *Journal of African Economies*, Oxford University Press, 12(4), 476–499, December.

Navaretti, G.B. and Venables, A.J. (2004), *Multinational Firms in the World Economy*, Princeton, NJ: Princeton University Press.

Neary, P.J. (2008), 'Trade costs and foreign direct investment', in Brakman, S. and Garretsen, J.H. (eds), *Foreign Direct Investment and the Multinational Enterprise*, CESifo Seminar Series. Cambridge, MA: The MIT Press.

Neary, P.J. (2009), 'Trade costs and foreign direct investment', *International Review of Economics and Finance*, 18(2), 207–218.

Neffke, F., Henning, M. and Boschma, R. (2011), 'How do regions diversify over time? Industry relatedness and the development of new growth paths in regions', *Economic Geography*, 87(3), 237–265.

Neffke, F., Henning, M., Boschma, R., Lundquist, K.J. and Olander, L.O. (2009), 'The dynamics of agglomeration externalities along the life cycle of industries', Working Paper, Utrecht University, Utrecht.

Nel, E. and Rogerson, C.M. (2009), 'Re-thinking spatial inequalities in South Africa: lessons from international experience', *Urban Forum*, 20(2), 141–155.

Nelson, R.R. (1991), 'Why do firms differ and how does it matter?', *Strategic Management Journal*, 14, 61–74.

Nelson, R.R. (1992), 'National innovation systems: a retrospective on a study', *Industrial and Corporate Change*, 1(2), 347–374.

Nelson, R.R. (ed.) (1993), *National Innovation Systems: A Comparative Analysis*, Oxford: Oxford University Press.

Nelson, R.R. (1994), 'Why do firms differ and how does it matter?', in Rumelt, R.P., Schendel, D.E. and Teece, D.J. (eds), *Fundamental Issues in Strategy. A Research Agenda*, Cambridge, MA: Harvard Business School Press, Chapter 9.

Nelson, R.R. (2003), 'On the uneven evolution of human know-how', *Research Policy*, 32(6), 909–922.

Nelson, R.R. and Nelson, K. (2002), 'Technology, institutions, and innovation systems', *Research Policy*, 31, 265–272.

Nelson, R.R. and Rosenberg, N. (1993), 'Technical innovation and national systems', in Nelson, R.R. (ed.), *National Innovation Systems: A Comparative Analysis*, Oxford: Oxford University Press.

Nelson, R.R. and Winter, S.G. (1977), 'In search of useful theory of innovation', *Research Policy*, 6(1), 36–76.

Nelson, R.R. and Winter, S.G. (1982), *An Evolutionary Theory of Economic Change*, Cambridge, MA: Harvard University Press.

Ni, P-F. and Kresl, P.K. (2010), *The Global Urban Competitiveness Report*, Cheltenham, UK and Northampton, MA, USA: Edward Elgar Publishing.

Niosi, J. and Bellon, B. (1996), 'The globalization of national innovation systems', in de la Mothe, J. and Paquette, G. (eds), *Evolutionary Economics and the New International Political Economy*, London: Pinter, pp. 138–159.

Nishiguchi, T. (1993), *Strategic Industrial Sourcing*, Oxford: Oxford University Press.

Nocke, V. and Yeaple, S. (2007), 'Cross-border mergers and acquisitions vs. greenfield foreign direct investment: the role of firm heterogeneity', *Journal of International Economics*, 72, 336–365.

Nooteboom, B. (1999), 'Innovation, learning and industrial organization', *Cambridge Journal of Economics*, 23, 127–150.

Nooteboom, B. (2000), *Learning and Innovation in Organizations and Economies*, Oxford: Oxford University Press.

North, D.C. (1981), *Structure and Change in Economic History*, New York: Norton.

Norton, R.D. (1979), *City Life Cycles and American Urban Policy*, New York: Academic Press.

Norton, R.D. (1987), 'The role of services and manufacturing in New England's economic resurgence', *New England Economic Indicators*, October.

Norton, R.D. and Rees, J. (1979), 'The product cycle and the spatial decentralization of American manufacturing', *Regional Studies*, 13(2), 141–151.

O'Brien, R. (1992), *Global Financial Integration: The End of Geography*, New York: Council on Foreign Relations Press.

O'Donnell, S. and Blumentritt, T. (1999), 'The contribution of foreign subsidiaries to host country national competitiveness', *Journal of International Management*, 5(3), 187–206.

O'Rourke, K.H. and Williamson, J.G. (1999), *Globalization and History: The Evolution of a 19th Century Atlantic Economy*, Cambridge, MA: MIT Press.

Obstfeld, M. and Taylor, A.M. (2004), *Global Capital Markets: Integration, Crisis, and Growth*, Cambridge: Cambridge University Press.

Odagiri, H. and Goto, A. (1993), 'The Japanese system of innovation: past, present, and future', in Nelson, R.R. (ed.), *National Innovation Systems: A Comparative Analysis*, New York: Oxford University Press.

OECD (2006), *Competitive Cities in the Global Economy*, Paris: Organisation for Economic Cooperation and Development.

OECD (2007a), *International Investment Perspectives: Freedom of Investment in a Changing World*, Paris: Organisation for Economic Cooperation and Development.

OECD (2007b), 'Tax effects on foreign direct investment: recent evidence and policy analysis', OECD Tax Policy Studies No.17, Paris: Organisation for Economic Cooperation and Development.

OECD (2008), *The Internationalisation of Business R&D: Evidence, Impacts and Implications*, Paris: Organisation for Economic Cooperation and Development.

OECD (2009a), *How Regions Grow*, Paris: Organisation for Economic Growth and Development.

OECD (2009b), *Regions Matter: Economic Recovery, Innovation and Sustainable Growth*, Paris: Organisation for Economic Growth and Development.

OECD (2009c), *Regions at a Glance*, Paris: Organisation for Economic Cooperation and Development.

OECD (2010), *Promotions Transparency and Exchange of Information*, Paris: Organisation for Economic Growth and Development.

OECD (2011), *Regions at a Glance*, Paris: Organisation for Economic Cooperation and Development.

OECD (2012), *Redefining Urban*, Paris: Organisation for Economic Cooperation and Development.

Ohlin, B. (1933), *Interregional and International Trade*, Cambridge, MA: Harvard University Press.

Ohmae, K. (1990), *The Borderless World: Management Lessons in the New Logic of the Global Market Place*, London: Harper Business.

Ohmae, K. (1995), *The End of the Nation-State: the Rise of Regional Economies*, New York: Simon and Schuster.

Ohmae, K. (1999), *The Borderless World: Power and Strategy in the Interlinked Economy*, London: Harper Business.

Oliver, N. and Wilkinson, B. (1988), *The Japanization of British Industry*, Oxford: Basil Blackwell.

Oliver, N. and Wilkinson, B. (1989), 'Japanese manufacturing techniques and personnel and industrial relations practice in Britain: evidence and implications', *British Journal of Industrial Relations*, 27(1), 73–91.

Openshaw, S. and Taylor, P.J. (1979), 'A million or so correlation coefficients: three experiments on the modifiable areal unit problem', in Wrigley, N. (ed.), *Statistical Applications in the Spatial Sciences*, London: Pion.

Osegowitsch, T. and Sammartino, A. (2008), 'Reassessing (home-) regionalisation', *Journal of International Business Studies*, 39(2), 184–196.

Owen-Smith, J. and Powell, W.W. (2004), 'Knowledge networks in the Boston biotechnology community', *Organization Science*, 15, 5–21.

Ozawa, T. (1992), 'Japanese MNCs as potential partners in Eastern Europe's economic reconstruction', Kbh: Handelshøjskolen i København, Institut for International Økonomi og Virksomhedsledelse.

Paci, R. (1997), 'More similar and less equal: economic growth in the European regions', *Review of World Economics* (Weltwirtschaftliches Archiv), Springer, 133(4), 609–634, December.

Paci, R. and Usai, S. (2009), 'Knowledge flows across European regions', *The Annals of Regional Science*, Springer, 43(3), 669–690, September.

Padilla-Perez, R. (2006), 'Foreign direct investment and regional technological capabilities: the case of the electronics industry in Mexico', SPRU: Science Policy Research Unit, Sussex University. Unpublished PhD dissertation.

Palander, T.F. (1935), *Beiträge zur Standortstheorie*, Uppsala: Almqvist & Wicksell.

Panagariya, A. (2005), 'India in the 1980s and the 1990s: a triumph of reforms', in Tseng, W. and Cowen, D. (eds), *India and China's Recent Experience with Reform and Growth*, Basingstoke: International Monetary Fund and Palgrave.

Paniccia, I. (1998), 'One, a hundred, thousands of industrial districts. Organizational variety in local networks of small and medium-sized enterprises', *Organization Studies*, 19(40), 667–699.

Parr, J.B., Hewings, G.D., Sohn, J. and Nazara, S. (2002), 'Agglomeration and trade: some additional perspectives', *Regional Studies*, 36(6), 675–684.

Partridge, M.D., Rickman, D.S., Ali, K. and Olfert, M.R. (2007), 'Employment growth in the American urban hierarchy: long live distance', Paper presented at the 47th Congress of the European Regional Science Association, Paris, August.

Patel, P. and Le Bas, C. (2005), 'Does internationalisation of technology determine technological diversification in large firms? An empirical study', *Revue d'Économie Industrielle*, Programme National Persée, 110(1), 157–174.

Patel, P. and Pavitt, K. (1991), 'Large firms in the production of the world's technology – an important case of non-globalization', *Journal of International Business Studies*, 22(1), 1–21.

Patel, P. and Vega, M. (1999), 'Patterns of internationalisation of corporate technology: location vs. home country advantage', *Research Policy*, 28(2–3), 145–155.

Patrucco, P.P. (2001), 'The production and communication of technological knowledge at the regional level: the European evidence', in Stanford-Smith, B. and Chiozza, E. (eds), *E-work and E-commerce: Novel Solutions and Practices for a Global Networked Economy*, vol. 1, Amsterdam: IOS Press.

Patrucco, P.P. (2003), 'Institutional variety, networking and knowledge exchange: communication and innovation in the case of the Brianza technological district', *Regional Studies*, 37(2), 159–172.

Patrucco, P.P. (2009), 'Collective knowledge production costs and the dynamics of technological systems', *Economics of Innovation and New Technology*, 18, 3, 295–310.

Pavitt, K. (1984), 'Sectoral patterns of technical change: towards a taxonomy and a theory', *Research Policy*, 13, 343–373.

Pavitt, K. (2000), *Technology, Management and Systems of Innovation*, Cheltenham, UK and Northampton, MA, USA: Edward Elgar Publishing.

Pavitt, K. and Patel, P. (1991), 'Technological strategies of the world's largest companies', *Science and Public Policy*, 18(6), 363–368.

Pavitt, K., Robson, M. and Townsend, J. (1989), 'Technological accumulation, diversification and organisation in UK companies, 1945–83', *Management Science*, 35, 81–99.

Pearce R.D. (1983), 'Industrial diversification amongst the world's leading multinational enterprises', in Casson, M.C. (ed.), *The Growth of International Business*, London: Allen and Unwin.

Pearce, R.D. (1999), 'Decentralised R&D and strategic competitiveness: globalised approaches to generation and use of technology in multinational enterprises', *Research Policy*, 28(2), 157–178.

Pearce, R.D. and Singh, S. (1992), *Globalizing Research and Development*, London: Macmillan.

Peck, J. and Sheppard, E. (2010), 'Worlds apart? Engaging with the World Development Report 2009: Reshaping Economic Geography', *Economic Geography*, 86, 331–340.

Penrose, E. (1956), 'Foreign investment and the growth of the firm', *Economic Journal*, 66, 220–235.

Penrose, E. (1959), *The Theory of the Growth of the Firm*, New York: John Wiley and Sons.

Penrose, E. (1968), *The Large International Firm in Developing Countries: The International Petroleum Industry*, London: Allen & Unwin.

Penrose, E. (1990), 'Dumping, unfair competition and multinational corporations', *Japan and the World Economy*, 2, 181–187.

Penrose, E (1995), *The Theory of the Growth of the Firm*, 3rd edn, Oxford: Oxford University Press.

Perez, C. (1983), 'Structural change and assimilation of new technologies in the economic and social systems', *Futures*, 15(5), 357–375.

Perez, C. (1985), 'Microelectronics, long waves and world structural change: new perspectives for developing countries', *World Development*, 13(3), 441–463.

Peri, G. (2005), 'Determinants of knowledge flows and their effect on innovation', *Review of Economics and Statistics*, 87(2), 308–322.

Perobelli, F.S., Haddad, E.A. and Domingues, E.P. (2008), 'Interdependence among the Brazilian states: an input-output approach', in Farshchi, M., Janne, O.E.M. and McCann, P. (eds), *Technological Change and Mature Industrial Regions: Firms, Knowledge, and Policy*, Cheltenham, UK and Northampton, MA, USA: Edward Elgar Publishing.

Perraton, J., Goldblatt, D., Held, D. and Megrew, A. (1997), 'The globalization of economic activity', *New Political Economy*, 2(2), 257–277.

Petit, M.L. and Sanna-Randaccio, F. (2000), 'Endogenous R&D and foreign direct investment in international oligopolies', *International Journal of Industrial Organization*, 18, 339–367.

Phelps, N.A. and Fuller, C. (2000), 'Multinationals, intracorporate competition, and regional development', *Economic Geography*, 76, 224–243.

Phene, A. and Tallman, S. (2002), 'The effects of regional clusters on knowledge stocks and flows: evidence from the biotechnology industry', University of Utah, mimeo.

Pietrobelli, C. (1994), 'National technological capabilities: an international comparison', *Development Policy Review*, 12(2), 115–148.

Pietrobelli, C. (2000), 'Technology transfer for developing countries', in Schroeer, D. and Elena, M. (eds), *Technology Transfer*, London: Ashgate, pp. 209–234.

Pietrobelli, C. and Rabellotti, R. (2010), 'Global value chains meet innovation systems: are there learning opportunities for developing countries?', IDB Publications 11098, Inter-American Development Bank.

Pike, A., Dawley, S. and Tomaney, J. (2010), 'Resilience, adaptation and adaptability', *Cambridge Journal of Regions, Economy and Society*, 3, 59–70.

Pike, A., Rodríguez-Pose, A. and Tomaney, J. (2006), *Local and Regional Development*, London: Routledge.

Pike, A., Rodríguez-Pose, A. and Tomaney, J. (2007), 'What kind of local and regional development and for whom?', *Regional Studies*, 41(9), 1253–1269.

Pine, B.J. II, Victor, B. and Boynton, A.C. (1993), 'Making mass customization work', *Harvard Business Review*, 71(5), 108–119.
Piore, M.J. (1994a), 'Corporate reform in American manufacturing and the challenge to economic theory,' in Allen, T.J. and Scott Morton, M.S. (eds), *Information Technology and the Corporation of the 1990s: Research Studies*, New York: Oxford University Press.
Piore, M.J. (1994b), 'Unions: a reorientation to survive', in Kerr, C. and Staudohar, P.D. (eds), *Labor Economics and Industrial Relations*, Cambridge, MA: Harvard University Press.
Piore, M.J. and Sabel, C.F. (1984), *The Second Industrial Divide: Possibilities for Prosperity*, New York: Basic Books.
Piscitello, L. (2011), 'Strategy, location, and the conceptual metamorphosis of the MNE', *Global Strategy Journal*, 1, 127–131, doi: 10.1002/gsj.14.
Pitelis, C.N. (1991), 'The effects of advertising (and) investment on aggregate profits', *Scottish Journal of Political Economy*, Scottish Economic Society, 38(1), 32–40, February.
Pitelis, C.N (ed.) (2002a), 'The growth of the firm: the legacy of Edith Penrose', Oxford: Oxford University Press.
Pitelis, C.N. (2002b), 'Stephen Hymer: life and the political economy of multinational corporate capital', *Contributions to Political Economy*, Oxford University Press, 21(1), 9–26, December.
Pitelis, C.N. (2005), 'Stephen Herbert Hymer and/on the (theory of) the MNE and international business', *International Business Review*, 15(2), 103–110.
Pitelis, C.N. (2007a), 'Edith Penrose and a learning-based perspective on the MNE and OLI', *Management International Review*, 47(2), 207–219.
Pitelis, C.N. (2007b), 'A behavioral resource-based view of the firm: the synergy of Cyert and March (1963) and Penrose (1959)', *Organization Science*, 18(3), 478–490.
Piva, M. and Vivarelli, M. (2009), 'Demand-pulled innovation under liquidity constraints', *Applied Economics Letters*, Taylor and Francis Journals, 16(3), 289–293.
Polanyi, M. (1967), *The Tacit Dimension*, New York: Anchor Books.
Pomeranz, K. (2000), *The Great Divergence: China, Europe and the Making of the Modern World Economy*, Princeton, NJ: Princeton University Press.
Ponds, R., van Oort, F.G. and Frenken, K. (2007), 'The geographical and institutional proximity of research collaboration', *Papers in Regional Science*, 86, 423–443.
Porter, M.E. (1980), *Competitive Strategy*, New York: Free Press.

Porter, M.E. (1985), *Competitive Advantage: Creating and Sustaining Superior Performance*, New York: Free Press.
Porter, M.E. (1990), *The Competitive Advantage of Nations*, New York: Free Press.
Porter, M.E. (1998a), 'Competition in global industries: a conceptual framework', *Hitotsubashi Business Review*, 35(4).
Porter, M.E. (1998b), *On Competition*, Boston, MA: Harvard Business School Press.
Posner, M.V. (1961), 'International trade and technical change', *Oxford Economic Papers*, 13, 323–341.
Puga, D. (2001), 'European regional policies in light of recent location theories', *Journal of Economic Geography*, 2, 373–406.
Quéré, M. (2007), 'Sophia-Antipolis as a "reverse" science park: from exogenous to endogenous development', in Frenken, K. (ed.), *Applied Evolutionary Economics and Economic Geography*, Cheltenham, UK and Northampton, MA, USA: Edward Elgar Publishing.
Rabbiosi, L. (2008), 'The impact of reverse knowledge transfer on the parent company's innovativeness. Which role for organizational mechanisms and subsidiary characteristics?', in Piscitello L. and Santangelo G.D. (eds), *Multinationals and Local Competitiveness*, Milan: Franco Angeli, Chapter 6.
Radosevic, S. (1999a), 'Patterns of innovative activities in countries of Central and Eastern Europe: an analysis based on comparison of innovation surveys', SPRU Electronic Working Paper Series 35, University of Sussex, SPRU – Science and Technology Policy Research.
Radosevic, S. (1999b), 'Transformation of science and technology systems into systems of innovation in central and eastern Europe: the emerging patterns and determinants', *Structural Change and Economic Dynamics*, Elsevier, 10(3–4), 277–320, December.
Radosevic, S. (1999c), *International Technology Transfer and Catch-up in Economic Development*, Cheltenham, UK and Northampton, MA, USA: Edward Elgar Publishing.
Ramachandran, K. and Ray, S. (2005), 'Formation of information technology clusters: how late movers follow models different from early movers', 35th EISB Conference, Barcelona, 12–14 September.
Rantisi, N.M. (2002), 'The local innovation system as a source of "variety": openness and adaptability in New York City's garment district', *Regional Studies*, 36(6), 587–602.
Rawstron, E.M. (1958), 'Three principles of industrial location', Transactions and Papers (Institute of British Geographers), 25, 135–142.
Redding, S. (2010), 'The empirics of new economic geography', *Journal of Regional Science*, 50, 297–311.

Redlich, F. (1962), 'Approaches to business history', *Business History Review*, XXXVI, 1, 61–86, Spring.
Reed, H. (1981), *The Preeminence of International Financial Centers*, New York: Praeger.
Reich, R. (2001), *The Future of Success: Work and Life in the New Economy*, London: William Heinemann.
ReVelle, C. (1986), 'The maximum capture or sphere of influence location problem—Hotelling revisited on a network', *Journal of Regional Science*, 26, 343–358.
Revilla Diez, J. and Berger, M. (2006), 'Do firms require an efficient innovation system for the development of technological and innovation capabilities? – empirical evidence from Singapore, Malaysia, and Thailand', *International Journal of Technology Management*, 36, 267–285.
Revilla Diez, J. and Berger, M. (2005), 'The role of multinational corporations in metropolitan innovation systems: empirical evidence from Europe and Southeast Asia', *Environment and Planning A*, 35, 1813–1835.
Revilla Diez, J. and Kiese, M. (2006), 'Scaling innovation in Southeast Asia: empirical evidence from Singapore, Penang (Malaysia) and Bangkok', *Regional Studies*, 40(9), 1005–1023.
Reyes, J., Schiavo, S. and Fagiolo, G. (2010), 'Using complex networks analysis to assess the evolution of international economic integration: the cases of East Asia and Latin America', *The Journal of International Trade and Economic Development: An International and Comparative Review*, 19(2), 215–239.
Richardson, H.W. (1978), *Urban Economics*, Hinsdale, IL: The Dryden Press.
Richardson, H.W. and Bae, C-H.C. (eds) (2005), *Globalization and Urban Development*, Heidelberg: Springer.
Rietveld, P. and Vickerman, R. (2004), 'Transport in regional science: the "death of distance" is premature', *Papers in Regional Science*, 83, 229–248.
Roberts, M., Deichmann, U., Fingleton, B. and Shi, T. (2010), 'On the road to prosperity? The economic geography of China's national expressway network', World Bank Policy Research Series WPS5479.
Roberts, P. (1993), 'Managing the strategic planning and development of regions: lessons from a European perspective', *Regional Studies*, 27, 759–768.
Robertson, P.L. and Langlois, R.N. (1994), 'Institutions, inertia, and changing industrial leadership', *Industrial and Corporate Change*, 3(2), 359–378.
Robertson, P.L. and Langlois R.N. (1995), 'Innovation, networks, and vertical integration', *Research Policy*, 24(4), 543–562.

Rodríguez-Pose, A. (1994), 'Socioeconomic restructuring and regional change: re-thinking growth in the European Community', *Economic Geography*, 70(4), 325–343.

Rodríguez-Pose, A. (1998a), *The Dynamics of Regional Growth in Europe: Social and Political Factors*, Oxford: Clarendon Press and New York: Oxford University Press.

Rodríguez-Pose, A. (1998b), 'Social conditions and economic performance: the bond between social structure and regional growth in Western Europe', *International Journal of Urban and Regional Research*, 22(3), 443–459.

Rodríguez-Pose, A. (1999), 'Innovation prone and innovation averse societies: economic performance in Europe', *Growth and Change*, 30(1), 75–105.

Rodríguez-Pose, A. (2002), *The European Union: Economy, Society, and Polity*, Oxford: Oxford University Press.

Rodríguez-Pose, A. (2010), 'Economic geographers and the limelight: the reaction to the World Development Report 2009', *Economic Geography*, 86, 361–370.

Rodríguez-Pose, A. (2011), 'economists as geographers and geographers as something else: on the changing conception of distance in geography and economics', *Journal of Economic Geography*, 11(2), 347–356.

Rodríguez-Pose, A. and Arbix, G. (2001), 'Strategies of waste: bidding wars in the Brazilian automobile sector', *International Journal of Urban and Regional Research*, 25(1), 134–154.

Rodríguez-Pose, A. and Bilbao-Osorio, B. (2004), 'From R&D to innovation and economic growth in the EU', *Growth and Change*, 35 (4), 434–455.

Rodríguez-Pose, A. and Crescenzi, R. (2008), 'Research and development, spillovers, innovation systems, and the genesis of regional growth in Europe', *Regional Studies*, 42, 51–67.

Rodríguez-Pose, A. and Fratesi, U. (2004), 'Between development and social policies: the impact of European structural funds in Objective 1 regions', *Regional Studies*, 38, 97–113.

Rodríguez-Pose, A. and Storper, M. (2006), 'Better rules or stronger communities? On the social foundations of institutional change and its economic effects', *Economic Geography*, 82(1), 1–25.

Rodríguez-Pose, A. and Zademach, H-M. (2006), 'Industry dynamics in the German merger and acquisitions market', *Tijdschrift voor Economische en Sociale Geografie*, 97(3), 296–313.

Rodrik, D. (2005), 'Growth strategies', in Aghion, P. and Durlauf, S.N. (eds), *Handbook of Economic Growth*, North Holland: Elsevier Academic Press, 1A, 967–1014.

Rodrik, D. (2006), 'Goodbye Washington consensus, hello Washington confusion? A review of the World Bank's economic growth in the 1990s: learning from a decade of reform', *The Journal of Economic Literature*, 44(4), 973–987.

Rodrik, D. (2007), *One Economics Many Recipes*, Cambridge, MA: The MIT Press.

Rodrik, D., Subramanian, A. and Trebbi, F. (2004), 'Institutions rule: the primacy of institutions over geography and integration in economic development', *Journal of Economic Growth*, 9(2), 131–165.

Romer, P.M. (1986), 'Increasing returns and long-run growth', *Journal of Political Economy*, 94, 1002–1037.

Romer, P.M. (1990), 'Endogenous technological change', *Journal of Political Economy*, 98(5), S71–S102.

Rosenberg, N. (1969), 'The direction of technological change: inducement mechanisms and focusing devices', *Economic Development and Cultural Change*, 18(1), Part 1, 1–24, October.

Rosenberg, N. (1976), *Perspectives on Technology*, Cambridge, UK: Cambridge University Press.

Rosenthal, S.S. and Strange, W.C. (2004), 'Evidence on the nature and sources of agglomeration economics', in Henderson, V. and Thisse, J-F. (eds), *Handbook of Urban and Regional Economics*, vol. 4, Amsterdam: North-Holland.

Rostow, W.W. (1959), 'The stages of economic growth', *Economic History Review*, 12(1), 1–17.

Roubini, N. and Mihm, S. (2010), *Crisis Economics: A Crash Course in the Future of Finance*, London: Penguin Press.

Rugman, A.M. (1979), *International Diversification and the Multinational Enterprise*, Lexington, KY: DC Heath.

Rugman, A.M. (1981), *Inside the Multinationals: The Economics of Internal Markets*, New York: Columbia University Press.

Rugman, A.M. (2000), *The End of Globalisation*, London: Random House.

Rugman, A.M. (2005), *The Regional Multinationals*, Cambridge: Cambridge University Press.

Rugman, A.M. and Verbeke, A. (2002), 'Edith Penrose's contribution to the resource-based view of strategic management', *Strategic Management Journal*, 23, 769–780.

Rugman, A.M. and Verbeke, A. (2003), 'Multinational enterprises and clusters: an organising framework', *Management International Review*, 43(3), 151–169.

Rugman, A.M. and Verbeke, A. (2004a), 'A final word on Edith Penrose', *Journal of Management Studies*, 41(1), 205–217.

Rugman, A.M. and Verbeke, A. (2004b), 'A perspective on regional and global strategies of multinational enterprises', *Journal of International Business Studies*, 35(1), 3–18.

Rugman, A.M. and Verbeke, A. (2007), 'Liabilities of regional foreignness and the use of firm-level versus country-level data: a response to Dunning et al. (2007)', *Journal of International Business Studies*, 38(1), 200–205.

Rugman, A.M. and Verbeke, A. (2008a), 'The theory and practice of regional strategy: a response to Osegowitsch and Sammartino', *Journal of International Business Studies*, Palgrave Macmillan, 39(2), 326–332, March.

Rugman, A.M. and Verbeke, A. (2008b), 'A regional solution to the strategy and structure of multinationals', *European Management Journal*, Elsevier, 26(5), 305–313, October.

Ruigrok, W. and Tulder van, R.J.M. (1995), *The Logic of International Restructuring*, London: Routledge.

Rumelt, R. (1984), 'Toward a strategic theory of the firm', in Lamb, R. (ed.), *Competitive Strategic Management*, Englewood Cliffs, NJ: Prentice-Hall, 556–570.

Sachs, J.D. (2003), 'Institutions don't rule: direct effects of geography on per capita income', Working Paper 9490, NBER, Cambridge, MA.

Saliola, F. and Zanfei, A. (2009), 'Multinational firms, global value chains and the organization of knowledge transfer', *Research Policy*, Elsevier, 38(2), 369–381, March.

Salop, S.C. (1979), 'Monopolistic competition with outside goods', *Bell Journal of Economics*, The RAND Corporation, 10(1), 141–156, Spring.

Sammarra, A. (2010), 'A "low road" to competitiveness in the global apparel industry: the case of the Vibrata Valley', in Belussi, F. and Sammarra, A. (eds), *Business Networks in Clusters and Industrial Districts*, London and New York: Routledge, 114–135.

Sanna-Randaccio, F. (2002), 'The impact of foreign direct investment on home and host countries with endogenous R&D', *Review of International Economics*, 10(2), 278–298.

Santangelo, G.D. (2000), 'Corporate strategic technological partnerships in the European information and communications technology industry', *Research Policy*, 29(9), 1015–1031.

Santangelo, G.D. (2001), 'The impact of the information and communications technology revolution on the internationalisation of corporate technology', *International Business Review*, Elsevier, 10(6), 701–726, December.

Sassen, S. (1991), *The Global City: New York, London, Tokyo*, Princeton, NJ: Princeton University Press.

Sassen, S. (1994), *Cities in a World Economy*, Thousand Oaks, CA: Pine Forge.

Sassen, S. (1995), 'On concentration and centrality in the Global City', in, Knox, P.L. and Taylor, P.J. (eds), *World Cities in a World System*, Cambridge: Cambridge University Press.

Sassen, S. (2001), *The Global City: New York, London, Tokyo*, 2nd edn, Princeton, NJ: Princeton University Press.

Sassen, S. (ed.) (2002), *Global Networks: Linked Cities*, London: Routledge.

Sassen, S. (2006), *Cities in a World Economy*, Thousand Oaks, CA: Pine Forge.

Saxenian, A. (1994), *Regional Advantage: Culture and Competition in Silicon Valley and Route 128*, Cambridge, MA and London: Harvard University Press.

Schama, S. (2006), *Simon Schama's Power of Art*, London: BBC Books.

Scheve, K.V. and Slaughter, M.J. (2007), 'A new deal for globalization', *Foreign Affairs*, 34–47, July–August.

Schienstock, G. (2009), 'Organisational innovations and new knowledge management practices, their diffusion and influence on firm performance: results from a Finnish firm survey', IAREG Working Paper, no. 1.2.d, accessed at: http://www.iareg.org/fileadmin/iareg/media/papers/WP_IAREG_1.2d.pdf.

Schonberger, R.J. (1996), *World Class Manufacturing: The Next Decade*, New York: Free Press.

Schumpeter, J. (1934), *The Theory of Economic Development*, Cambridge, MA: Harvard University Press.

Schumpeter, J. (1939), *Business Cycles: A Theoretical, Historical and Statistical Analysis of the Capitalist Process*, New York: McGraw-Hill.

Schumpeter, J. (1942), *Capitalism, Socialism and Democracy*, New York: Harper Row.

Schwerin, J. (2004), 'The evolution of the Clyde region's shipbuilding innovation system in the second half of the nineteenth century', *Journal of Economic Geography*, 4(1), 83–101.

Schwerin, J. and Werker, C. (2003), Learning innovation policy based on historical experience, *Structural Change and Economic Dynamics*, 14, 385–404, December.

Scott, A.J. (1988), *New Industrial Spaces*, London: Pion.

Scott, A.J. (1991), 'Flexible production systems: analytical tasks and, theoretical horizons – a reply to Lovering', *International Journal of Urban and Regional Research*, 15, 130–134.

Scott, A.J. (1998), *Regions and the World Economy*, Oxford: Oxford University Press.

Scott, A.J. (ed.) (2001), *Global City-Regions: Trends, Theory, Policy*, Oxford: Oxford University Press.

Scott, A.J., Agnew, J., Soja, E.W. and Storper, M. (2001), 'Global city-regions', in Scott, A.J. (ed.), *Global City-Regions: Trends, Theory, Policy*, Oxford: Oxford University Press, pp. 11–32.

Seely Brown, J. and Duguid, P. (2001), 'Knowledge and organisation: a social-practice perspective', *Organisation Science*, 12(2), 198–213.

Senker, J. (1995), 'Tacit knowledge and models of innovation', *Industrial and Corporate Change*, 2, 425–447.

Shapiro, J.M. (2006), 'Smart cities: quality of life, productivity, and the growth effects of human capital', *Review of Economics and Statistics*, 88(2), 324–335.

Shaver, J.M. and Flyer, F. (2000), 'Agglomeration economies, firm heterogeneity, and foreign direct investment in the United States', *Strategic Management Journal*, 21, 1175–1193.

Shearmur, R. (2011), 'Innovation, regions and proximity: from neo-regionalism to spatial analysis', *Regional Studies*, 45(9), 1225–1243.

Shimizutani, S. and Todo, Y. (2007), 'What determines overseas R&D activities? The case of Japanese multinational firms', Discussion Papers 07010, Research Institute of Economy, Trade and Industry (RIETI).

Simmie, J. (1997), 'The origins and characteristics of innovation in highly innovative areas – the case of Hertfordshire', in Simmie, J. (ed.), *Innovation, Networks and Learning Regions*, London: Jessica Kingsley.

Simmie, J. (1998), 'Reasons for the development of "islands of innovation": evidence from Hertfordshire', *Urban Studies*, 35(8), 1261–1289.

Simmie, J. (2003), 'Innovation and urban regions as national and international nodes for the transfer and sharing of knowledge', *Regional Studies*, 37(6&7), 607–620.

Simmie, J. (2004), 'Innovation and clustering in the globalized international economy', *Urban Studies*, 41(5–6), 1095–1112.

Simmie, J. and Sennet, J. (1999), 'Innovative clusters: theoretical explanations and why size matters', *National Institute Economic Review*, 4/99(170), 87–98.

Simmie, J., Siino, C., Zuliani, J.M., Jalabert, G. and Strambach, S. (2004), 'Local innovation system governance and performance: a comparative analysis of Oxfordshire, Stuttgart and Toulouse', *International Journal of Technology Management*, 28(3–6), 534–559.

Simon, H.A. (1947), *Administrative Behavior*, 2nd edn, New York: The Free Press.

Simon, H.A. (1952a), 'A behavioural model of rational choice', *Quarterly Journal of Economics*, 69(1), 99–118.

Simon, H.A. (1952b), 'A formal theory of interaction in social groups', *American Sociological Review*, 17, 202–211.

Simon, H.A. (1959), 'Theories of decision-making in economics and behavioral science', *American Economic Review*, 49, 253–283.

Slangen, A.H.L. and Beugelsdijk, S. (2010), 'The impact of institutional hazards on foreign multinational activity: a contingency perspective', *Journal of International Business Studies*, doi:10.1057/jibs.2010.1.

Solow, R.M. (1956), 'A contribution to the theory of economic growth', *Quarterly Journal of Economics*, 70, 65–94.

Spence, M. (2011), 'The impact of globalization on income and employment. The downside of integrating markets', *Foreign Affairs*, 90(4), 28–41.

Steger, M.B. (2003), *Globalization: A Very Short Introduction*, Oxford: Oxford University Press.

Steingart, G. (2008), *The War for Wealth: The True Story of Globalization and Why the Flat World is Broken*, New York: McGraw Hill.

Steinmueller, W.E. (1995a), 'Technology infrastructure in information technology industries', Research Memoranda 007, Maastricht: MERIT, Maastricht Economic Research Institute on Innovation and Technology.

Steinmueller, W.E. (1995b), 'The U.S. software industry: an analysis and interpretative history', Research Memoranda 006, Maastricht: MERIT, Maastricht Economic Research Institute on Innovation and Technology.

Steinmueller, W.E. (2000), 'Will new information and communication technologies improve the "codification" of knowledge?', *Industrial and Corporate Change*, Oxford University Press, 9(2), 361–376, June.

Steinmueller, W.E. (2001), 'ICTs and the possibilities for leapfrogging by developing countries', *International Labour Review*, 140(2), 193–210.

Steinmuller, W.E. (2005), 'Technical compatibility standards and the co-ordination of the industrial and international division of labour', Conference: Advancing Knowledge and the Knowledge Economy, Washington, DC.

Stiglitz, J. (2002), *Globalization and its Discontents*, London: Penguin.

Storper, M. (1985), 'Oligopoly and the product cycle: essentialism in economic geography', *Economic Geography*, 61(3), 260–282.

Storper, M. (1992), 'The limits of globalisation: technology districts and international trade', *Economic Geography*, 68, 60–92.

Storper, M. (1997), *The Regional World: Territorial Development in a Global Economy*, New York: Guilford Press.

Storper, M. (1998), *The Regional World: Territorial Development in a Global Economy*, 2nd edn, New York: Guilford Press.

Storper, M. (2009), 'Regional context and global trade', *Economic Geography*, 85(1), 1–20.
Storper, M. (2011), 'Why do regions develop and change? The challenge for geography and economics', *Journal of Economic Geography*, 11, 333–346.
Storper, M. and Venables, A.J. (2004), 'Buzz: face-to-face contact and the urban economy', *Journal of Economic Geography*, 4, 351–370.
Storper, M. and Walker, R. (1989), *The Capitalist Imperative: Territory, Technology, and Industrial Growth*, New York and Oxford: Basil Blackwell.
Strange, S. (1992), 'States, firms, and diplomacy', *International Affairs*, 68(1), 1–15.
Streeck, W. (1991), 'On the institutional conditions of diversified quality production', in Matzner, E. and Streeck, W. (eds), *Beyond Keynesianism: Socioeconomics of Production and Full Employment*, Aldershot, UK and Brookfield, VT, USA: Edward Elgar Publishing, 21–61.
Streeten, P. (1996), 'Governance of the Global Economy', paper presented at the International Conference on 'Globalization and Citizenship', Geneva, United Nations Research Institute for Social Development, December, 1996.
Sturgeon, T.J. (2000), 'How SiliconValley came to be', in Kenney, M. (ed.), *Understanding SiliconValley; The Anatomy of an Entrepreneurial Region*', Stanford, CA: Stanford University Press, 15–47.
Suarez-Villa, L. and Walrod, W. (1997), 'Operational strategy, R&D, and intra-metropolitan clustering in a polycentric structure: the advanced electronics industries of the Los Angeles basin', *Urban Studies*, 34(9), 1343–1380.
Sullivan, D. and Daniels, J. (2008), 'Innovation in international business research: a call for multiple paradigms', *Journal of International Business Studies*, 39(6), 1081–1090.
Sutton, J. (1991), *Sunk Cost and Market Structure: Price Competition, Advertising, and the Evolution of Concentration*, Cambridge, MA: The MIT Press.
Swan, T.W. (1956), 'Economic growth and capital accumulation', *Economic Record*, 32, 334–361.
Swann, G.M.P. (1997), 'Television: looking at the adverts', *Economic Review*, 15(1), 20–23.
Swann, G.M.P. (1998), 'Rapid technological change, shorter business horizons and under-investment', in Coombs, R., Green, K., Richards, A. and Walsh, V. (eds), *Technological Change and Organisation*, Cheltenham, UK and Lyme, NH, USA: Edward Elgar Publishing, 108–123.
Swann, G.M.P. (2006), *Putting Econometrics in its Place: A New Direction*

in Applied Economics, Cheltenham, UK, and Northampton, MA, USA: Edward Elgar Publishing.
Swann, G.M.P. (1993), 'Clusters in high-technology industries', *The Business Economist*, 25, 27–36.
Swann, G.M.P. and Prevezer, M. (1996), 'A comparison of the dynamics of industrial clustering in computing and biotechnology', *Research Policy*, 25, 1139–1157.
Tabellini, G. (2010), 'Culture and institutions: economic development in the regions of Europe', *Journal of the European Economic Association*, 8, 677–716.
Tavares, A.T. and Teixeira, A. (2006), 'Conclusions and policy issues', in Tavares, A.T. and Teixeira, A. (eds), *Multinationals, Clusters and Innovation*, Basingstoke: Palgrave Macmillan.
Taylor, A. (2000), 'Bumpy roads for global auto makers', *Fortune*, 18, p. 284.
Taylor, P.J. (2004), *World City Network: A Global Urban Analysis*, London: Routledge.
Taylor, P.J. and Aranya, R. (2008), 'A global "urban roller coaster"? Connectivity changes in the world city network, 2000–2004', *Regional Studies*, 42(1), 1–16.
Taylor, P.J. and Walker, D.R.F. (2001), 'World cities: a first multivariate analysis of their service complexes', *Urban Studies*, 38(1), 23–47.
Taylor, P.J., Walker, D.R., Catalano, G. and Hoyler, M. (2002a), 'Diversity and power in the world city network', *Cities*, 19(4), 231–241.
Taylor, P.J., Catalano, G. and Walker, D.R.F. (2002b), 'Measurement of the world city network', *Urban Studies*, 39(13), 2367–2376.
Taylor, P.J., Evans, D.M. and Pain, K. (2008), 'Application of the interlocking network model to mega-city regions: measuring polycentricity within and beyond city-regions', *Regional Studies*, 42(8), 1079–1093.
Taylor, P.J., Ni, P., Derudder, B., Hoyler, M., Huang, J. and Witlox, F. (eds) (2011), *Global Urban Analysis: A Survey of Cities in Globalization*, London: Earthscan.
Taylor, R.S. (1986), *Value-Added Processes in Information Systems*, Norwood, NJ: Ablex.
Taylor, R.S. (1987), 'New thought in the twenties: the case of Springfield, Illinois', *Historian*, 49, 329–347, doi: 10.1111/j.1540-6563.1987.tb01919.x.
Teece, D.J. (1977), 'Technology transfer by multinational firms: the resource costs of transferring technological know-how', *Economic Journal*, 87(2), 242–261.
Teece, D.J. (1980), 'The diffusion of an administrative innovation', *Management Science*, 26(5), 464–470, May.

Teece, D.J. (2000), *Managing Intellectual Capital: Organizational, Strategic, and Policy Dimensions*, Oxford: Oxford University Press.

Teece, D.J. and Pisano, G. (1994), 'The dynamic capabilities of firms: an introduction', *Industrial and Corporate Change*, 3(3), 537–556.

Teece, D.J., Pisano, G. and Shuen, A. (1997), 'Dynamic capabilities and strategic management', *Strategic Management Journal*, 18(1), 509–533.

Ter Wal, A.L.J. (2010), 'Cluster emergence and network evolution: a longitudinal analysis of the inventor network in Sophia-Antipolis', *Regional Studies*.

Thissen, M. and Van Oort, F. (2010), 'European place-based development policy and sustainable economic agglomeration', *Tijdschrift voor Economische en Sociale Geografie*, 101(4), 473–480.

Tödtling, F. and Trippl, M. (2005), 'One size fits all?: Towards a differentiated regional innovation policy approach', *Research Policy*, 34(8), 1203–1219..

Tolentino, P.E. (2002), 'Hierarchical pyramids and heterarchical networks: organisational strategies and structures of multinational corporations and its impact on world development', *Contributions to Political Economy*, 21, 69–89.

Tomiura, E. (2007), 'Foreign outsourcing, exporting, and FDI: a productivity comparison at the firm level', *Journal of International Economics*, 72, 113–127.

Torre, A. and Gilly, J.P. (2000), 'On the analytical dimension of proximity dynamics', *Regional Studies*, 34(2), 169–180.

Torre, A. and Rallet, A. (2005), 'Proximity and localization', *Regional Studies*, 39(1), 47–59.

Trevor, M. and Christie, I. (1988), '*Manufacturers and Suppliers in Britain and Japan: Competitiveness and the Growth of Small Firms*', London: Policy Studies Institute.

Tsekouras, G. (1998), 'Integration, organisation and management: investigating capability development', PhD thesis, SPRU, University of Sussex.

Tuan, C. and Ng, L.F.Y. (2003), 'A miracle via agglomerations: the HK-PRD connection', Monograph UNDP#CPR/91/161, The Chinese University of Hong Kong, Hong Kong, in Chinese.

Tybout, J.R. (2003), 'Plant and firm level evidence on new trade theories', in Choi, E.K. and Harrigan, J. (eds), *Handbook of International Trade*, Oxford: Basil Blackwell, 388–414.

Ulltveit-Moe, K.H. (2008), 'Live and let die – industrial policy in a globalised world', Expert Report No 22 to Sweden's Globalization Council, Västerås: Edita.

United Nations Conference on Trade And Development (UNCTAD)

(2000), World Investment Report: Cross-Border Mergers and Acquisitions and Development, New York and Geneva: UNCTAD.
UNCTAD (2003), World Investment Report: FDI Policies for Development: National and International Perspectives, United Nations Conference on Trade and Development, United Nations, New York and Geneva.
UNCTAD (2004), World Investment Report: The Shift Toward Services, United Nations Conference on Trade and Development, United Nations, New York and Geneva.
UNCTAD (2005), World Investment Report: Transnational Corporations and the Internationalization of R&D, United Nations Conference on Trade and Development, United Nations, New York and Geneva.
UNCTAD (2006), World Investment Report: FDI from Developing and Transition Economies: Implications for Development, United Nations Conference on Trade and Development, United Nations, New York and Geneva.
UNCTAD (2007), World Investment Report: Transnational Corporations, Extractive Industries and Development, United Nations Conference on Trade and Development, United Nations, New York and Geneva.
UNCTAD (2009), World Investment Report 2009: Transnational Corporations, Agricultural Production and Development, United Nations, New York and Geneva.
UNCTAD (2010), World Investment Report 2010: Investing in a Low-Carbon Economy, United Nations, New York and Geneva.
UNCTAD (2011), World Investment Report 2011: Non-Equity Modes of International Production and Development, United Nations, New York and Geneva.
UNCTAD (2012), World Investment Report 2012: Towards a New Generation of Investment Policies, United Nations Conference on Trade and Development, United Nations, Geneva.
United Nations Population Fund (UNFPA) (2007), 'State of the world's population: unleashing the potential of urban growth', New York: UNFPA.
UNFPA (2008), 'State of the world's population: unleashing the potential of urban growth', New York: UNFPA.
UN-HABITAT (2008), State of the World's Cities Report 2008/09, United Nations-HABITAT Programme, Nairobi.
United Nations (2008), World Urbanization Prospects, United Nations, New York, accessed at http://esa.un.org/unup/p2k0data.asp.
Utterback, J.M. and Abernathy, W.J. (1975), 'A dynamic model of process

and product innovation', *The International Journal of Management Science*, 3(6), 639–656.
Uyarra, E. (2010), 'Conceptualizing the regional roles of universities, implications and contradictions', *European Planning Studies*, 18(8), 1227–1246.
Uyarra, E. and Flanagan, K. (2010a), 'From regional systems of innovation to regions as innovation policy spaces', *Environment and Planning C: Government and Policy 2010*, 28, 681–695.
Uyarra, E. and Flanagan, K. (2010b), 'Understanding the innovation impacts of public procurement', *European Planning Studies*, 18(1), 123–145.
Vanhoudt, P., Mathä, T. and Smid, B. (2000), 'How productive are capital investments in Europe?', *EIB Papers*, 5(2), 81–106.
Van Oort, F.G. (2004), *Urban Growth and Innovation. Spatially Bounded Externalities in the Netherlands*, Aldershot: Ashgate.
Venables, A.J. (1999a), 'The international division of industries: clustering and comparative advantage in a multi-industry model', *Scandinavian Journal of Economics*, Wiley Blackwell, 101(4), 495–513, December.
Venables, A.J. (1999b), 'Regional integration agreements – a force for convergence or divergence?', Policy Research Working Paper Series 2260, The World Bank.
Venables, A.J. (2006), 'Shifts in economic geography and their causes', Federal Reserve Bank of Kansas City, *Economic Review*, 91(4), 61–85.
Venables, A.J. (2007), 'European integration: a view from spatial economics', in Fujita, M. (ed.), *Regional Integration in East Asia: From the Viewpoint of Spatial Economics*, Basingstoke: Palgrave.
Venables, A.J. (2010), 'Economic geography and African development', *Papers in Regional Science*, 89(3), 469–483.
Verbeke, A., Li, L. and Goerzen, A. (2009), 'Toward more effective research on the multinationality-performance relationship', *Management International Review*, 49, 149–162.
Vernon, R. (1957), The international patent system and foreign policy: [study] of the Committee on the Judiciary, United States Senate, Eighty-fifth Congress, first session, pursuant to S. Res. 55. Washington, DC: US Government. Printing Office.
Vernon, R. (1959), 'The changing economic function of the central city', New York: Area Development Committee of CED.
Vernon, R. (1960), *Metropolis*, Cambridge, MA: Harvard University Press.
Vernon, R. (1963), *Metropolis 1985*, New York: Doubleday Anchor Books.
Vernon, R. (1966), *The Myth and Reality of our Urban Problems*, Cambridge, MA: Harvard University Press.

Vernon, R. (1974), *Big Business and the State: Changing Relations in Western Europe*, Cambridge, MA: Harvard University Press.
Vernon, R. (1979a), *Storm over the Multinationals: The Real Issues*, London: Macmillan.
Vernon, R. (1979b), *Opportunities and Challenges for Multinational Firms from Developing Countries*, Seoul, Korea: Korea International Economic Institute.
Vernon, R. (1991a), 'The longer view the coming global metropolis cities of the next century', *Journal of the American Planning Association*, 57(1), 3–6.
Vernon, R. (1991b), 'Sovereignty at bay: twenty years after', *Millennium Journal of International Studies*, 20(2), 191–195.
Verspagen, B. (2000), 'Economic growth and technological change: an evolutionary interpretation', Working Paper series 00.12, Eindhoven Center for Innovation Studies (ECIS).
Verspagen, B. and Caniëls, M.C.J. (2001), 'Barriers to knowledge spillovers and regional convergence in an evolutionary model', *Journal of Evolutionary Economics*, 11(3), 307–329.
Verspagen, B. and Schoenmakers, W. (2004), 'The spatial dimension of patenting by multinational firms in Europe', *Journal of Economic Geography*, Oxford University Press, 4(1), 23–42, January.
Vertova, G. (2002), 'A historical investigation of the geography of innovative activities', *Structural Change and Economic Dynamics*, 13(3), 259–283.
Vertova, G. (2006), *The Changing Economic Geography Of Globalization: Reinventing Space*, Abingdon and New York: Routledge.
Veugelers, R. and Cassiman, B. (1999), 'Innovative strategies and know-how flows in international companies: some evidence from Belgian manufacturing', Open Access publications from Katholieke Universiteit, Leuven.
Veugelers, R. and Cassiman, B. (2004), 'Foreign subsidiaries as a channel of international technology diffusion: some direct firm level evidence from Belgium', *European Economic Review*, 48, 455–476.
Vicente, J., Dalla Pria, Y. and Suire, R. (2007), 'The ambivalent role of mimetic behavior in proximity dynamics: evidence from the "Silicon Sentier"', in Surinach, J. (ed.), *Knowledge Externalities, Innovation Clusters and Regional Development*, Cheltenham, UK and Northampton, MA, USA: Edward Elgar Publishing.
Vilanova, M. and Leydesdorff, L. (2001), 'Why Catalonia cannot be considered as a regional innovation system', *Scientometrics*, 50(2), 215–240.
von Hippel, E. (1994), 'Sticky information and the locus of problem solving: implications for innovation', *Management Science*, 40, 429–439.

von Tunzelmann, G.N. (1995), *Technology and Industrial Progress*, Aldershot, UK and Brookfield, VT, USA: Edward Elgar Publishing.

von Tunzelmann, N. (2003), 'Historical coevolution of governance and technology in the industrial revolutions', *Structural Change and Economic Dynamics*, 14, 365–384.

von Tunzelmann, N. (2009a), 'Competencies vs. capabilities: a reassessment', *Economia Politica, Journal of Analytical and Institutional Economics*, 3, 435–464.

von Tunzelmann, N. (2009b), 'Regional capabilities and industrial regeneration', in Farshchi, M., Janne, O. and McCann, P. (eds), *Technological Change and Mature Industrial Regions: Firms, Knowledge and Policy*, Cheltenham, UK and Northampton, MA, USA: Edward Elgar Publishing, pp.11–28.

von Tunzelmann, N. and Wang, Q. (2003), 'An evolutionary view of dynamic capabilities', *Economie Appliquée*, 6, 33–64.

von Tunzelmann, N. and Wang, Q. (2007), 'Capabilities and production theory', *Structural Change and Economic Dynamics*, 18, 192–211.

Walker, R. (1998), 'The geographical organization of production systems', *Environment and Planning: Society and Space*, 6, 377–408.

Warf, B. (1995), 'Telecommunications and the clustering geographies of knowledge transmission in the late 20th century', *Urban Studies*, 32(2), 361–378.

Weber, A. (1909), *Theory of the Location of Industries*, Chicago: The University of Chicago Press.

Wernerfelt, B. (1984), 'A resource-based view of the firm', *Strategic Management Journal*, 5, 171–180.

Wickham, J. and Vecchi, A. (2008), 'Local firms and global reach: business air travel and the Irish software cluster', *European Planning Studies*, 16(5), 693–710.

Wilkins, D.B. (2001), 'Why global law firms should care about diversity: five lessons from the American experience', in Drolshammer, J. and Pfeifer, M. (eds), *The Internationalization of the Practice of Law*, Kluwer Law International.

Wilkins, M. (1994), 'Comparative hosts', *Business History*, 36(1), 18–50.

Wilkins, M. (2001), 'The history of multinational enterprise', in Rugman, A.M. and Brewer, T.L. (eds), *The Oxford Handbook of International Business*, Oxford: Oxford University Press, 3–35.

Wilkins, M. (2001), 'The history of multinational corporations', in Rugman, A.M. and Brewer, T.L. (eds), *The Oxford Handbook of International Business*, Oxford: Oxford University Press.

Williamson, J.G. (1965), 'Regional inequality and the process of national

development: a description of the patterns', *Economic Growth and Cultural Change*, 13, 1–84.
Williamson, O.E. (1975), *Markets and Hierarchies*, New York: Free Press.
Williamson, O.E. (1979), 'Transaction-cost economics: the governance of contractual relations', *Journal of Law and Economics*, 22(2), 233–261.
Williamson, O.E. (1985), *The Economic Institutions of Capitalism*, New York: Free Press.
Williamson, O.E. (1987), *The Economic Institutions of Capitalism: Firms, Markets, Relational Contracting*, New York: Free Press.
Winter, S.G. (1984), 'Schumpeterian competition in alternative technological regimes', *Journal of Economic Behavior and Organization*, 5, 287–320.
Winter, S.G. (1987), 'Knowledge and competence as strategic asset', in Teece, D.J. (ed.), *The Competitive Challenge: Strategies for Industrial Renewal*, Cambridge, MA: Ballinger Publishing.
Winter, S.G. (1993), 'On Coase, competence and the corporation', in Williamson, O.E. and Winter, S.G. (eds), *The Nature of the Firm*, Oxford: Oxford University Press.
Winter, S.G. (2003), 'Understanding dynamic capabilities', *Strategic Management Journal*, 24, 991–995.
Wolfe, D.A. and Gertler, M.S. (2004), 'Clusters from the inside and out: lessons from the Canadian study of cluster development', *Urban Studies*, 41(5-6), 1071–1093, May.
World Bank (1998), 'Corruption and development', PREM Notes No. 4, May, Washington, DC.
World Bank (2003a), *Doing Business 2008*, Washington, DC: The World Bank.
World Bank (2003b), *Sustainable Development in a Dynamic World: Transforming Institutions, Growth, and Quality of Life*, Washington, DC: The World Bank.
World Bank (2005), *World Development Report 2005: A Better Investment Climate for Everyone*, Washington, DC: The World Bank.
World Bank (2007), *2007 World Development Indicators*, Washington, DC: The World Bank.
World Bank (2008a), *World Trade Indicators 2008: Benchmarking Policy and Performance*, Washington, DC: The World Bank.
World Bank (2008b), *2008 World Development Indicators*, Washington, DC: The World Bank.
World Bank (2008c), *World Development Report 2008: Agriculture for Development*, Washington, DC: The World Bank.
World Bank (2009), *World Development Report 2009: Reshaping Economic Geography*, Washington, DC: The World Bank.

World Gazetteer (n.d.), World Gazetteer (available at: http://www.world-gazetteer.com).

Xu, C. (2011), 'The fundamental institutions of China's reforms and development', *Journal of Economic Literature*, 49(4), 1076–1151.

Yeaple, S.R. (2003), 'The role of skill endowments in the structure of U.S. outward foreign direct investment', *The Review of Economics and Statistics*, The MIT Press, 85(3), 726–734.

Yeaple, S.R. (2009), 'Firm heterogeneity and the structure of US multinational activity', *Journal of International Economics*, 78, 206–215.

Young, S. and Tavares, A.T. (2004), 'Centralization and autonomy: back to the future', *International Business Review*, 13(2), 215–237.

Young, S., Hood, N. and Peters, E. (1994), 'Multinational enterprises and regional economic development', *Regional Studies*, 28(7), 657–677.

Zademach, H-M., and Rodríguez-Pose, A. (2009), 'Crossborder M&As and the changing economic geography of Europe', *European Planning Studies*, 17, 765–789.

Zaheer, S. (1995), 'Overcoming the liability of foreignness', *Academy of Management Journal*, 38, 341–363.

Zander, I. (1997), 'Technological diversification in the multinational corporation: historical trends and future prospects', *Research Policy*, 26, 209–227.

Zander, I. (1999a), 'Where to the multinational? Evolution of technological capabilities in the multinational network', *International Business Review*, 8, 261–291.

Zander, I. (1999b), 'How do you mean "global"? An empirical investigation of innovation networks in the multinational corporation', *Research Policy*, 28, 195–214.

Zanfei, A. (2000), 'Transnational firms and changing organisation of innovative activities', *Cambridge Journal of Economics*, 24(5), 515–542.

Zhao, M. (2006), 'Conducting R&D in countries with weak intellectual property rights protection', *Management Science*, 52(8), 1185–1199.

Zollo, M. and Winter, S.G. (2002), 'Deliberate learning and the evolution of dynamic capabilities', *Organization Science*, 13, 339–351.

Zook, M.A. (2005), *The Geography of the Internet*, Oxford: Blackwell.

Index